武汉大学2020年规划核心教材建设项目

Basic Engineering Geology
工程地质基础

王　涛　主编

［英］萨霍斯·瓦西利斯
［尼泊尔］夏姆·桑达尔·哈德卡　副主编

中国建筑工业出版社

图书在版编目(CIP)数据

工程地质基础 = Basic Engineering Geology：英文 / 王涛主编；（英）萨霍斯·瓦西利斯，（尼泊尔）夏姆·桑达尔·哈德卡副主编．—北京：中国建筑工业出版社，2022.1

ISBN 978-7-112-27060-6

Ⅰ. ①工… Ⅱ. ①王… ②萨… ③夏… Ⅲ. ①工程地质-英文 Ⅳ. ①P642

中国版本图书馆CIP数据核字（2021）第275799号

本书是在国际化教学背景下编写的英文版教材。教材根据最新的水利、土木相关专业的课程设置和教学要求编写，并力求采用最专业的英文来表述工程地质基础理论知识。全书共分11章，系统讲述了矿物与岩石、地质年代、地质构造、地质图、地下水、河流地质与地貌、岩体的工程地质力学特性、坝基的工程地质研究、岩质边坡的工程地质研究、地下洞室及隧洞的工程地质研究。

本书可作为水利、土木专业的本科生以及采矿、石油、地质、地球物理、测绘、资源与环境等相关专业学生的教材；同时，本书可供工程地质和岩土力学领域研究生及研究人员参考使用。

责任编辑: 辛海丽
责任校对: 芦欣甜

Basic Engineering Geology
工程地质基础

王　涛　主　编
［英］萨霍斯·瓦西利斯　副主编
［尼泊尔］夏姆·桑达尔·哈德卡

*

中国建筑工业出版社出版、发行(北京海淀三里河路9号)
各地新华书店、建筑书店经销
北京鸿文瀚海文化传媒有限公司制版
北京中科印刷有限公司印刷

*

开本：787毫米×1092毫米　1/16　印张：21¼　字数：331千字
2022年1月第一版　2022年1月第一次印刷
定价：**68.00**元

ISBN 978-7-112-27060-6
(38813)

Preface

Engineering geology can be defined as the application of geology in engineering practice. It involves the geological factors that affect the siting, design, construction, operation, and maintenance of an engineering project. Therefore, it draws on the knowledge of many geological disciplines, such as mineralogy, petrology, stratigraphy, geomorphology, tectonic geology, and hydrogeology. In addition, the knowledge of rock mechanics and soil mechanics is an indispensable part that bridges geology and engineering.

I have been teaching the course of engineering geology in the School of Water Resources and Hydropower Engineering (SWRHE) at Wuhan University (WHU) since 1996 and started to offer courses of Engineering Geology in English in 2016. In the process of teaching practice, I deeply realized that it is necessary to have a suitable textbook in English. On the one hand, under the background of international teaching, Textbooks in English are the prerequisite for the All-English teaching program. On the other hand, with the rapid development of engineering geology and related disciplines domestically and internationally, new methods, theories, concepts, and new norms emerge one after another. In contrast, most textbooks in Chinese are updated very slowly in these aspects. To bridge these gaps, I started to compile this textbook under the sponsorship of the WHU 2020 Core Textbooks Plan Project.

Overall, this book is based on my many years of teaching experience, using the relevant course syllabus as guidelines, with efforts to incorporate the most recent scientific discoveries, large-scale engineering practices, and the latest relevant domestic and international standards. Although initially intended to offer a textbook for undergraduate students majoring in hydropower and civil engineering, it can also be used as a textbook for both undergraduate and graduate students in the major of mining, geology, petroleum, geophysics, surveying, and mapping. In other words, all students who are engaged in earth-science-related programs may benefit from reading this book. In addition, this textbook can be used as a helpful reference for researchers in related areas.

The textbook consists of 11 chapters. Chapters 1~5 introduce the basic geological concepts. Chapters 6、7 focus on groundwater and fluvial geomorphology. Chapter 8 describes the fundamental engineering geomechanics of rock mass. Chapters 9~11 present some application examples of engineering geology. The author Tao Wang from WHU is responsible for the technical content arrangement and outline of the textbook. Vasilis Sarhosis (Associated Professor, Faculty of Engineering and Physical Sciences) from The University of Leeds, Shyam Sundar Khadka (Assistant Professor, Head of Department of Civil Engineering) from Kathmandu University, Nepal, Guan Rong, Kai Su, Ji He, Qiao Wang, Yao Yue, Yinlong Jin, Quan Liu, Yonggang Cheng, Jing Lu from WHU, Yanhui Han from Aramco Americas, and a few other teachers also participated in the writing, compilation or review. The author is very grateful to the graduate students at WHU, including Bowen Fan, Kai Yang, Suifeng Wang, Ming Lei, Yang Zhao, Songcheng Liu, Congpeng Zhang, Qian Liu, Wentao Xu, Xianyu Zhao, Faisal Waqar, and Ali Sajid, for their contributions in editing and reviewing this textbook.

I would like to express my gratitude to the authors of the references cited in this textbook. Special appreciation goes to Professor R.T. Stacey at the University of the Witwatersrand in South Africa for providing his precious research results in this book. The writing of this textbook was funded by the "Double First-Rate" discipline construction project of WHU, and supported by the faculty and students of Undergraduate School and SWRHE. The students of the international Engineering of 2019 class from Hongyi School, and students of 2018 and 2019 classes from SWRHE, who are majoring in Port, Waterway and Coastal Engineering, and Hydropower Engineering, have put forward many suggestions for revising the textbook. I would like to express my heartfelt thanks together.

Due to the rapid development of engineering geology and the wide range of fields involved, it is inevitable that a lot of essential materials should be included but are missed in this edition. I will improve this textbook in its future editions based on the feedback from readers. Email:htwang@whu.edu.cn.

前　言

工程地质学可以定义为地质学在工程实践中的应用，即它涉及的是那些影响工程的选址、设计、施工和维护的地质因素。因此，它需要借鉴许多地质学学科的知识，如矿物学、岩石学、地层学、地貌学、构造地质学和水文地质学。此外，工程地质学还涉及岩石力学和土力学等领域的知识。

自 1996 年起，我就负责武汉大学水利水电学院的工程地质课程教学工作，从 2016 年又开始了水利工程全英文实验班的工程地质教学任务。在教学实践过程中，我深刻地意识到出版一本完善的英文版教材是非常必要的。一方面在国际化教学背景下，英文版教材是全英文教学的必备条件；另一方面当今国内外工程地质及相关学科发展迅速，新方法、新理论、新概念和新的规范层出不穷，而传统的中文教材在这些方面更新速度过慢。基于此，在武汉大学 2020 年规划核心教材项目的支持下，我们开始编写本教材。

本教材立足于长期的教学积累，遵从国内的相关教学大纲，力争在吸收该领域国内外大师们思想精髓的基础上，融入最新的科学发现、大型的工程实践和最新的相关标准。教材最初是面向土木、水利专业的本科生，但同时可供采矿、地质、石油、地球物理、测绘、资源与环境等相关专业学生使用，只要是从事与地球相关学习的同学都可以本教材中受益。另外，本教材也可作为相关领域研究人员的参考书。

全书共 11 章，其中第 1~5 章为基础地质知识部分，第 6、7 章为地下水和河流地貌部分，第 8 章为岩体工程地质力学部分，第 9~11 章为工程地质应用部分。全书的章节安排和统稿由武汉大学王涛负责，参加本书编写的有英国利兹大学工程和物理科学学院副教授萨霍斯・瓦西利斯（Vasilis Sarhosis），尼泊尔加德满都大学助理教授、土木工程系系主任夏姆・桑达尔・哈德卡（Shyam Sundar Khadka），武汉大学荣冠、苏凯、何吉、王桥、岳遥、金银龙、刘全、程勇刚、陆晶以及韩彦辉（美国沙特阿美）等老师。非常感谢参加编写的武汉大学研究生范博文、杨凯、王穗丰、雷鸣、赵杨、刘松成、张丛彭、刘骞、徐文韬、赵先宇、Faisal Waqar 和 Ali Sajid 等同学。

我们谨向本书所引用参考文献的作者们表示感谢，他们的成果为本书的撰写提供了基础。特别感谢南非金山大学（The University of the Witwatersrand）斯特西（R.T. Stacey）教授提供了他早期珍贵的研究成果作为本教材的素材。在本教材编写过程中，得到了武汉大学“双一流”学科建设项目的资助；得到了武汉大学本科生院、武汉大学水利水电学院师生们的支持；武汉大学弘毅学堂 2019 级国际工程专业、水利水电学院港口航道与海岸工程专业和水利水电工程专业（2018级和2019级）的学生对教材提出了很多修订意见，在此一并致以衷心的感谢。

由于工程地质学发展很快，涉及的领域也很广，因此教材中会出现很多需要持续改进的地方，敬请读者批评指正！电子邮箱：htwang@whu. edu. cn。

Contents

1
Introduction

1.1 Engineering geology and its tasks

Engineering geology is a branch of geology, which studies geological problems related to engineering construction. Engineering geology is an interdisciplinary science formed by the interpenetration and intercrossing of engineering and geology, which is engaged in studying the interrelationship between human engineering activities and the geological environment. It is an applied science serving engineering construction and belongs to the category of applied geology(Bell, 2007; Shi & Yan, 2017; Tang, 2008).

1.1.1 Geology and engineering geology

Geology is the field of knowledge concerning the present and past morphology and structure of the earth, its environments, and the fossil record of its inhabitants.

It is an enormous subject that comprises all the rocks, soils, and water bodies on the planet and the development of life. The study aids orderly and efficient exploration of the crust for oil and mineral deposits; helps to explain the mysteries connected with our development as beings and the nature of our planet; permits us to mitigate the hazards of earthquakes, volcanic eruptions, floods, and so forth; and is helpful in many other fields. It is taught in most of the world's great universities and colleges, and occupies the minds of a multitude of researchers.

By comparison, its subfield, engineering geology has simple objectives and a small compass; it exists solely to serve the art and science of engineering by describing the structure and attributes of the rocks connected with engineering works(Goodman, 1993). It is concerned with mapping and characterizing all the materials proximate to a project. It is also part of engineering geology to identify and evaluate natural hazards like landslides and earthquakes that may affect the success of an engineering project. It is concerned with the storage and movement of water through void spaces within rocks and soils and the associated consequences for an engineering project and the natural regime.

All kinds of human engineering activities will react to the geological environment, change the natural geological conditions, affect the stability and routine use of buildings, and even threaten human life and the living environment. For example, land subsidence caused by massive pumping of groundwater in coastal cities will lead to seawater intrusion, destruction, loss of municipal transportation facilities'utility, deterioration of underground water quality, etc. The construction of large reservoirs changes the hydrologic and hydrogeological conditions in the upper and lower reaches of the river. It causes problems such as bank reconstruction, flood, siltation, earthquake, and even deterioration of the ecological environment. We should fully anticipate the impact of constructing notable projects on the geological environment to take corresponding countermeasures.

1.1.2 Basic tasks of engineering geology

The engineering activities of human beings and the geological environment are in the contradictions of mutual contact and mutual restriction. It has become the essential task of engineering geology to study the relationship between geological environment and human engineering activities and promote the transformation and solution of contradictions between them. The primary mission of engineering geology construction is to provide geological data needed for the project planning, design, building, and solving various geological problems encountered in engineering. At the same time, it is necessary to demonstrate the changes in the geological conditions and propose corresponding measures for the rational use of the geological environment to ensure that the buildings are safe, reliable, and economically reasonable (Záruba & Mencl, 1976). To be specific, the main tasks of engineering geology at present include:

(1) Studying the influence of human activities on the geological environment and the effect of geological disasters;

(2) Evaluating and predicting the space, time, and strength between the geological conditions and engineering construction through the theory and method of mechanics, mathematics, and system engineering;

(3) Evaluating engineering geological conditions, to grasp the material migration rule of geological environment, and to demonstrate the technical possibility of construction and operation of the construction project as well as the safety and economic rationality of the implementation under the current engineering geological conditions;

(4) Assessing the engineering geological environment and putting forward measures to prevent and control geological disasters and protect the geological environment;

(5) Developing engineering geological survey and geological disaster monitoring technology and methods to provide tools and means for the solution of new problems and new phenomena;

(6) Conducting the classification of engineering geology, which can provide a scientific basis for the scientific layout of basic engineering and rational utilization and protection of the geological environment.

The engineering geologist can accomplish the above tasks by working closely with the engineering planning, design, and construction. From the above tasks, it is necessary to clarify the meaning of engineering geological conditions and engineering geological problems.

Engineering geology plays a vital role in national economic construction. If engineering geology work is done well, engineering design and construction will be smooth, engineering construction safety operation will be guaranteed, engineering geological disasters can be prevented, and the geological environment will be protected.

1.1.3 Engineering geological conditions and problems

Engineering geological conditions refer to the synthesis of geological factors related to engineering construction. These factors include geotechnical type, engineering properties, geological structure, geomorphology, hydrogeology, engineering dynamic geology and natural building materials. It is a comprehensive concept so a certain factor can not be summarized as engineering geological conditions, but only a certain factor of engineering geological conditions. Engineering geological conditions directly affect the safety, economy and normal use of engineering buildings. So before the construction of any type of building, the first is to find out the engineering geological conditions of the construction site, which is the basic task of engineering geological survey.

Because the geological environments of different areas are not the same, the geological factors that affect the engineering buildings vary in primary and secondary aspects. We should make a specific analysis of the local engineering geological conditions, make clear the primary and secondary, and further point out the favorable and unfavorable aspects of the engineering buildings. To understand engineering geological conditions, we must start from introductory geology and then understand the geological development history of the study area, the characteristics of various elements, and the regularity of their combination.

Engineering geological problems refer to the contradiction between engineering geological conditions and engineering buildings. Engineering geological conditions exist objectively in nature. Whether they can meet the needs of engineering construction must be related to the type, structure, and scale of engineering buildings. Engineering buildings of different styles, designs, and scales have additional requirements for the geological environment due to different working modes and loads on geological bodies. So engineering geological problems are complex and diverse. For example, industrial and civil architecture's main engineering geological problems are foundation bearing capacity and deformation. The main engineering geological problems of the underground cavern are surrounding rock stability and water inrush. The main engineering geological problem of open-pit mining is the stability of the mining slope. In hydropower engineering, the most important thing to pay attention to is the seepage and leakage of the dam foundation. The concrete gravity dam is the problem of the anti-sliding stability, and the arch dam is the problem of the anti-sliding stability of the abutment. The analysis and evaluation of engineering geological problems are the critical tasks of engineering geological engineers.

1.2 Research contents of engineering geology

The tasks of engineering geology determine its research contents (Tang, 2008), which include many aspects, including the following elements:

(1) Geotechnical engineering properties. No building on earth can be built without rock and soil mass. Whether analyzing engineering geological conditions or evaluating engineering geological problems, it is necessary to study the engineering properties of rock and soil first. The research contents include the geotechnical engineering geological properties, formation and change law of rock-soil, the testing technology and method of various parameters, the type and distribution law of rock and soil mass, and the improvement of its harmful properties. The research on this aspect is carried out by geotechnical engineering science, a branch of engineering geology.

(2) Engineering dynamic geology effects. The crust surface is affected by various natural forces, including the earth's internal force and external force and human engineering activities, affecting the stability and normal use of buildings. This kind of geological effect that has an impact on engineering construction is called engineering dynamic geology. The geological phenomena caused by natural forces are called physical geological phenomena, and those caused by human engineering activities are called engineering geological phenomena. The research contents of engineering dynamic geology, another branch of engineering geology, are to study

the formation mechanism, scale, distribution, development, and evolution law of engineering dynamic geology, the related engineering geological problems. Then evaluate them qualitatively and quantitatively, and effectively prevent and cure them.

(3) Theory and technique of engineering geological survey. To find out the engineering geological conditions of construction sites, demonstrate the engineering geological problems, make the engineering geological evaluation correctly, and provide the geological data needed for the design, construction, and use of buildings, it is necessary to carry out engineering geological survey. Buildings with different types, structures, and scales have additional requirements for engineering geological conditions and engineering geological problems, so the selection of survey methods, the arrangement of work principles, and the use of workload are also different. To ensure the safety and routine use of all kinds of buildings, it is necessary to study the possible engineering geological problems first in detail and depth and arrange the investigation work on this basis. Various survey specifications or work manuals applicable to different types of engineering construction should be formulated as the guide of survey work to ensure survey quality. A branch of specialized engineering geology researches this aspect.

(4) Regional engineering geology. Engineering geological conditions are different in different regions due to various natural geological conditions. It is the content to understand the formation and distribution of regional engineering geological conditions and predict these conditions'change under the influence of human engineering activities and make engineering geological zoning maps. Regional engineering geology is a branch of this research. It can be seen that engineering geology is a subject with a robust application. It plays a critical role in engineering construction, serves a wide range of objects, and has rich research content.

1.3 Research methods of engineering geology

The research method of engineering geology is adapted to its research content. There are natural history analysis, mechanics, model simulation, and engineering geology analogy methods.

1.3.1 Natural history analysis

Natural history analysis is the most basic research method in engineering geology. Geological bodies and various geological phenomena, which are the research objects of engineering geology, are formed in natural geological history and continue to evolve with the change of conditions. Therefore, when conducting engineering geological research on dynamic geology or construction sites, it is necessary to do essential geological work well first, find out natural geological conditions and various geological phenomena as well as their relations, and predict the trend of their development and evolution. Only in this way can we find out the engineering geological conditions in the studied area, which serve as the basis for further study of engineering geological problems.

For example, to study slope deformation and failure problems, you should start from the morphological research, determine the slope deformation and destruction of type, size and boundary conditions, analyze slope deformation and destruction mechanism and various controlling factors to show the spatial distribution pattern, analyze its formation, development, evolution and development stage, and then reveal its internal law from space distribution and time series. Predicting its changes under human engineering activities lays a foundation for further engineering geological evaluation of slope stability.

For the problem of dam foundation stability against sliding, we must find out the formation characteristics of dam foundation rock mass, investigate the geological structures and groundwater condition, pay attention to the existence of the weak surfaces in the rock mass, speculate the possible slip plane and the cut surface and their relationships with engineering force. The engineering geological behaviors of slip surfaces are studied to provide a basis for further study of the anti-sliding stability of dam foundations.

However, only the natural history method can not completely meet the requirements of engineering geological evaluation because it ultimately belongs to qualitative research. In order to study a specific engineering geological in-depth problem, a quantitative research method must be adopted. Mathematical and mechanical analysis methods and model simulation test methods are quantitative research methods.

1.3.2 Mechanical analysis

Mechanical analysis is based on natural history analysis. For an engineering geological problem or dynamic geological phenomenon, according to the determined boundary conditions and calculation parameters, a quantitative calculation is carried out using theoretical or empirical formulas. For example, the rigid body limit equilibrium theory method used in slope stability calculation concentrates various forces on the possible sliding surface and anti-sliding force on the premise that the slope is a rigid body. The stability coefficient of the surface is worked out as the basis of quantitative evaluation. In order to make clear the boundary conditions and reasonably select the calculated parameters, it is necessary to carry out engineering geological exploration and test, which sometimes costs vast capital and human resources. Therefore, except for large or important buildings, general buildings are often calculated using a mechanical analogy.

Due to the complexity of natural geological conditions, it is often necessary to simplify the conditions appropriately in the calculation and simplify the spatial problems into plane problems to deal with. Generally, the geological model is established first and then abstracted into mechanical and mathematical models, which are substituted into the calculation parameters. Due to the development of modern computing technology, various mathematical and mechanical computing models are increasingly used in engineering geology.

1.3.3 Model simulation test method

The model simulation test method can help us explore the law of natural geological action, reveal the mechanical mechanism of engineering dynamic geological action or engineering geological problems, and the whole process of evolution to make a correct engineering geological evaluation. Simple mathematical expressions can express some natural laws or the interaction between buildings and geological environments. However, some mathematical expressions are complicated to solve or even cannot be expressed by mathematical expressions because it is difficult to find the rules of their action. In this case, the model simulation test is efficient. In addition to engineering mechanics, rock mechanics, soil mechanics, hydraulics, groundwater dynamics, and other theories, the dimensional principle and similarity principle must be followed to conduct model simulation tests.

The difference between model and simulation tests is whether the fundamental law based on the test is consistent with the basic law of actual action. For example, a dam foundation seepage test with a seepage channel is a model test because the test is based on Darcy's law, the same fundamental law that controls dam foundation seepage in practice. If such an experiment is carried out using a network method, it is a simulation experiment because it is based on Ohm's law in electricity. Ohm's law and Darcy's law are similar in form but fundamentally different. Standard model tests in engineering geology include surface flow and groundwater seepage, slope stability, foundation stability, anti-sliding stability of hydraulic structures, and surrounding rock stability of underground chambers. The commonly used simulation tests include photoelastic and photoplastic, and electrical network simulations to simulate groundwater seepage (Tang, 2008).

1.3.4 Engineering geology analogy method

Engineering geology analogy is another standard engineering geology research method, which can be used for qualitative evaluation or semi-quantitative evaluation. It is to apply the evaluation experience of the engineering geological problems of the existing buildings to the similar buildings planned to be built with roughly the same natural geological conditions. This approach is based on similarity: the natural geological conditions, the way the building works, and the predicted engineering geological problems should be roughly the same or similar. The experience of the researcher often limits it. Because the natural geological conditions can not be precisely the same, and the analogy often simplifies the conditions, this method is relatively rough, generally suitable for small projects or preliminary evaluation.

The four research methods mentioned above have their characteristics, which should complement each other and be applied comprehensively. Natural history analysis is the most fundamental research method, which is the basis of other research methods.

1.4 The relationship between engineering geology and other disciplines

Engineering geology involves a wide range of knowledge, which must be based on the knowledge of many disciplines. The sub-disciplines of geology, such as mineralogy, petrology, structural geology, stratigraphy, quaternary geology, geomorphology, and hydrogeology, are closely related to engineering geology. Engineering geology research cannot be carried out without the basic knowledge of the above disciplines. The practice has proved that many problems in engineering geological surveys are caused by the lack of basic geological knowledge (Zhang et al.,2005).

In engineering geology, the theories and methods of various geological branches are often applied. However, engineering geology is for the service of engineering construction, and its research purpose is obvious, so the research purpose and method are different from other branches of geology. Dynamic geological process, for example, is a dynamic geology and engineering geology research object. However, the former is mainly qualitative to study morphology, distribution, conditions, etc., while the latter is not only to qualitatively research, but also to further study the formation mechanism, quantitatively study the evolution of its occurrence and development rule and the influence degree of the buildings in engineering, as well as effective prevention and control measures, etc. In order to evaluate engineering geologic quantitatively, engineering geology needs mathematics and mechanics knowledge as its foundation. Therefore, advanced mathematics, applied mathematics, engineering mechanics, elasticity, soil mechanics, and rock mechanics are closely related to engineering geology. A large number of computational problems in engineering geology are studied in soil mechanics and rock mechanics. In the broad concept of engineering geology, soil mechanics and rock mechanics are often included. Soil mechanics and rock mechanics are two subjects to study engineering geology of soil and rock mass. However, from the point of mechanics, they belong to a branch of mechanics. Combining geology with mathematics and mechanics, and based on many engineering practices, Gu Dezhen put forward the concept of rock mass structure and founded the discipline of " Engineering Geology mechanics of Rock Mass", and Sun Guangzhong established the theoretical system of rock mass structure mechanics(Gu, 1979; Sun, 1988).

During the 1960s, 30 years after soil mechanics was established as a science, engineering geology emerged as a new discipline in the broad field of geotechnics, almost simultaneously with rock mechanics. By the end of that decade, it became accepted worldwide that geotechnics, as a branch of engineering, embraced soil mechanics, rock mechanics, and engineering geology (Figure 1-1). It was not possible to define the boundaries clearly between them (IAEG,2014).

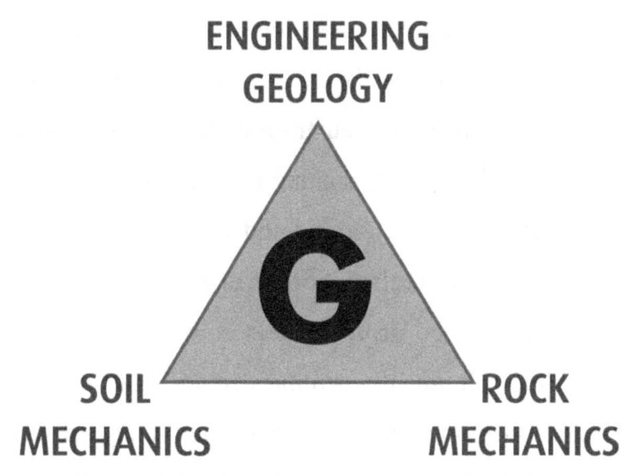

Figure 1-1 Geotechnics and embraced sciences

Professionals in engineering geology are generally well established and accepted today. In their activities, they are frequently required to interact with specialists in other disciplines (Rengers et al., 2008) to solve critical engineering geotechnical works and natural hazards (Figure 1-2). Their contribution starts at the feasibility stage and goes through design (ground model, the definition of parameters, the conception of solutions, and their dimensioning in some cases), construction, supervision, and operation.

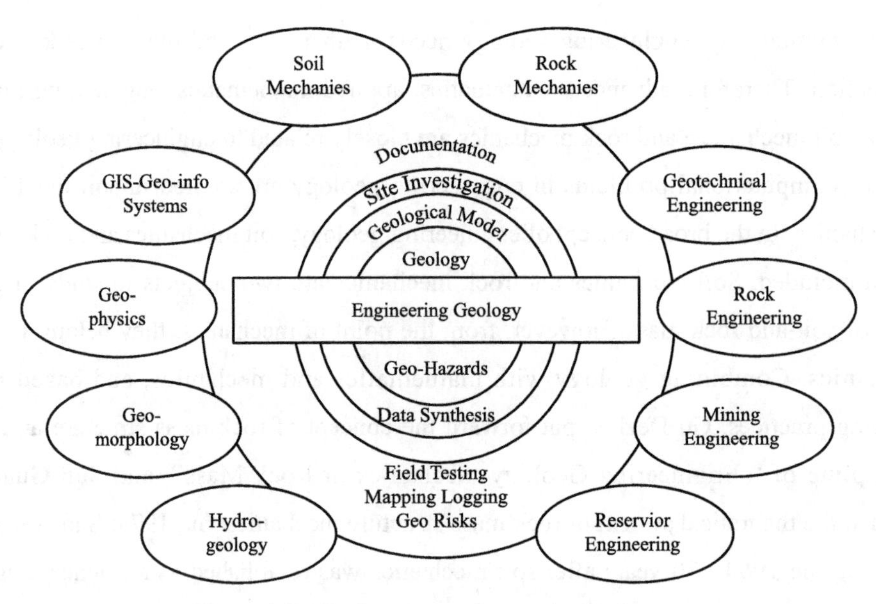

Figure 1-2 Engineering geology and related areas

Site investigation and testing methods are readily available. The improvements experienced in the last years have been mainly related to informatics tools and their upgrades (remote sensing advances, GIS, GPS, Beidou, telematics, 3D modeling, and so on). They are increasingly used to provide more accurate and reliable interpretations to achieve higher quality outputs. Laser, radar, GRACE, and other techniques developed for other scientific purposes are frequently used in engineering geological investigations of particular problems (Oliveira, 2010).

In addition, engineering geology work relating to many recent topics (mainly connected to the environment, such as waste disposal, ground and groundwater pollution, land subsidence, land-use planning, seismic zonation, reclamation of degraded areas from surface mining and quarrying) calls for very experienced professionals able to interact with practitioners from other disciplines. More and more frequently, similar demands also occur in the rehabilitation of essential engineering works constructed decades ago and are today experiencing the aging of the structures and their foundations (IAEG,2014).

1.5 The characteristics and learning requirements of this course

As an engineer of engineering construction, it is necessary to have specific knowledge of engineering geology, not only to be able to read and use the results of engineering, geological survey data, but also to understand, analyze and deal with the relevant geological problems, to make the right design and construction scheme. Pan Jiazheng, a famous expert in hydropower engineering, once said: "A geologist who does not know design cannot become an excellent engineering geologist, just as a designer who does not know engineering geology cannot become an excellent designer. " This is a critical understanding that can be summarized from many production practices and engineering accidents. Some engineering geological problems will inevitably be neglected, or even a wrong design or construction scheme will be made without necessary engineering geological knowledge (Cui & Zhu, 2014).

The practical teaching of this course is significant. The indoor experiments and the field course are also essential teaching sections in addition to classroom teaching. In particular, the field course plays a vital role in this course. It is more appropriate to call it "field teaching" than field practice. It is not only verifying, consolidating, and deepening the teaching content from class, but also a knowledge that cannot be taught in class, which can only be learned by the teacher's explanation infield, and students' observation, analysis, and practical operation. The field course is vital to cultivating students' independent observation, thinking, analysis, and practical operation. Engineering geology teaching is incomplete without this critical practical segment. Therefore, enough attention should be paid to field study when making teaching plans and studying the syllabus.

Reference

[1] BELL F G. Engineering geology[M]. 2nd ed. Amsterdam: Elsevier, Butterworth-Heinemann, 2007.

[2] CUI Guanying, ZHU Jixiang. Hydraulic engineering geology[M]. 4th Ed. Beijing: China Water and Power Press, 2014. (In Chinese)

[3] GOODMAN R E. Engineering geology-rock in engineering construction[M]. Chichester: John Wiley & Sons, 1993.

[4] GU Dezhen. Fundamentals of engineering geology mechanics of rock mass[M]. Beijing: Science Press, 1979. (In Chinese)

[5] The International Association for Engineering Geology and the Environment (IAEG). The international association for engineering geology and the environment 50 years[M/OL]. Beijing: Science Press, 2014[2021-12-07]. https://www.iaeg.info/book/#.

[6] OLIVEIRA R. Geologia de Enegnharia. Génese, evolução e situação actual[J/OL]. Revista Luso-Brasileira de Geotecnia, 2010, 118, 3-13[2021-12-07]. https://vdocuments.net/reader/full/revista-geotecnia-118.

[7] RENGERS N, BOCK H, MANOLIU I, RADULESCU N. Competency-orientated curricula

development in Geo-engineering with particular reference to Engineering Geology[C]// MANOLIU I, RADULESCU N. International Conference on Education and Training in Geo-Engineering Sciences, Constantza, Romania, 2-4 June 2008. 1st ed. London: CRC Press, Taylor & Francis Group, 2008: 101-110.

[8] SHI Bin, YAN Changhong. Engineering geology[M]. Beijing: Science Press, 2017. (In Chinese)

[9] SUN Guangzhong. Rock mass structural mechanics[M]. Beijing: Science Press, 1988. (In Chinese)

[10] TANG Huiming. Fundamentals of engineering geology[M]. Beijing: Chemical Industry Press, 2008. (In Chinese)

[11] ZARUBA Q, Mencl V. Engineering geology[M]. Amsterdam: Elsevier, 1976.

[12] ZHANG Xiangong, WANG Sijing, LI Zhiyi. Introduction to engineering geology[M]. Beijing: Seismological Press, 2005. (In Chinese)

2
Minerals and Rocks

2.1 Introduction

The Earth is an ellipsoid that moves along a specific orbit in the universe, and its outer layer is surrounded by atmosphere and water. The only rocks geologists can study directly in place are those of the crust, and Earth's crust is nothing more than a thin skin of rock, making up less than 1% of Earth's total volume. Mantle rocks brought to the Earth's surface basalt flows and diamond-bearing gabbro pipes, as well as the tectonic adherence of lower parts of the oceanic lithosphere to the continental crust, give geologists a glimpse of what the underlying mantle is. Meteorites also provide clues about the possible composition of the core of Earth. But to learn more about Earth's deep interior, geologists must study it indirectly, mainly by using the tools of geophysics—that is, seismic waves and the measurement of gravity, heat flow, and Earth magnetism.

The evidence from geophysics suggests that Earth is divided into three primary compositional layers:

(1) the crust on Earth's surface;

(2) the rocky mantle beneath the crust;

(3) the metallic core at the center of Earth.

The plate tectonics study has shown that the crust and uppermost mantle can be further divided into the brittle lithosphere and the ductile or plastic asthenosphere. The study of seismic refraction and reflection enabled scientists to plot the three main zones of the Earth's interior (Figure 2-1). The crust is the external layer of rock, which forms a thin coat on Earth's surface. Below the crust lies the mantle, a chunky shell of rock that separates the crust above from the core below. The core is the cardinal zone of the Earth, and it is probably metallic and the source of Earth's magnetic field.

Studies of seismic waves have shown: (1) that the crust is thinner beneath the oceans than beneath the continents (Figure 2-2); (2) that seismic waves travel faster in the oceanic crust than in the continental crust. Because of this velocity difference, it is assumed that the two types of crust are made up of different kinds of rock.

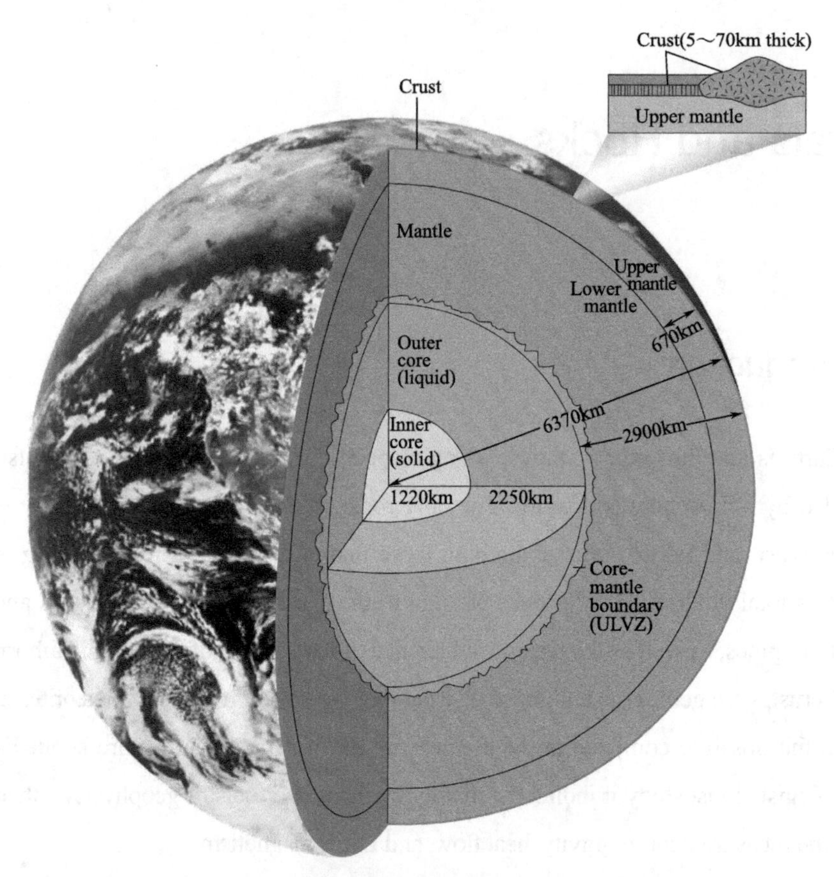

Figure 2-1 Earth's interior. Seismic waves show the three main divisions of Earth: the crust, the mantle, and the core (Photo by NASA)

Seismic P-waves cross through the oceanic crust at about 7km/s, which is also the speed they travel through basalt and gabbro. Samples of rocks taken from the seafloor by oceanographic ships confirm that the upper part of the oceanic crust is basalt and indicate that the lower part is gabbro. The oceanic crust averages 7km in thickness, varying from 5 to 8 km (Table 2-1).

Table 2-1 Characteristics of oceanic crust and continental crust

	Oceanic Crust	Continental Crust
Average thickness	7km	20~70km (thickest under mountains)
Seismic P-wave velocity	7km/s	6km/s (higher in lower crust)
Density	3.0gm/cm^3	2.7gm/cm^3
Probable composition	Basalt underlain by gabbro	Granite, other plutonic rocks, schist, gneiss (with sedimentary rock cover)

Seismic P-waves travel more slowly through the continental crust—about 6km/s, the same speed they move through granite and gneiss. Continental crust is commonly called "granitic," but the term should be put in quotation marks because most rocks exhibited on land are not granite. The continental crust is greatly varied and complex, consisting of a crystalline basement made up of granite, other plutonic rocks, gneisses, and schists, all capped by a sheet of sedimentary rocks. Since a single rock term cannot properly describe the crust that varies immensely in make-up, some geologists use the term felsic (rocks high in feldspar and silicon) for the continental crust and mafic (rocks high in magnesium and iron) for oceanic crust.

Continental crust is much thicker than oceanic crust, averaging 30 to 50km in thickness, though it varies from 20 to 70km. Seismic waves show that the crust is thickest under geologically young mountain ranges, such as the Andes and Himalayas, bulging downward as a mountain root into the mantle (Figure 2-2). The continental crust is also less dense than the oceanic crust, which is important in plate tectonics (Table 2-1).

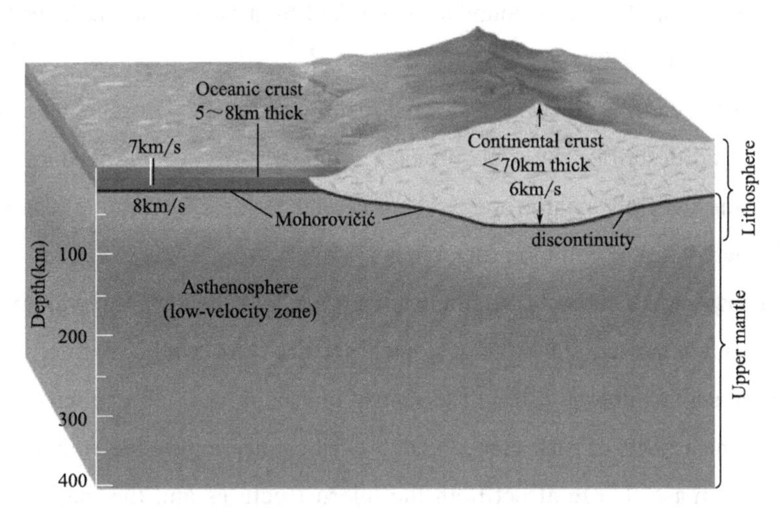

Figure 2-2 Thin oceanic crust has a P-wave velocity of 7km/s, whereas thick continental crust has a lower velocity. Mantle velocities are about 8km/s. The oceanic and continental crust and the upper mantle build the lithosphere. The asthenosphere underlies the lithosphere and is prescribed by a decrease in P-wave velocities

The boundary that separates the crust from the mantle beneath is called the Mohorovicic discontinuity (Moho for short). Note from Table 2-2 that the mantle lies closer to Earth's surface beneath the ocean than beneath continents. The idea behind an ambitious program called Project Mohole (begun during the early 1960s) was to use specially equipped ships to drill through the oceanic crust and obtain samples from the mantle. Although the project was abandoned because of high costs, ocean-floor drilling has become routine since then, but not the great depth necessary to sample the mantle. Perhaps the original concept of drilling to the mantle through oceanic crust will be revived in the future.

Because of the way seismic waves pass through the mantle, geologists think that it, like the crust, is made of solid rock. The melted rock's localized magma chambers may occur as isolated pockets of liquid in both the crust and the upper mantle, but most of the mantle seems to be solid. Because P-waves travel at about 8km/s in the upper mantle, it appears that the mantle is a different type of rock from either the oceanic crust or continental crust. The best hypothesis that geologists can make about the composition of the upper mantle is that it consists of ultramafic rock such as peridotite. Ultramafic rock is a dense igneous rock made up chiefly of ferromagnesian minerals such as olivine and pyroxene. Some ultramafic rocks contain garnet, and feldspar is extremely rare in the mantle.

The crust and uppermost mantle jointly form the lithosphere, the external casing of the Earth that is relatively vigorous and brittle. The lithosphere constitutes the plates of plate-tectonic theory. The lithosphere averages about 70km thick underneath oceans and maybe 125~250km thick beneath continents. Its lower boundary is marked by a quaint mantle layer where seismic waves slow down (Figure 2-2).

The crust of the earth is composed of three major rock types. This chapter covers the three groups of rocks and describes the rocks in each group. It provides an account of the various minerals, including their physical properties and crystal forms. The chapter also elucidates the identifying characteristics of common rock types.

Rock composition is a subject commonly covered under petrology, mineralogy, which forms an essential part of the curriculum in the geological sciences (Attewell & Farmer, 1976). Detailed addressing of the subject area is difficult to justify in a book devoted explicitly to engineering geology. Still, some aspects of rock makeup have a critical bearing on the engineering properties of rocks and rock mass. These affect both the microstructures and the massive forms of the rocks. The essential belongings derive from the specific physical properties of some minerals, particularly some clay minerals, which can absorb vast quantities of water and have structures that tend to offer the material anisotropic. In the current chapter, the composition of rocks and the basic physical properties and texture of the more essential minerals are considered and defined.

2.2 Rock-forming minerals

2.2.1 What is a mineral?

Most rocks (and rock particles) comprise a mechanically-bonded aggregate of mineral particles. A mineral is a naturally occurring solid formed by geologic processes with a crystalline structure and a definable chemical composition. Almost all minerals are inorganic, and rocks are mixtures of minerals, and most rocks are composed of at least two minerals. Mineralogy is the study of chemistry, atomic structure, physical properties, and the genesis of minerals.

The International Mineralogical Association (IMA) provides the following definition for a mineral: "A mineral is an element or chemical compound that is normally crystalline, and that has been formed as a result of geological processes." According to IMA, there are currently more than 4000 known minerals that occur in various types of rocks. These minerals are formed in varied environments and different chemical and physical conditions such as pressure and temperature. The leading group of mineral formations is found in igneous rocks, in which minerals occur as crystalline products formed from the magmatic melt. In sedimentary rocks, minerals are the result of weathering, erosion, and sedimentation processes. In metamorphic rocks, new minerals are formed at the expense of the existing ones due to increasing temperature and pressure or both on some pre-existing rock types. Minerals are also formed by a hydrothermal process in which they are chemically precipitated from the solution. However, in all three types of rocks, the minerals are interwoven with rocks as interlocking fabrics. Hydrothermal solutions tend to follow fracture zones with open spaces that develop minerals with suitable crystal forms.

2.2.2 Crystal geometry of minerals

The crystal form of a mineral is the outward expression of the internal molecular structure, which is dependent on the state of equilibrium due to the interatomic forces. Crystal geometry or crystal morphology deals with the symmetry, faces, and forms of a crystal. A crystal possesses the following three types of symmetry shown in Figure 2-3.

(1) Plane of symmetry: It divides a crystal into two halves such that one half is the mirror image of the other. A crystal may have one or many planes of symmetry or none.

(2) Center of symmetry: It is the central point about which every face and edge of a crystal is matched by one parallel to it on the opposite side of the crystal.

(3) Axis of rotation symmetry: It is a line or an axis through a crystal about which the crystal can be rotated to bring it into an identical position several times in the course of one revolution.

1. Crystal system

The crystal system is a category of crystals regarding the position of their crystal faces and

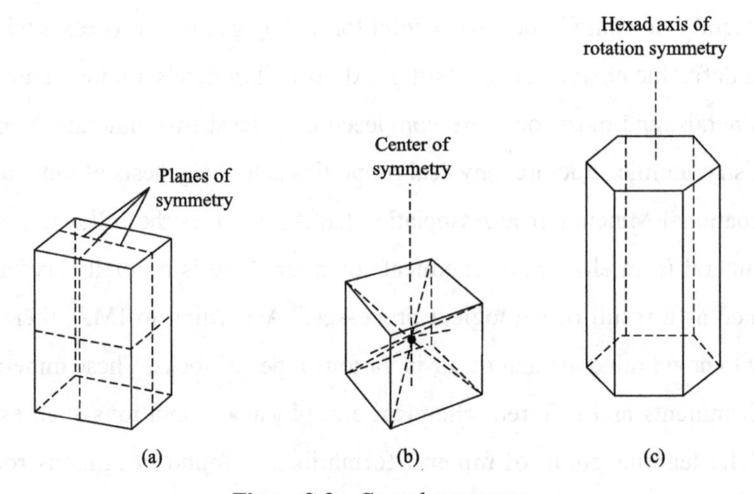

Figure 2-3 Crystal symmetry

(a) plane of symmetry;(b) center of symmetry; (c) axial of rotation symmetry

the relationship of the intercepts that the planes containing the faces make with three (or four) axes intersecting at an origin (Figure 2-4). In all, 32 combinations of symmetry elements are recognized under seven crystal systems depending upon the possible planes of symmetry and axes of symmetry. They are as follows:

(1) Cubic (isometric): Three orthogonal tetrad axes of equal length, a_1, a_2, a_3;

(2) Tetragonal: Three orthogonal axes, two horizontal diads of equal length and one vertical tetrad, a_1, a_2, c;

(3) Orthorhombic: Three orthogonal diad axes of unequal lengths, a, b, c;

(4) Hexagonal: Four axes, three horizontal diads of equal length 120° apart and one vertical hexad at right angles, a_1, a_2, a_3, c;

(5) Trigonal: Four axes, three horizontal diads of equal length 120° apart and one vertical triad at right angles, a_1, a_2, a_3, c;

(6) Monoclinic: Three unequal axes, one vertical, one horizontal diad, and a third making an oblique angle with the plane containing the other two, a, b, c;

(7) Triclinic: Three unequal axes, none at right angles, a, b, c.

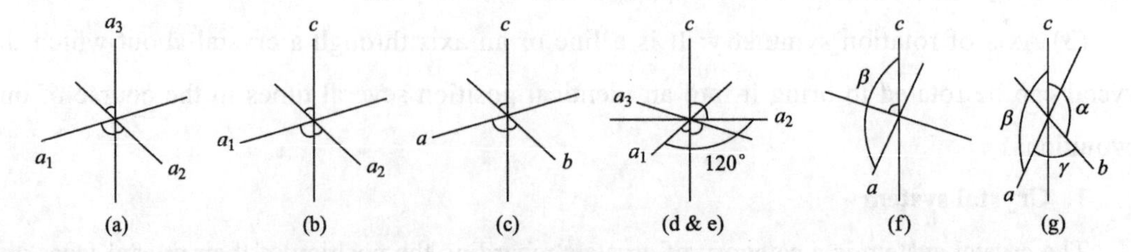

Figure 2-4 Crystal systems showing intercepts of the crystal faces with three or four axes at the origin

Different crystal forms can be referred to as the same set of crystal axes and hence belong to the same crystal system. For example, cube, rhomb dodecahedron, and octahedron are different crystals of the same cubic system.

Each mineral crystallizes in one of the seven crystal systems. Figure 2-5 shows the outlines of these crystal forms with names of the common minerals, that is, isometric (fluorite), tetragonal (zircon), hexagonal (beryl), trigonal (tourmaline), orthorhombic (olivine), monoclinic (hornblende), and triclinic (albite/feldspar). A mineral showing formation of crystals is termed crystalline. If no crystal of any geometric shape is visible in the constituent minerals of rock in its thin section under the microscope, it is called amorphous. If only some tiny crystals are detected in the thin rock section under the microscope, it is termed cryptocrystalline (Subinoy, 2013).

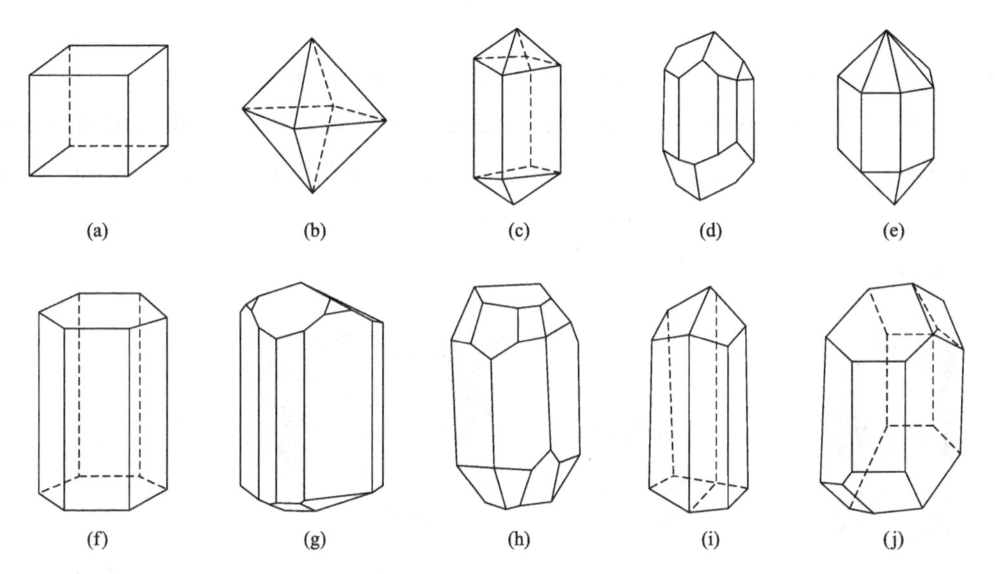

Figure 2-5 Different systems of crystal forms with a representative mineral of each system
(a) & (b) isometric (fluorite); (c) & (d) tetragonal (zircon); (e) & (f) hexagonal (beryl); (g) trigonal (tourmaline); (h) orthorombic (olivine); (i) monoclinic (hornblende); (j) triclinic (albite)

2. Twinning of crystal

Two or more crystals of the same species can occur intergrown or in contact together, remarkably symmetric; such crystals are called twins. Twinning is found in different crystals of many minerals such as calcite, fluorite, gypsum, feldspar, staurolite, and rutile (Figure 2-6).

2.2.3 Form

1. Habit

If a mineral specimen is well crystallized, meaning that it shows well-developed crystal faces, then the crystal form can be used to help identify the mineral, and we can say that the

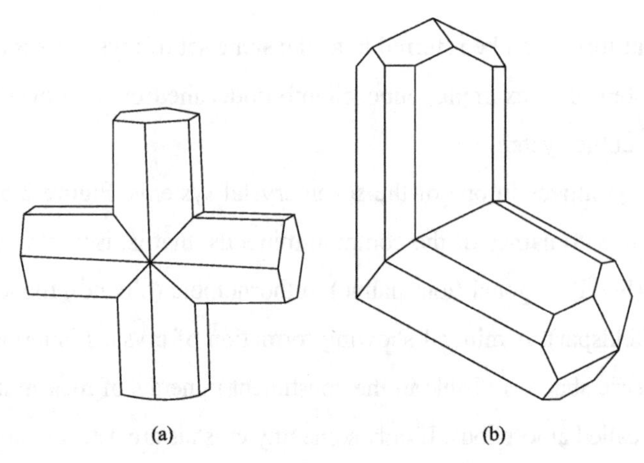

Figure 2-6 Twinning of crystals

(a) staurolite; (b) rutile

crystal form is diagnostic in our mineral identification. The external crystal form is an outward expression of the ordered internal atomic arrangement (Klein & Philpotts, 2017). Here we concentrate on properties that can be assessed without specific knowledge of crystallography. Different adjectives describe various habits, and several of these are described here and illustrated with photographs in Figure 2-7.

Figure 2-7 Photographic illustrations of many of the mineral habits

(a) Prismatic, as in diopside crystals; (b) Columnar, as in a pink tourmaline crystal; (c) Acicular (or needlelike), as in this tufted aggregate of mesolite fibers (mesolite is a member of the zeolite group of minerals); (d) A tabular crystal of barite; (e) A bladed crystal of kyanite; (f)Fibrous celestite; (g) A dendritic pattern of manganese oxide minerals on a flat surface of siltstone; (h) Foliated muscovite as a rosette of crystals

(1) Prismatic means that the mineral has an elongate habit with the bounding faces forming a prismlike shape, as is common in members of the pyroxene and amphibole groups of silicates.

(2) Columnar exhibits rounded columns, as is common in tourmaline.

(3) Acicular means "needlelike," as is common for natrolite, a member of the zeolite group.

(4) Tabular describes crystal masses that are flat like a board, as commonly seen in barite.

(5) Bladed refers to crystal shapes that are elongate and flat, as in a knife blade. Kyanite shows this commonly.

(6) Fibrous refers to threadlike masses, as exhibited by chrysotile, the most common mineral included in the commercial term asbestos.

(7) Dendritic describes minerals that show a treelike branching pattern, as is common in manganese oxide minerals. The term is derived from the Greek word dendron, meaning "tree."

(8) Foliated refers to a stack of thin leaves or plates that can be separated from each other, as in mica and graphite.

(9) Massive describes a mineral specimen that is totally devoid of crystal faces.

2. State of aggregation

Most mineral specimens, unless unusually well crystallized, appear as aggregates of smaller grains. Some additional adjectives best describe such occurrences (photographs of the appearance of several of these are given in Figure 2-8).

Figure 2-8 Illustrations of many of the states of aggregation

(a) Granular grains of yellow-green olivine in a volcanic bomb of dunite (a rock type consisting essentially of olivine); (b) Finely banded as in this polished slab of agate; (c)Botryoidal chalcedony is also known as grape agate; (d) Botryoidal goethite; (e) Reniform hematite; (f) A geode with an outer lining of slightly blue chalcedony and a cavity lined with purple quartz crystals; (g) Oolitic limestone; (h) Pisolitic bauxite

(1) Granular applies to rock and mineral specimens consisting of approximately equal-sized mineral grains, as in the rock dunite, composed essentially of granular olivine grains.

(2) Compact describes a specimen that is so fine-grained that the state of aggregation is not

apparent, as in samples of clay minerals or chert (flint).

(3) Banded is said of a mineral specimen that shows bands of different colors or textures, but that may or may not differ in mineral composition. A banded agate may show various differently colored bands, but each band is composed of the same silica, SiO_2, known as chalcedony. In contrast, banded iron-formations commonly show banding, on a millimeter scale, of two or three different minerals, chert (light-colored) and hematite (red), and/or magnetite (black).

(4) Mammillary is from the Latin word mamma, meaning "breast," and describes minerals that occur as smoothly rounded masses resembling portions of spheres. Examples are goethite and hematite.

(5) Botryoidal is from the Greek word botrys, meaning "band or cluster of grapes" or "having a surface of spherical shapes." In a botryoidal appearance, the rounded prominences are generally smaller than those described as mammillary. Familiar as the outer surface of chalcedony, a microcrystalline variety of quartz, SiO_2.

(6) Reniform describes the surface of a mineral aggregate that resembles that of a kidney. Seen in hematite. Derived from the Latin word renis, meaning "kidney."

(7) Stalactitic is a term used for a mineral that is made up of forms like small stalactites. Some limonite occurs this way, as does rhodochrosite.

(8) Geode is a rock cavity partly filled with minerals. Standard mineral fill is well-crystallized quartz or the purple variety of quartz, amethyst. Agates may be partly or wholly filled with rock cavities, and they may show attractive color banding in the chalcedony, microcrystalline SiO_2 that lines the outer cavity.

(9) Oolitic describes the occurrence of mineral grains in rounded masses the size of fish roe. Derived from the Greek word oön, meaning "egg." This structure may occur in iron ore, made up mainly of hematite, known as oolitic iron ore, or in sedimentary rocks known as oolitic limestone, consisting of millimeter-sized spheres of rhythmically precipitated calcite.

(10) Pisolitic applies to rounded mineral grains the size of a pea. From the Greek word pisos, meaning "pea." Bauxite, a primary ore of aluminum, is commonly pisolitic. Pisolitic grains are larger than those described as oolitic.

2.2.4 Physical properties of minerals

All minerals are recognized by their physical characteristics. It includes color, luster, form, hardness, cleavage, fracture, tenacity, and specific gravity. Not all of these properties would necessarily be needed to identify anyone mineral; two or three of them taken together may be sufficient, apart from optical properties. Other characters such as fusibility, fluorescence, magnetism, and electrical conductivity are also helpful in some cases as means of identification. They will be referred to as they arise in the descriptions of mineral species. In a few instances, taste (e.g., rock-salt) and touch (e.g., talc, feels soapy) are valuable indicators.

1. Specific gravity

It is easy to tell that a brick is heavier than a loaf of bread just by hefting each of them. The brick has a higher density, weight per given volume, than the bread. Density is commonly expressed as specific gravity, the ratio of the mass of a substance to the mass of an equal volume of water.

Liquid water has a specific gravity of 1 (Ice, being lighter, has a specific gravity of about 0.9) Most common silicate minerals are about two and a half to three times as dense as equal volumes of water: quartz has a specific gravity of 2.65; the feldspars range from 2.56 to 2.76. Unique scales are needed to determine specific gravity precisely. However, a person can easily distinguish by hand very dense minerals such as galena (a lead sulfide with a specific gravity of 7.5) from the much less dense silicate minerals.

Gold, with a specific gravity of 19.3, is much denser than galena. Because of its high density, gold can be collected by "panning". While the lighter clay and silt particles in the pan are sloshed out with the water, the gold lags in the bottom of the pan.

2. Color

It indicates the appearance of the minerals in reflected light or transmitted light for translucent minerals. Many minerals show different types of body colors. Quartz, for instance, is commonly white, but it may also be grey, yellow, or red. Iridescence is the play of colors due to surface or internal interference. Labradorite (a feldspar of plagioclase group) exhibits internal iridescence, whereas sphalerite often shows the surface effect.

3. Streak

It refers to the color of the mineral powder, which is different from the body color. It is the color produced when a mineral is rubbed against unglazed porcelain.

Some minerals have a distinctive color, for example, the green color of chlorite, but most naturally occurring minerals contain traces of substances that modify their color. Thus quartz, is colorless when pure, maybe white, grey, pink, or yellow when certain chemical impurities or included particles are present. Much more consistent is the color of a mineral in the powdered condition, known as the streak. This may be produced by rubbing the mineral on a piece of unglazed porcelain, called a streak plate, or other rough surfaces.

Streak is practical, for example, in distinguishing the various oxides of iron - hematite (Fe_2O_3)

gives a red streak, limonite (hydrated Fe_2O_3) a brown, and magnetite (Fe_3O_4) a grey streak (Blyth et al. 2005).

4. Luster

Luster is the appearance of a mineral surface in reflected light. It may be described as metallic, as in pyrite or galena, glassy or vitreous, as in quartz; resinous or greasy, as in opal; pearly, as in talc; or silky, as in fibrous minerals such as asbestos and satin-spar (fibrous gypsum). Minerals with no luster are described as dull.

5. Cleavage

Many minerals tend to split effortlessly in specific regular directions and yield smooth plane surfaces called cleavage planes when thus broken. These directions depend on the arrangement of the atoms in a mineral and are parallel to definite crystal faces. Perfect, good, distinct, and imperfect are terms used to describe the quality of mineral cleavage. Mica, for example, has a perfect cleavage through which it can be split into very thin flakes; feldspars have two sets of good cleavage planes, and Calcite has three directions of cleavage.

6. Fracture

A fracture is developed in a mineral by a sharp blow with a hard object. The resultant surface is characteristic of the mineral and maybe conchoidal, agged, fibrous, or irregular. Thus, quartz develops a conchoidal fracture that appears like a concave surface. The nature of a broken surface of a mineral is known as a fracture, the break being irregular and independent of cleavage. It is sometimes characteristic of a mineral and, also, a fresh fracture shows the actual color of a mineral Fracture is described as conchoidal, when the mineral breaks with a curved surface, e.g., in quartz and flint; as even, when it is nearly flat; as uneven, when it is rough; and as hackly when the surface carries small sharp irregularities (Figure 2-9). Most minerals show an uneven fracture.

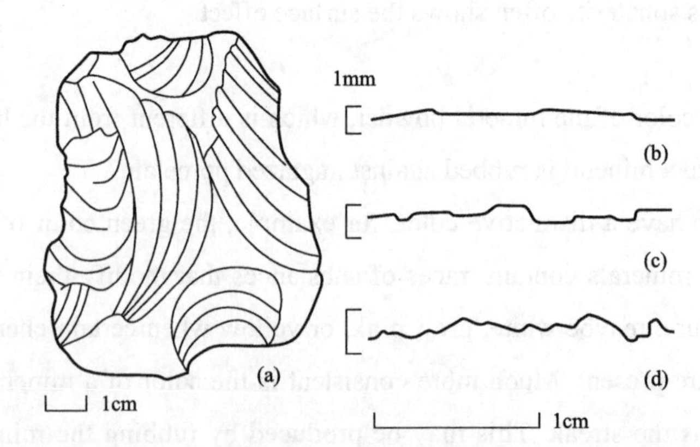

Figure 2-9 Mineral fracture

(a) Conchoidal, which can occur on a smaller scale than shown; (b) Even; (c) Uneven; (d) Hackly

7. Hardness

Hardness, or abrasion resistance, is measured relative to a standard scale, often minerals, known as Mohs' scale of hardness (Table 2-2). These minerals are chosen so that their hardness increases in the order 1 to 10. Hardness is tested by attempting to scratch the minerals of the scale with the specimen under examination. For example, a mineral that scratches calcite, but not fluorspar, is said to have a hardness between 3 and 4, or $H = 3\sim4$. Talc and gypsum can be scratched with a fingernail, and a steel knife will cut apatite (5) and perhaps feldspar (6), but not quartz (7). Soft glass can be scratched by quartz. The hardness test, in various forms, is simple, easily made, and valuable; it is a ready means for distinguishing, for example, between quartz and calcite.

Table 2-2 Mohs' scale of hardness

Hardness	Minerals	Character
1	Talc	Soapy to greasy feel
2	Gypsum	Can be scratched by finger nails
3	Calcite	Can be scratched by a copper coin
4	Fluorspar	Can be scratched by a penknife
5	Apatite	Can be scratched by window glass
6	Orthoclase feldspar	Can be scratched by a steel file
7	Quartz	Equivalent hard object porcelain
8	Topaz	Can be scratched by corundum but not by quartz
9	Corundum	Cannot be scratched by any mineral except diamond
10	Diamond	Cannot be scratched by corundum

This scale is based on the relative abrasiveness of minerals, increasing orders being determined in terms of the ability of one mineral to scratch another. It is a helpful scale for determining the relative toughness of minerals but might seem to have minor use in determining the mechanical properties of rocks since strength does not always correlate directly with hardness. On the other hand, the Schmidt hammer hardness test is being increasingly used for rapidly indexing intact rock via a correlation with its compressive or tensile strength.

Indentation values are obtained by an instrument that measures the force necessary to make a slight indentation into a substance.

Rather than carrying samples of the ten standard minerals, a geologist doing fieldwork usually relies on common objects to test for hardness (Figure 2-10). A fingernail normally has a hardness of approximately 2.5. If you can scrape the smooth surface of a mineral with your fingernail, the mineral's hardness must be less than 2.5. A copper coin has a hardness between

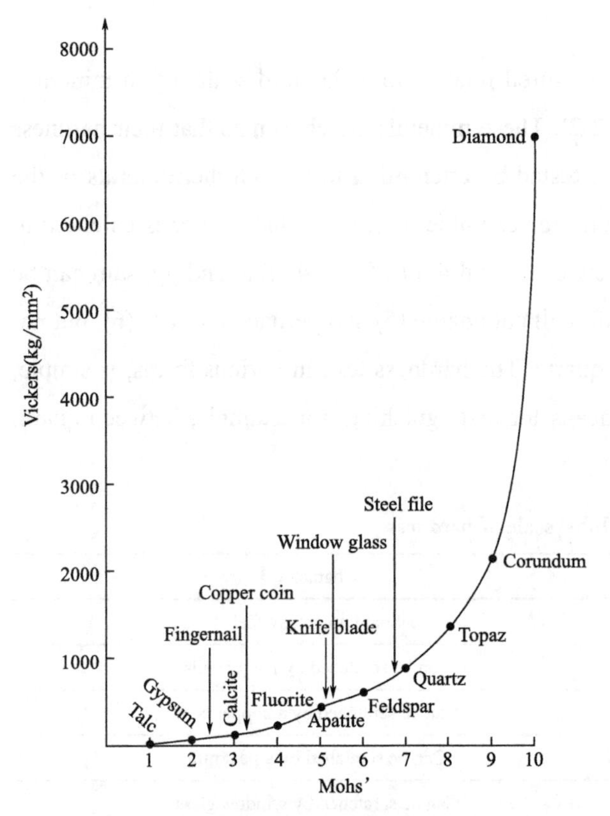

Figure 2-10 Mohs' hardness scale plotted against actual hardness values (kg/mm^2)

3 and 4; however, most pennies' brown, oxidized surface is much softer, so check for an identation into the coin. A knife blade or a steel nail usually has a hardness a little greater than 5, but it depends on the particular steel metal used. A geologist uses a knife blade to differentiate between softer minerals, such as calcite and similarly appearing harder minerals, such as quartz. Ordinary window glass, usually somewhat harder than a knife blade (although some glass, such as that including lead, is much softer), can be used in the same manner as a knife blade for hardness tests. A file (one made of tempered steel for filing metal) can be used in a hardness of between 6 and 7. A porcelain plate also has a hardness of roughly 6.5.

A team of materials scientists from Yanshan University in China has discovered the critical proportion of crystallized and amorphous carbon needed to create glass with remarkable properties that won't weaken under intense pressure. This new type of glass can even scratch diamonds (Zhang et al.,2020).

8. Other properties

Taste and magnetic properties are a diagnosis of a few minerals. Mineral associations are also used: some minerals commonly occur together, whereas others are never instituting together because they are unstable as a chemical combination and would react to create another mineral.

Nearly all identification of minerals in hand specimens in the field are made with the condition that the specimen being examined is not a rare mineral, but is one of a dozen or so ordinary rock-forming minerals, or one of a couple of dozen minerals commonly found in the sheetlike veins that cut rocks (McLean & Gribble, 2005). The distinction between common quartz and one particular rare mineral in a hand specimen is minor and readily missed, but identification mistakes are presumably as rare as the mineral. Using such simple techniques can only be favorable in certain circumstances, such as differentiating feldspar from a fine-grained rock containing feldspar.

Three or four properties are usually sufficient for particular mineral identification, and there is little point in determining the others. For example, a mineral with a metallic luster, three cleavages all at right angles, gray color, and a black streak is almost certainly the ordinary lead ore, galena.

2.2.5 Minerals and rock

All rocks are composed of minerals, homogeneous substances with fixed chemical composition, crystal forms, and other distinctive characteristics such as color, luster, and hardness. The chemical elements and the crystal forms determine the properties of the minerals, and the chemical elements depend on the composition of the rock in the liquid state.

It is essential to keep the distinction between elements, minerals, and rocks clear when learning geology. Rocks are composed of minerals, and minerals are composed of atoms of elements bonded together in an orderly crystalline structure.

Rocks are defined as naturally formed aggregates of minerals or mineral-like substances. The granite in Figure 2-11, therefore, is a rock that is made up of the minerals quartz, plagioclase

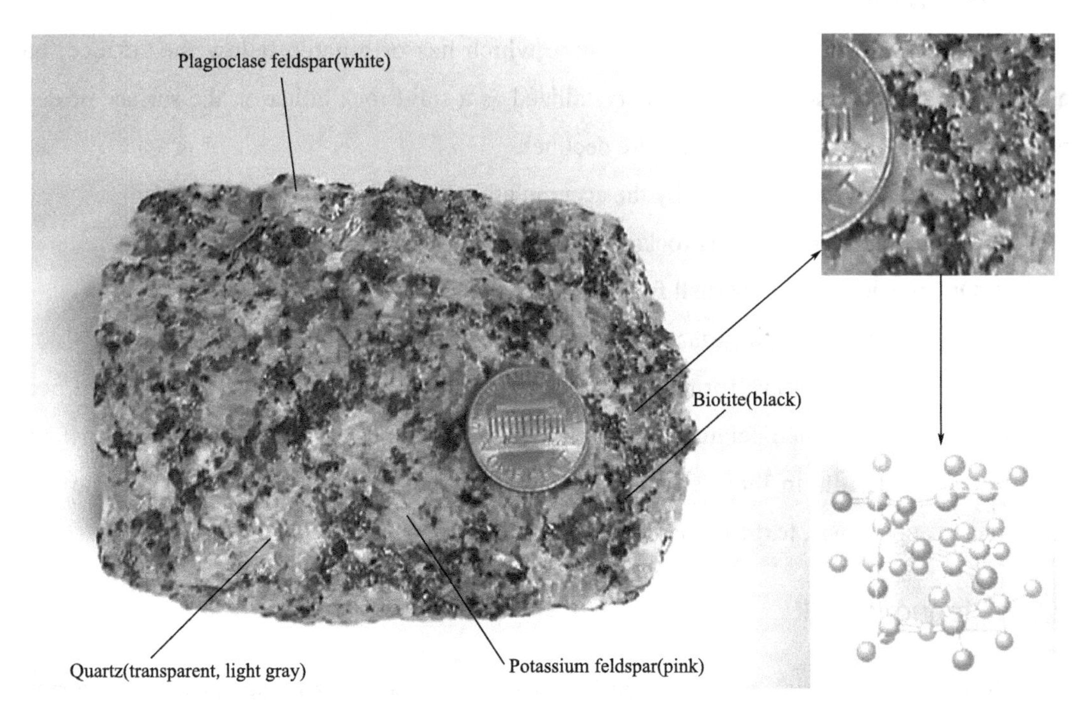

Figure 2-11 The rock granite comprises of the minerals quartz, potassium feldspar, plagioclase feldspar, and biotite mica. The mineral quartz (SiO_2) is made up of atoms of the elements silicon and oxygen bonded together

feldspar, potassium feldspar, and biotite. A rock can also be composed of a single mineral. For example, limestone is composed of the mineral calcite. The reason that limestone is a rock and not defined simply as the mineral calcite is that the limestone is made up of multiple crystals of calcite either grown in an interlocking pattern or cemented together. Although limestone is made up of a single mineral type, it is still an aggregate of many mineral grains. Some rocks can be comprised of nonmineral substances. For example, coal is made of partially decomposed organic matter. Obsidian is made of silica glass which is not crystalline, and therefore not a mineral.

2.3 Rocks

2.3.1 The nature of rocks

Rocks are aggregates of one or more minerals. The properties of the rock are defined by the minerals in it (particularly those key minerals, which individually make up more than 95% of its volume) and by how the minerals are disposed of relative to each other (that is, the texture of the rock). An individual rock type or specimen is always depicted in terms of its mineral composition and texture, and both are used to classify rocks. According to their manner of formation, rocks are of three main types:

Igneous rocks are formed from magma, which has originated below the surface, has ascended towards the surface, and has crystallized as a solid rock either on the surface or deep within the Earth's crust as its temperature declines.

Sedimentary rocks are formed by the accumulation and compaction of:

(1) Fragments of pre-existing rocks that have been shattered by erosion.

(2) Organic debris such as shell fragments or dead plants.

(3) Material dissolved in surface waters (rivers, oceans, etc.) or groundwater.

Metamorphic rocks are formed from pre-existing rocks of any type, which have been endured to increases of temperature (T) or pressure (P), or both, such that the rocks transform. This transform results in the metamorphic rock being different from the original parental substance in appearance, texture, and mineral composition.

2.3.2 Igneous rock

Molten rock material, generated within or below the Earth's crust, reaches the surface from time to time and flows out from volcanic orifices as lava. Similar material may be injected into the rocks of the crust, giving rise to a variety of igneous intrusions which cool slowly and solidify; many of which were formed during the past geological ages are now exposed to view after the removal of their covering rocks by denudation. These solidified lavas and intrusions constitute igneous rocks. The molten material from which igneous rocks have solidified is called magma. Natural magmas are hot, viscous siliceous melts. The chief elements present are silicon and oxygen and the metals potassium, sodium, calcium, magnesium, aluminum, and iron (in the order of their chemical activity). These main constituents are small amounts of many other elements and gases such as CO_2, SO_2, and H_2O. Magmas are thus complex bodies, and the rocks derived from them have a wide variety of compositions. When cooled quickly, magma solidifies as a rock-glass without crystals, while rock-forming minerals crystallized from it when they cooled slowly.

1. Forms of extrusive igneous rocks

Extrusive igneous rocks usually occur in cups, plates, and basins of volcanic lava (Figure 2-12). Volcanic necks are the product of an old volcanic rim composed of several layers (outflow) of solidified lava. This occurs when the acidic, highly viscous, and therefore poorly formed mobile lava solidifies around a volcanic crater forms a conical hill or dome. A volcanic plate or lava cover presents a body of extensive propagation and relatively small thickness, formed in spout or outbursts of low viscous voluble lava on a large area around the volcano. The basin (flow) of lava is formed by cooling lava flows like a fiery river pouring down the volcano's slopes. Volcanic rocks are also formed by outpourings of lava on the seabed, typically within the volcanic mass. The pulsating pouring of lava and mixing with seawater create a spherically or cushion shape-structure, known as "pillow lavas".

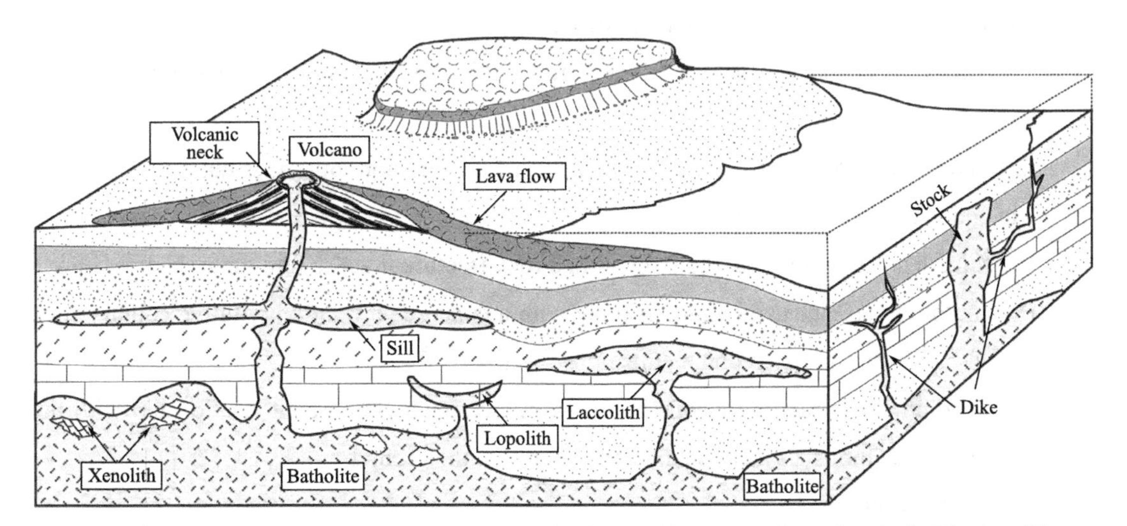

Figure 2-12 Conceptual diagram showing the major forms of igneous rocks such as batholiths, lopolith, xenolith, laccolith, sill, dyke, stock, volcano, volcanic neck and lava flow (After Haldar & Tišljar,2013)

2. Forms of intrusive (plutonic) igneous rocks

It is established that most intrusive igneous rocks are formed by the cooling and crystallization of magma at depths of 1.5~20km.

The slow cooling of magma, deep in the lithosphere, under the surface of the Earth, created a vast body of intrusive igneous rocks of irregular shape, whose propagation is several thousand kilometers with an unknown base in depth. Such massive, intrusive bodies are called batholiths (Figure 2-12). There are often smaller or larger enclaves, xenoliths of surrounding rocks at the edges of a batholith, which are incorporated in the magma and partially altered or completely metamorphosed under the influence of high-temperature fluids from the magma. Stocks are smaller irregular bodies with 10km in maximum dimension and are associated with batholiths. The term massive refers to round and irregular intrusive body of larger size.

The batholith, stock, and massive occur by crystallization in the depths of the lithosphere can reach the Earth's surface by a variety of tectonic movements, erosion, and denudation processes. The batholiths, stocks, and massive granodiorite, diorite, peridotite, gabbro, and granite are often found on the Earth's surface or shallow depth.

Magma can be injected into the surrounding sedimentary rock layers, raising the layers above it, creating a more petite igneous body. The newly created body has a shape like a dome or mushroom and is well known as laccolith. The length of laccolith usually does not exceed a few hundred meters to several kilometers, similar to the lopolith, which is smaller, lenticular in shape with a depressed central region (Figure 2-12).

(1) Batholith, large plutons which, by definition, cover at least 100km^2. They are found in the core of tectonically deformed mountain belts.

(2) Sill, a sheet-like body formed by magma injection between parallel layers of pre-existing bedded rock.

(3) Dike, similar to sills in that they are a sheet-like intrusion, but dikes cut across bedding layers in country-rock are discordant.

(4) Vein, deposits of minerals found within a rock fracture that are foreign to the host rock.

(5) Laccolith is an intrusive igneous body with a concordant relation with the country-rock (native rock) and appears like a dome at its upper part.

(6) Lopolith, an intrusive igneous body that maintains a concordant relationship with the country-rock and exhibits a saucer-shaped appearance at the ground.

3. Composition

Many different minerals occur in igneous rocks, but only about eight (Table 2-3) are generally present as essential constituents of a rock. Which of the eight are present is controlled primarily by the composition of the magma. Each mineral starts to crystallize at a particular temperature and continues to form throughout a limited temperature range as the magma cools. More than one mineral is usually formed at any one time. Since the crystals formed early to have higher specific gravity than the remaining liquid of the magma, they settle downwards. Alternatively, the two fractions, crystals, and liquid may be separated by some other process. As time progresses, different minerals crystallize from the magma. Eventually, this process gives rise to a sequence of minerals of different compositions, from high temperature, high specific gravity minerals at the bottom to low temperature, low specific gravity minerals at the top. Such magma is said to be differentiated.

When consolidated, the highest temperature layer of minerals at the bottom, consisting of olivine, calcium-rich plagioclase, and often augite, will form an ultrabasic igneous rock. At relatively low temperatures, the last magmatic liquid solidifies into the rock containing quartz, orthoclase, sodium-rich plagioclase, and micas. Because of its relative richness in silica, it is called an acid rock (McLean & Gribble,2005). These relationships are given in Table 2-4.

Table 2-3 Mineral crystallization from a magma

Mineral	Ultrabasic	Basic	Intermediate	Acid
quartz			■	■
orthoclase			■	■
plagioclase	■ Ca-rich	■	■	■ Na-rich
muscovite				■
biotite			■	■
hornblende		■	■	■
augite	■	■	■	
olivine	■	■		
	early(high temperature)		late(low temperature)	

Note: ■ Contains

Table 2-4 Minerals present in the four main groups of igneous rock

Rock composition	Amount of SiO_2 (%)	Minerals
acid	65	quartz, orthoclase, Na-plagioclase, muscovite, biotite (±hornblende)
intermediate	55~65	plagioclase, biotite, hornblende, quartz, orthoclase (±augite)
basic	45~55	Ca- plagioclase, augitr(±olivine, ±hornblende)
ultrabasic	45	Ca- plagioclase, olivine(±augite)

Igneous rock names are based on texture (notably grain size) and mineralogical composition (chemical composition). Mineralogically (and chemically) equivalent rocks are granite-rhyolite, diorite-andesite, and gabbro-basalt. The relationships between igneous rocks are shown in Figure 2-13.

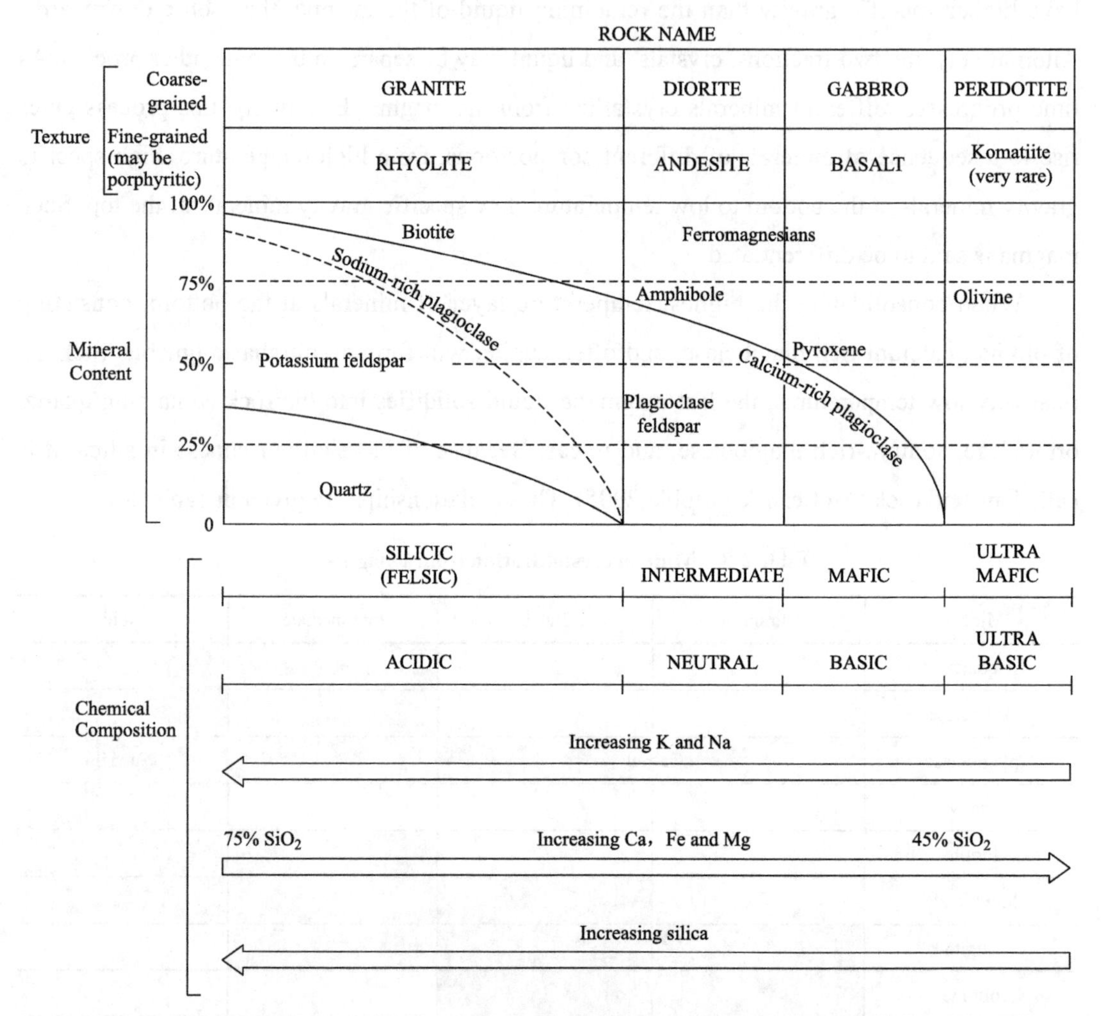

Figure 2-13 Classification chart for the most common igneous rocks (Modified from Carlson et al.,2011)

4. Texture

Texture refers to a rock's appearance with regard to the size, shape, and arrangement of its grains or other constituents. Most (but not all) igneous rocks are crystalline; that is, they are made of interlocking crystals (of, for instance, quartz and feldspar). The most significant aspect of texture in igneous rocks is grain (or crystal) size.

Extrusive rocks typically are fine-grained (Aphanitic) rocks, in which most of the grains are smaller than 1 millimeter. If they are crystals, the grains are small because magma cools rapidly at the Earth's surface, so they have less time to form. Some intrusive rocks are also fine-grained;

these occur as smaller bodies that solidified near the surface upon intrusion into relatively cold country rock (probably within a couple of kilometers of the Earth's surface). Basalt, andesite, and rhyolite are the common fine-grained igneous rocks.

Igneous rocks that formed at substantial depth—usually more than several kilometers—are called plutonic rocks (after Pluto, the Roman god of the underworld). In general, these rocks are coarse-grained, reflecting the slow cooling and solidification of magma. Coarse-grained (or coarsely crystalline) rocks are defined as those in which most grains are larger than 1 millimeter. The crystalline grains of plutonic rocks are commonly interlocked in a mosaic pattern (Figure 2-14). An extremely coarse-grained (grains over 5cm) igneous rock is called a pegmatite.

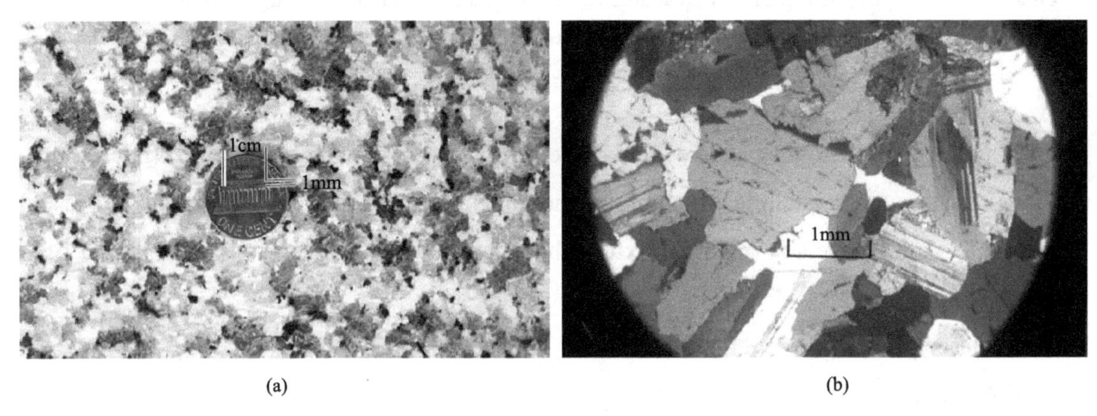

(a) (b)

Figure 2-14 The crystalline grains of plutonic rocks

(a) Coarse-grained texture feature of a plutonic rock. Feldspars are white and pink. Although this quartz is translucent, it appears light gray. Biotite mica is black;(b) A similar rock is seen through a polarizing microscope

The crystals or grains of most fine-grained rocks are considerably smaller than 1mm and cannot be distinguished by the unaided eye. So, for practical purposes, if you can discern the individual grains, regard the rock as coarse-grained; if not, consider it fine-grained.

Some rocks are porphyritic, and large crystals are enclosed in a groundmass of fine-grained crystals or glass. A chocolate bar containing almonds has the appearance of porphyritic texture. If the groundmass is fine-grained, extrusive rock names are used. For instance, Figure 2-15 shows porphyritic andesite. Porphyritic extrusive rocks are generally defined as having begun to crystallize slowly underground, followed by the eruption and rapid solidification of the remaining magma at the Earth's surface. Some porphyritic rocks have a coarse-grained groundmass in which the individual grains are over 1mm. The larger crystals enclosed in the groundmass are much bigger, usually two or more centimeters across. Porphyritic granite is an example.

Igneous rocks may be classified according to their chemical/mineral composition as felsic, intermediate, mafic, and ultramafic. By texture or grain size: intrusive rocks are coarse-grained

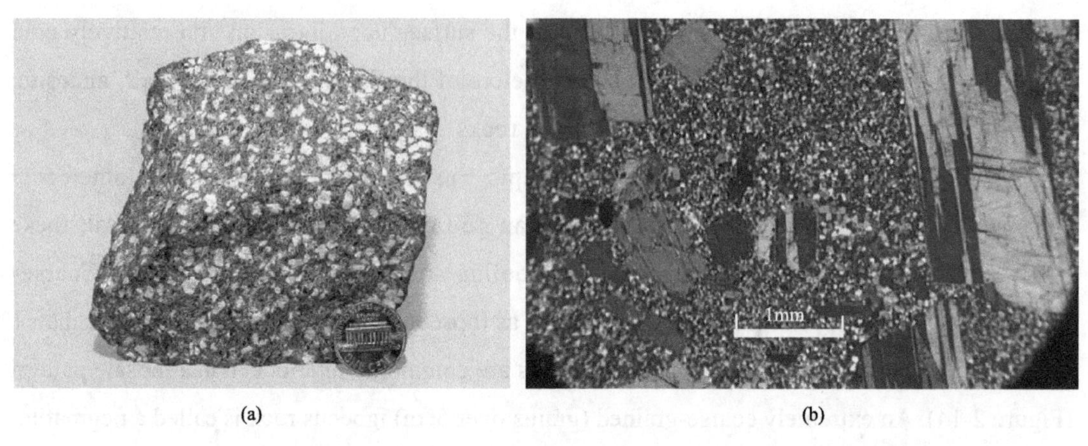

Figure 2-15 Porphyritic andesite. A great number of fine grains surround a few large crystals (phenocrysts)
(a) Hand specimen, Grains in the groundmass are too fine to see; (b) Photomicrograph (using polarized light) of the same rock, The black-and-white striped phenocrysts are plagioclase, and the green ones are ferromagnesian minerals

(all crystals are visible to the naked eye), while extrusive rocks may be fine-grained (microscopic crystals) or glass (no crystalline structure; i.e., no minerals). Volcanic rocks, especially felsic and intermediate, often have a porphyritic texture characterized by visible crystals floating in a fine-grained groundmass.

5. Description of typical igneous rocks

1) Granite

Granite is a medium-to-coarse grained, light-colored rock with white or pink tint according to the color of the feldspars. When examined under the microscope, the rock is found to essentially contain quartz, feldspar (orthoclase, plagioclase, and microcline), and ferromagnesian minerals such as hornblende and mica. The accessory minerals include apatite, rutile, zircon, and magnetite. Granite with large crystals of potash feldspar embedded in comparatively fine-grained groundmass is known as porphyritic granite. Granitic rock in which feldspar is abundant is called monzonite. If plagioclase content is more than orthoclase, it is known as granodiorite. The common name 'granite'is used in engineering geological description without making distinctions such as monzonite or granodiorite. The rock is rugged and durable and possesses high strength; it is suitable for construction material. Fresh granite outcrop provides excellent foundation conditions for engineering structures. However, when weathered and decomposed, porphyritic granite with large feldspars may cause weakness in the foundation (Subinoy, 2013).

Granites and granodiorites are the most familiar rocks of the plutonic association, and they are characterized by a coarse-grained, holocrystalline, granular texture. Although the term granite lacks precision, normal granite has been defined as a rock in which quartz forms more than 5% and less than 50% of the quarfeloids (quartz, feldspar, feldspathoid content), potash feldspar constitutes 50% to 95% of the total feldspar content, the plagioclase is sodi-calcic, and the mafites form more than 5% and less than 50% of the total constituents (Bell, 2007).

2) Pegmatite

Pegmatite is a very coarse-grained rock consisting mainly of feldspar, quartz, and mica. It occurs as a vein or dyke or as a large rock body intrusive into the granite. Minerals found in large crystals in pegmatite result from the cooling of residual magmatic fluid. Large books of muscovite used in the industry are obtained from pegmatite. The decomposition of feldspar in pegmatite forms kaolin, which is very soft and, when present in a foundation, is considered a weak feature. Aplite containing aggregates of sugary-grained quartz and feldspar are commonly associated with pegmatite.

3) Tuff and volcanic breccia

The ashes ejected with fragments from volcanoes are deposited on the surface, and consolidation of these materials by natural processes form the rock called tuff. It is a light-colored soft rock in which volcanic ash content is dominant. After mixing with many angular and uneven fragments from the surface, it forms volcanic breccia or agglomerate.

4) Syenite

Syenite is a medium-grained, light-colored, and even-textured rock consisting essentially of feldspar, hornblende, and mica, or pyroxene. This rock is of deep-seated (plutonic) origin containing intermediate silica content, and is high in alkali. A common variety of syenite is nepheline syenite, consisting of a significant range of nepheline. Syenite appears similar to granite in hand specimen but is not a typical rock.

5) Diorite

Diorite is grey to greenish-grey in color. Diorite is a medium-to-coarse grained and even-textured rock. It consists mainly of plagioclase feldspar and ferromagnesian minerals. The percentage of ferromagnesian minerals present in the rock equals or exceeds the feldspar content, hence the dark color. If the rock contains sufficient content of quartz, it is called quartz-diorite. Diorite is a commonly occurring rock and is used for foundation and construction purposes.

6) Gabbro

Gabbro is a coarse-grained basic igneous rock consisting essentially of feldspar and ferromagnesian minerals such as pyroxene. It is similar to diorite in appearance but contains more significant amounts of ferromagnesian minerals. If the hypersthene content in gabbro exceeds that of augite, it is called norite. If the mineral contents in gabbro are essentially plagioclase and labradorite with very little augite, the rock is anorthosite. When fresh, the rock chips can be used for the construction of roads and other engineering purposes. After decomposition, it develops weakness due to alteration of olivine to serpentine and chlorite, and such altered or weathered gabbro is not to be used for any engineering work.

7) Basalt

Basalt is a fine-grained, dark-colored basic igneous rock that originates from the cooling of the volcanic lava flow. When examined under the microscope, it is found to contain feldspars (plagioclase and oligoclase), pyroxene (augite), olivine, and iron oxide with an ophitic texture primarily. Glass, which is present in some varieties, makes the rock vulnerable to chemical weathering. A vesicular variety is known as amygdaloidal basalt containing secondary minerals such as chalcedony, opal, agate, jasper, and zeolite. Basalt with silica minerals in high content may cause an alkali-aggregate reaction, so using it as a concrete aggregate should be avoided. The weathering product of basalt is laterite, which is widely used in engineering constructions of roads and buildings.

8) Obsidian, pitchstone, and pumice

These are very dark-colored glassy igneous rocks formed by the rapid cooling of ejected lavas from volcanoes. The mineral grains cannot be distinguished in these glassy rocks. If the rock is vitreous in luster and bright looking, it is called obsidian. When the luster is dull and pitchy, it is pitchstone. The pumice has an earth texture and dull look. It is spongy and very porous. These rocks occur only in restricted areas, and if present in an engineering project site, their weathering characteristics are to be critically considered (Subinoy 2013).

9) Rhyolite, dacite, trachyte, andesite, and phonolite

These volcanic rocks are commonly known as felsites. They are very dense and homogeneous and have a flint-like or glassy appearance. The main constituent minerals of these different rock types are as follows: rhyolite—alkaline feldspars and quartz; dacite—lime-soda feldspar and quartz; trachyte—alkaline feldspar without or with very little quartz; andesite—soda-lime feldspar with little or no quartz; and phonolite—alkaline feldspar with nepheline. All these volcanic rocks are very fine-grained or glassy and light to pale grey except andesite, which is dark-colored. They are of common occurrence from volcanic eruptions as lavas on the surface and partial intrusion through cracks in the earth's crust.

2.3.3 Sedimentary Rocks

Sediments form a relatively skinny surface layer of the Earth's crust, covering the igneous or metamorphic rocks that underlie them. This sedimentary shield is discontinuous and averages about 0.8km in thickness. Still, it locally reaches a thickness of 12km or more in the long orogenic belts that are the sites of previous geosynclines. It has been assessed that the sedimentary rocks constitute little more than 5% of all crustal rocks (to a depth of 16km); within this percentage, the proportions of the three main sedimentary types are shales and clays (4%), sandstones (0.75%), and limestones (0.25%). Among other category of smaller amounts are rocks composed of organic remains, such as coals and lignites, and those formed by chemical sedimentation.

1. Development

The components of sediments become hardened into sedimentary rocks such as sandstone, quartzite, limestone and shale by changes that commence soon after the sediment has accumulated. Water percolating through the voids (or pores) between the particles of sediment carries mineral matter which coats the grains and acts as a cement that binds them together. Such processes are known as cementation; they may eventually fill the pores and are responsible for converting many coarse-grained sediments to rock. The conversion of muddy sediment to rock is mainly achieved by the very small particles of silt and clay of which they are mainly composed being pressed together by the weight of overlying sediment, interstitial water being squeezed out and mineral matter precipitated in the microscopic network of pores. In the course of time, mud will become a coherent mass of clay, mudstone, or shale. This process is called compaction, it being more than simple consolidation, and it affects the muddy sediments to a much greater degree than the sands. During compaction, while much pore-contained water in the mud is pressed out, some water with its dissolved salts may remain in the sediment; it is known as connate water (connate means "born with").

The general term diagenesis denotes the processes outlined above, which convert sediments into sedimentary rocks. Diagenetic procedures include cementation, compaction, solution, and redeposition of material to produce rocks. Other changes, such as replacement, take effect on particular rocks (e.g., dolomitization). Much salt in marine sediments is removed by leaching. All these changes take place near the Earth's surface at average temperatures. When fully-formed rocks come again into the zone of weathering, perhaps after a long burial history, soluble substances are removed, and insoluble particles are released to begin a new cycle of sedimentation in rivers and the sea.

2. Composition

The raw materials from which the sedimentary rocks have been formed include accumulations of loose sand and muddy detritus, derived from the breakdown of older rocks and brought together and sorted by water or wind. Some sediments are formed mainly from the remains of animals and plants that lived in rivers, estuaries, on deltas, along coast-lines, and in the sea. Shelly and coral limestones, coal, and many sedimentary iron ores are composed of such remains. Sediments may also be formed by water evaporation and precipitation of the soluble minerals within it, as in playa lakes. When seawater evaporates, its components precipitate valuable deposits of chemicals such as sodium chloride (rock salt, halite). The chemical properties of seawater are due to dissolved matter brought by rivers and contributed by volcanic eruptions and to the presence of marine organisms and sediments.

3. Texture

The size of grains is an essential textural feature of a terrigenous rock, as an indication of the distance between its source and depositional areas and an easily observed property that may be used to distinguish and classify the rock. The coarsest particles are deposited nearest to the source area, and most of the finest particles are carried in suspension to greater distances before they settle. These clay particles, or rock flour (which represents minute particles of the common rock-forming minerals), are deposited only when the current slackens to nearly zero. In some cases, the salinity of the water of an estuary or the open sea makes the minute specks of clay clot (flocculate) into larger particles.

The degree of roundness of grains is related to the totality of abrasion suffered during transport, and hence to the distance traveled from their source before deposition. The roundness is associated with the sharpness or curvature of the sides and corners of grains. It is also reliant on the size and hardness of the grains and the intensity of the impact of one against another. No matter how far they move, sand grains lack the required momentum when they crash with water to build ideal rounds, and it is only when dry grains are blown by the wind that quite rounded grains are generated. Bigger fragments, however, may be well rounded following transportation by water for distances of less than 150km. The degree of roundness is a critical attribute in the sand used to make concrete or other engineering intentions. Sand grains deposited from ice usually are more angular than those river deposits, and the most rounded grains generally occur in the dunes.

A property associated with the roundness of grains is sphericity, which defines the degree to which a particle or grain approaches the shape of a sphere (McLean et al. 2005). Equidimensional particles have a more fantastic prospect of becoming spherical during transportation than other shapes of particles. Sphericity is controlled by directions of weakness, such as bedding planes or fractures. It is also related to size, in that the larger the grain above 8mm, the lower the sphericity. The relationship of rounding to particles less than 2mm of high and low sphericity is given in Figure 2-16.

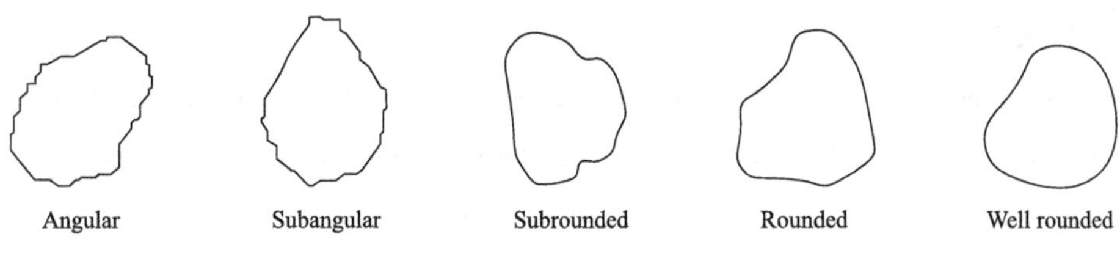

Figure 2-16 Sand-sized particles showing variation in roundness and sphericity

A positive correlation exists between sphericity and roundness. Still, other factors, notably particle size, planes of weakness, and particle composition, may have a marked influence on the final shape of the particle when it is finally deposited. Settling velocity will also influence the shape of the deposit.

Most terrigenous rocks contain grains of different sizes (Figure 2-17). The relative homogeneity of rock is expressed as its degree of sorting, a well-sorted rock in one which consists of similarly sized particles. In contrast, a poorly sorted rock has a wide range of particle sizes (grades). It should be noted that, in engineering practice, soil such as gravel, containing a wide range of sizes (degrees), is said to be well graded, and that "well-graded" is opposite in meaning to "well sorted".

Figure 2-17 Conglomerate, named from the Latin conglomeratus for "heaped, rolled or pressed together." Form in alluvial fans, stream beds, and pebble beaches

Chemical sedimentary rocks usually have a crystalline texture. However, some are formed of fragments, and their textures depend on the sizes, shapes, and arrangement of these fragments. If the rock has been created from organic debris, the fragments may comprise particles of shell or wood, but the texture can be characterized in equal terms as are used for other fragmented rocks.

4. Classification

Sandstone and shale (clastic), and limestone (non-clastic) are the most common sedimentary rocks. It has been established that 99% of the sedimentary rocks belong to these types. A simplified classification is derived based on the composition and size of the rock fragments or grains in clastic sediments and the content of minerals in non-clastic types (Table 2-5).

Table 2-5 Classification of sedimentary rocks (Zhu & Cui, 2017; Subinoy, 2013)

Group	Size of particle/rock fragments	Type of constituent materials/rock fragments	Name of rock
Clastic	>2mm	Gravel	Conglomerate/Breccia
	0.5~2mm	Coarse Sands	Sandstone
	0.25~0.5mm	Medium Sands	Sandstone/Arkose
	0.005~0.25mm	Fine Sands	Sandstone/Siltstone
	<0.005mm	Clay	Claystone/Shale
	Unassorted large to small sizes	Rock fragments in clayey matrix	Tillite/ Till
Non-clastic	Crystalline	>50% calcite	Limestone
		>50% dolomite	Dolomite
	Organic matter	Plant remains	Peat/Coal/Lignite

5. Description of typical sedimentary rocks

1) Sandstones

Sandstones are the most dominant type of sedimentary rock. They generally occur as layered or bedded strata and vary grain size from coarse-sand to fine-sand (Table 2-5). They contain quartz in varied proportions up to 95%, with other minerals such as feldspar, iron oxide, mica, and chloride occurring in subordinate amounts. Sandstones vary widely concerning their matrix, texture, porosity, and strength. If the matrix is siliceous, the rock will be hard, compact, and possesses high strength. Sandstone with argillaceous and calcareous matrix loses its strength if it remains saturated with water for a prolonged period.

2) Arkose

Arkose is light grey to pale pink and consists of coarse and angular grains of quartz and feldspar in nearly equal proportion. The grains are well-sorted and derived from the disintegration of granitic rock. The rock also contains mica, iron oxide, carbonate, and clay in subordinate amounts. If the feldspar content is more than 25%, the rock is termed feldspar-rich sandstone. The occurrence of feldspar-rich arkose-type sandstone reveals extreme cold climate conditions that had prevailed during the formation.

3) Siltstone

Siltstone is a light-colored rock consisting mainly of sand particles, and it also contains tiny amounts of coarser sand and clay. Siltstone has a gritty feel, which is an identifying character in a hand specimen. Siltstones are generally interbedded with shale and sandstone.

4) Shale

Shale is a soft sedimentary rock with thin layering or lamination along which it breaks easily. This easily breakable characteristic of shale is called fissility. Due to high compaction, some varieties of shale, such as slate, attain hardness. Shales vary widely in color, from light grey to dark grey and white, yellow, brown, and black. The rock contains both silt and clay particles, with clay-size particles (2μm or less) being the maximum. If the clay percentage predominates and there is no layering, the rock is called claystone or mudstone. Shale and claystone contain clay minerals, which play a significant role in engineering geological investigations.

5) Conglomerate

This rock is formed by consolidating rounded boulders with the siliceous, calcareous, or argillaceous types of cementing materials. If the matrix is siliceous, it is hard and of high strength. But when the rock contains other matrix types such as clay, it becomes porous and breaks easily when saturated with water. Depending upon the dominance in size of these rounded rock fragments, the conglomerates are termed pebbly conglomerate or pebblestone, cobbley conglomerate or cobblestone, and boulder conglomerate (Table 2-5). When most boulders in a sedimentary deposit are angular and uneven, the rock is termed breccia. When boulders in large numbers remain embedded in clay, the resulting residue is called tillite. Conglomerate, breccias, and tillite are not suitable as foundation rocks as they crumble under pressure in saturated conditions and permit water leakage through them.

6) Limestone

Limestone is the primary type of non-clastic sedimentary rock. Calcite is the dominant constituent of limestone with dolomite and aragonite in subordinate amounts with the occasional presence of a few quartz grains. When dolomite (mineral) content is more, the rock is called dolomite (rock). The rock is found in different shades of white, grey, yellow, brown, and black. Fossil shells are frequent in many varieties of limestone, for example, coral limestone and foraminiferal limestone. Limestone dissolves into solution forming large caverns, which are of great concern when present in a reservoir project. If hard and free from cavities, the rock is suitable for the foundation.

2.3.4 Metamorphic Rocks

Metamorphic rocks are formed from preexisting rocks through the action of heat and pressure. This process of the transformation of one rock type into another is called metamorphism. The word "Metamorphism" comes from Greek, where "meta" means after, and "morph" means form, so metamorphism means the after formation. Metamorphism most often occurs deep within the earth. Under increased temperature and pressure, the minerals of preexisting rocks become unstable and recrystallize in a solid-state to become new minerals. The study of metamorphic rocks yields valuable information about metamorphic conditions on rock and the geologic history of a region.

Metamorphism is the term used to denote the transformation of rocks into new types by the recrystallization of their constituents. The original rocks may be igneous, sedimentary, or ones that have already been metamorphosed. Changes in them result from the addition of heat or the operation of pressure. Heat and pressure are the agents of metamorphism which impart energy to the rocks, sufficient to mobilize the constituents of minerals and reassemble them as new minerals whose composition and crystal lattice are in equilibrium with existing conditions. (Figure 2-18). Metamorphic minerals grow in solid rock, their development being aided by solvents, especially water expelled from remaining pores and the dehydration of clay minerals.

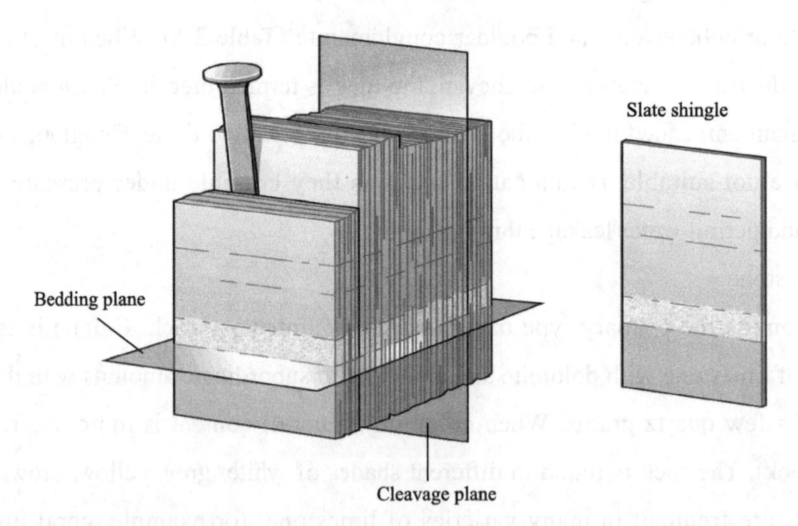

Figure 2-18 Cleavage in slate. Note the bedding plane is not necessarily parallel to the cleavage

1. Types of metamorphism

1) Contact metamorphism

The intrusion of a hot igneous mass such as granite or gabbro produces an increase of temperature in the surrounding rocks (Figure 2-19). This increase promotes the recrystallization of some or all of the components of the rocks affected, the most marked changes occurring near the contact with the igneous body. When recrystallization can develop uninhibited by an external stress acting on the rocks, the new minerals grow haphazardly in all directions and the metamorphosed rock acquires a granular fabric, which is known as the hornfels texture.

During the metamorphism, there may also be a transfer of material at the contact. When hot gases from the igneous mass penetrate the country rocks, the process is known as pneumatolysis (Blyth et al., 2005). The country rocks are not melted, but hot emanations such as carbonic acid, SO_2, water vapor, and volatile compounds of boron and fluorine percolate into them and result in the formation of new minerals. Temperatures may range from about 500 to 800℃ during pneumatolysis, and associated hydrothermal emanations may carry ore-metals such as Sn, Zn, and Fe, which are deposited as mineral veins in fissures in the country-rocks.

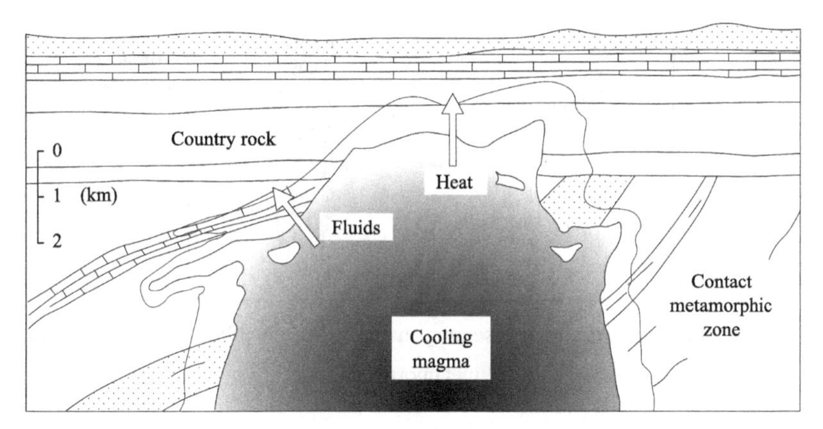

Figure 2-19 Contact metamorphism

2) Regional metamorphism

The operation of stresses and rise of temperature results in recrystallization, with the formation of new minerals, many of which grow in their length or flat cleavage surfaces at the direction of the maximum compressive stress (Figure 2-18). High temperatures and stresses are produced in orogenic belts of the crust, and regional metamorphic rocks are found in these great fold-belts, where they have become exposed after denudation. Many components of the rocks have acquired a largely parallel orientation, which gives the rocks characteristic textures: the oriented texture produced by platy or columnar minerals is known as schistosity (Figure 2-20), and an alteration of schistose layers with others less schistose gives the banded texture known as foliation. Argillaceous rocks under the influence of moderate to low temperature and high stress develop slaty cleavage (Blyth & Freitas, 2005).

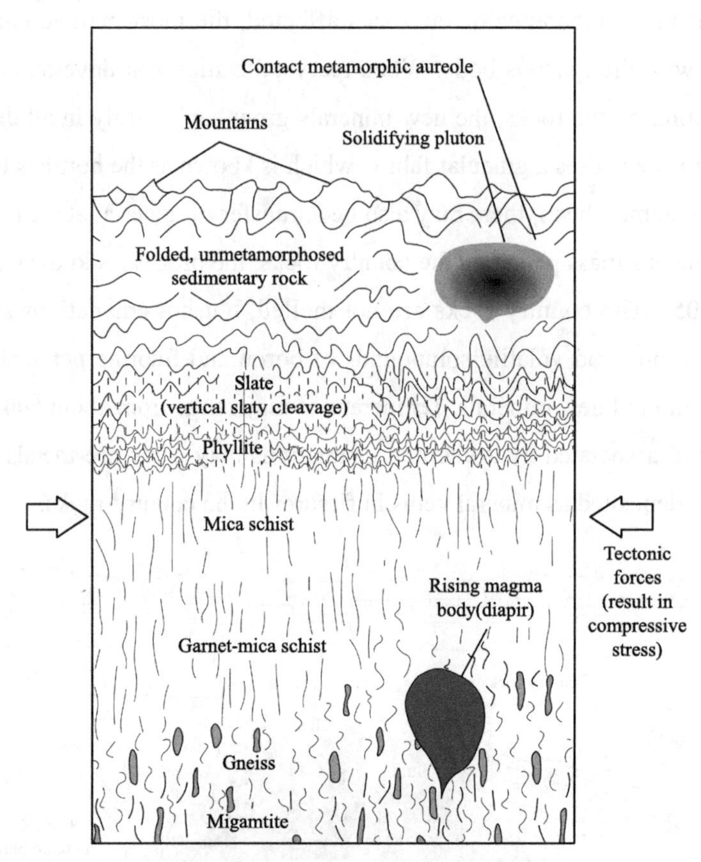

Figure 2-20 Schematic cross-section representing an approximately 30km portion of Earth's crust during metamorphism

3) Burial Metamorphism

When sedimentary rocks are buried to depths of several hundred meters, temperatures greater than 300℃ may develop in the absence of differential stress. New minerals grow, but the rock does not appear to be metamorphosed. The main minerals produced are the Zeolites. Burial metamorphism overlaps, to some extent, with diagenesis and grades into regional metamorphism as temperature and pressure increase.

4) Dynamic Metamorphism

This type of metamorphism is due to mechanical deformation, like when two bodies of rock slide past one another along a fault zone. Heat is generated by the friction of sliding along the zone, and the rocks tend to be crushed and pulverized due to the sliding. Dynamic metamorphism is not very common and is restricted to a narrow area along which the sliding occurred. The rock that is produced is called mylonite.

2. Texture

The term texture of the metamorphic rock is used to describe the size, shape, and arrangement of particles within a rock. Most igneous and sedimentary rocks comprise mineral grains that have a random direction and thus come out the same when viewed from any orientation. By contrast, metamorphic rocks containing platy minerals (micas) and elongated minerals (amphiboles) generally display some favored direction. The mineral grains show a parallel to subparallel alignment. Like many pencils, rocks enclosing elongated minerals that are oriented parallel to each other will appear distinct when viewed from the side than when viewed head-on. A rock that displays a preferred orientation of its minerals is said to hold foliation.

The term foliation refers to any planar (nearly flat) organization of mineral grains or structural features within a rock. Although foliation may take place in some sedimentary and even a few types of igneous rocks, it is an elementary feature of regionally metamorphosed rocks. Rock units that have been firmly deformed, primarily by folding. In metamorphic environments, foliation is eventually driven by compression stresses that shorten rock units, resulting mineral grains in preexisting rocks to produce parallel, or almost parallel, alignments. Samples of foliation include the parallel alignment of platy minerals, flattened pebbles, and improvisatory banding. The divergence of dark and light minerals generates a laminated occurrence; and rock cleavage where rocks can be easily divided into tabular slabs. These diversified types of foliation can form in many different ways, including:

(1) Rotation of platy and/or elongated mineral grains into a parallel or nearly parallel orientation.

(2) Recrystallization that produces new minerals with grains that exhibit a preferred orientation.

(3) Mechanisms that change spherically shaped grains into elongated shapes that are aligned in a preferred orientation.

The rotating of existing mineral grains is the easiest of these mechanisms to visualize. Figure 2-21 demonstrates the process by which platy or elongated minerals are turned. Note that the new alignment is approximately perpendicular to the direction of maximum shortening. Although the physical rotation of platy minerals contributes to the development of foliation in low-grade metamorphism, other mechanisms dominate in more extreme environments.

Recrystallization is the creation of new mineral grains out of old ones. When crystallization occurs as the rock is being subjected to different stresses, any elongated and platy minerals that form tend to recrystallize perpendicular to the direction of maximum stress. Thus, the newly formed mineral grains will possess a parallel alignment, and the metamorphic rock containing them will exhibit foliation. Mechanisms that change the shapes of existing grains are essential for developing preferred orientations in rocks that hold minerals such as quartz, calcite, and olivine—minerals that usually generate roughly spherical crystals.

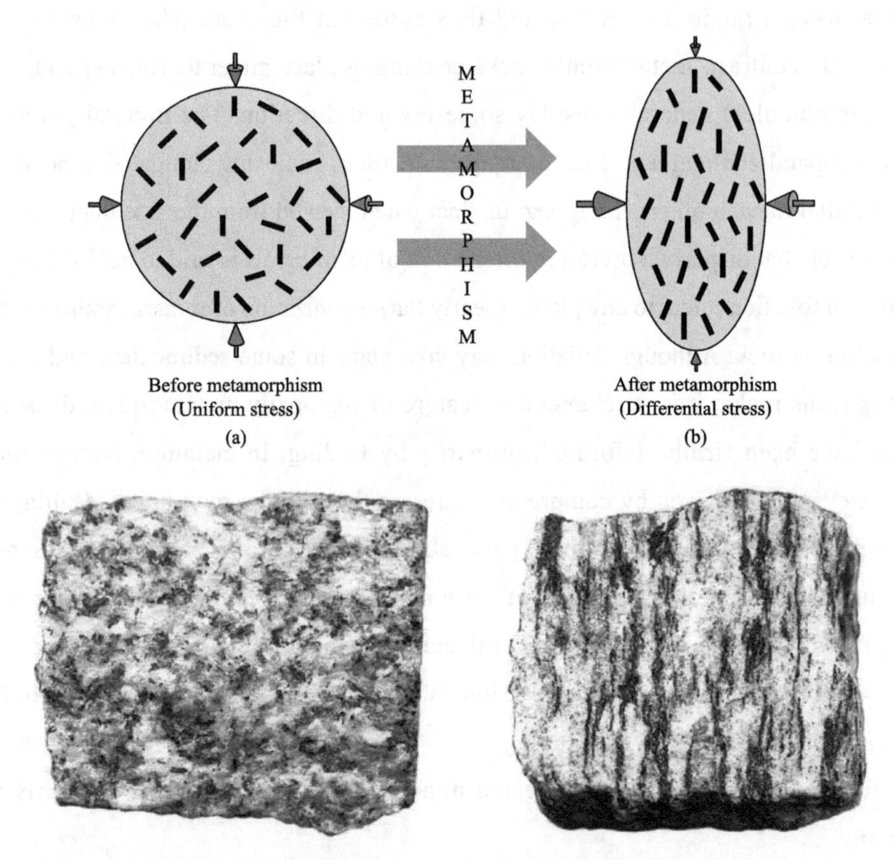

Figure 2-21 Mechanical rotation of platy or elongated mineral grains

(a) Existing mineral grains keep their random orientation if force is uniformly applied; (b) As differential stress causes rocks to flatten, mineral grains rotate toward the plane of flattening (Lutgens et al., 2018)

These processes operate in metamorphic environments where differential stresses exist. A change in grain shape can occur as units of a mineral's crystalline structure slide relative to one another along discrete planes. The type of gradual solid-state flow involves slippage that disrupts the crystal lattice as atoms shift positions. This process consists of the breaking of existing chemical bonds and the formation of new ones. The shape of a mineral may also change as ions move from a highly stressed location along the margin of the grain to a less-stressed position on the exact grain.

Mineral matter dissolves where grains are in contact with each other (areas of high stress) and precipitates in pore spaces (areas of low pressure). As a result, the mineral grains tend to elongated in the direction of maximum stress. Hot, chemically active fluids aid this mechanism.

1) Foliated Textures

Various types of foliation exist, depending mainly upon the grade of metamorphism and the mineral content of the parent rock. We will look at three: rock or slaty cleavage, schistosity, and gneissic texture.

ROCK OR SLATY CLEAVAGE. Rock cleavage refers to closely spaced, flat surfaces along which rocks split into thin slabs when hit with a hammer. Rock cleavage develops in various metamorphic rocks but is best displayed in slates, which exhibit an excellent splitting property called slaty cleavage.

Depending on the metamorphic environment and the composition of the parent rock, rock cleavage develops in several ways. In a low-grade metamorphic environment, rock cleavage is known to develop where beds of shale (and related sedimentary rocks) are strongly folded and metamorphosed to form slate. The process begins as platy grains are kinked and bent, generating microscopic folds with roughly aligned limbs (sides). This new alignment is enhanced with further deformation as old grains break down and recrystallize preferentially in the direction of the newly developed orientation. In this manner, the rock creates narrow parallel zones where mica flakes are concentrated. These features take turns with zones containing quartz and other mineral grains that do not exhibit a pronounced linear orientation. It is along these very narrow zones of platy mineral that slate splits.

Because slate typically forms during the low-grade metamorphism of shale, evidence of the original sedimentary bedding planes is often preserved. However, the slate's cleavage orientation usually develops at an oblique angle to the original sedimentary layers. Thus, unlike shale, which splits along bedding planes, slate often splits across them. Other metamorphic rocks, such as schists and gneisses, may also split along planar surfaces and exhibit rock cleavage.

SCHISTOSITY. Under higher temperature-pressure regimes, the minute mica and chlorite crystals in slate begin to grow. When these platy minerals are large enough to be perceptible to the unaided eye and exhibit a planar or layered structure, the rock shows a type of foliation called schistosity. The rocks having this texture are described as schists. In addition to platy minerals, schists often contain deformed quartz and feldspar crystals that appear flat or lens-shaped and are hidden among the mica grains.

GNEISSIC TEXTURE. During high-grade metamorphism, ion migration can result in the segregation of minerals, as shown in Figure 2-22. Notice that the dark biotite crystals and light silicate minerals (quartz and feldspar) have separated, giving the rock a banded appearance called gneissic texture. A metamorphic rock with this texture is called gneiss (pronounced "nice"). Although foliated, gneisses will not usually split as easily as slates and some schists.

2) Non-Foliated Textures

Not all metamorphic rocks show a foliated texture and those that are not referred to as

Figure 2-22 This rock displays a gneissic texture

nonfoliated. Nonfoliated metamorphic rocks typically develop in minimal deformation, and the parent rocks are made up of minerals that exhibit equidimensional crystals, such as quartz or calcite. For example, when a fine-grained limestone (made of calcite) is metamorphosed by the intrusion of a hot magma body, the small calcite grains recrystallize to form larger interlocking crystals. The resulting rock, marble, exhibits large, equidimensional grains that are randomly oriented, similar to coarse-grained igneous rock.

Another texture familiar to metamorphic rocks consists of substantial grains, called porphyroblasts, surrounded by a fine-grained matrix of other minerals. Porphyroblastic textures develop in many rock types and metamorphic environments when minerals in the parent rock recrystallize to form new minerals. During recrystallization, certain metamorphic minerals, including garnet, staurolite, and andalusite, often develop a small number of very large crystals. By contrast, minerals such as muscovite, biotite, and quartz typically form a large number of tiny grains. As a result, when metamorphism generates the minerals garnet, biotite, and muscovite in the same setting, the rock will contain large crystals (porphyroblasts) of garnet embedded in a finer-grained matrix of biotite and muscovite Figure 2-23).

3. Classification

As we noted before, the kind of metamorphic rock is determined by the metamorphic environment (primarily the particular combination of pressure, stress, and temperature) and by the chemical constituents of the parent rock. Many kinds of metamorphic rocks exist because of the many possible combinations of these factors. These rocks are classified based on broad

Figure 2-23 Garnet-mica schist. The dark red garnet crystals (porphyroblasts) are embedded in a matrix of fine-grained micas

similarities. The relationship of texture to rock name is summarized in Table 2-6. First, consider the texture of a metamorphic rock and determine whether is foliated or non-foliated (Figure 2-24).

If the rock is non-foliated, it is named based on its composition. The two most common non-foliated rocks are marble and quartzite, respectively, of calcite and quartz. If the rock is foliated, you need to decide the kind of foliation to name the rock. For example, a schistose rock is called schist. But this name informs us nothing about what minerals are in this rock, so we need to outline the composition—for example, garnet-mica schist.

Figure 2-24 Photomicrographs taken through a polarizing microscope of metamorphic rocks

(a) Nonfoliated rock; (b) Foliated rock. Multicolored grains are biotite mica; gray and white are mostly quartz

Table 2-6 Classification and Naming of Metamorphic Rocks (Plummer et al., 2016)

<table>
<tr><td colspan="4">Non-foliated</td></tr>
<tr><td colspan="4">Name Based on Mineral Content of Rock</td></tr>
<tr><td>Usual Parent Rock</td><td>Rock Name</td><td>Predominant Minerals</td><td>Identifying Characteristics</td></tr>
<tr><td>Limestone</td><td rowspan="2">Marble</td><td rowspan="2">Calcite</td><td>Coarse interlocking grains of calcite (or, less commonly, dolomite)</td></tr>
<tr><td>Dolomite</td><td>Calcite (or dolomite) has rhombohedral cleavage; hardness intermediate between glass and fingernail. Calcite effervesces in weak acid</td></tr>
<tr><td>Quartz Sandstone</td><td>Quartzite</td><td>Quartz</td><td>Rock composed of small interlocking granules of quartz. Has a sugary appearance and vitreous luster; scratches glass</td></tr>
<tr><td>Shale</td><td>Hornfels</td><td>Fine-grained micas</td><td rowspan="2">A fine-grained, dark rock that generally will scratch glass. May have a few coarser minerals present</td></tr>
<tr><td>Basalt</td><td>Hornfels</td><td>Fine-grained</td></tr>
<tr><td colspan="4">Foliated</td></tr>
<tr><td colspan="4">Name Based Principally on Kind of Foliation Regardless of Parent Rock. Adjectives Describe the Composition (e.g., biotite-garnet schist)</td></tr>
<tr><td>Texture</td><td>Slate</td><td>Clay and other sheet silicates</td><td>A very fine-grained rock with an earthy luster. Splits easily into thin, flat sheets</td></tr>
<tr><td>Intermediate between slaty and schistose</td><td>Phyllite</td><td>Mica</td><td>Fine-grained rock with a silky luster. Generally, splits along wavy surfaces</td></tr>
<tr><td>Schistose</td><td>Schist</td><td>Biotite and muscovite amphibole</td><td>Composed of visible platy or elongated minerals that show planar alignment. A wide variety of minerals can be found in various types of schist (e.g., garnet-mica schist, hornblende schist, etc.)</td></tr>
<tr><td>Gneissic</td><td>Gneiss</td><td>Feldspar, quartz, amphibole, biotite</td><td>Light and dark minerals are found in separate, parallel layers or lenses. Commonly, the dark layers include biotite and hornblende; the light-colored layers are composed of feldspars and quartz. The layers may be folded or appear contorted</td></tr>
</table>

4. Description of typical metamorphic rocks

1) Gneiss

Gneiss is a medium to coarse-grained and banded or foliated metamorphic rock. Observed under the microscope, it comprises mainly quartz, feldspar, hornblende, micas, and accessory minerals such as garnet, tourmaline, apatite, zircon, sphene, and magnetite. Alternating bands of light and dark minerals are typical in gneiss. The dark bands consist primarily of biotite and amphibole and the white bands of quartz and feldspar. The characteristic banding of this rock is due to the platy minerals such as micas and hornblende that are arranged in thin planes or bands along which the rock splits easily. The gneisses are designated as biotite gneiss, hornblende gneiss, or granite gneiss, depending upon the mineralogical composition. The rocks are generally derived from igneous rocks, such as granites, but may also be from sedimentary rocks. The gneiss is a rugged rock and is suited for the foundation of engineering structures.

2) Schist

Schist consists mainly of quartz, feldspar, muscovite, and biotite. Various minerals such as garnet, staurolite, kyanite, sillimanite, epidote, chlorite, and talc are also present in the rock. These rocks are named according to the presence of the flaky or platy minerals, for example, mica schist, talc schist, and chlorite schist. The rocks are of varied colors, such as white, brown, yellow, and black. These are derived from the metamorphism of igneous and sedimentary rocks. In schist, the micas and other platy minerals lie with their cleavage planes parallel to each other to provide schistosity fabric. The rocks are generally of low strength and split easily along the planes of schistosity. However, it can be used for the construction of houses, walls, and so on.

3) Phyllite

Phyllite is a grey-colored rock characterized by a glistering sheen derived from planes of partings or surfaces lined with fine-grained micas. Two types of phyllite are common—arenaceous (quartzitic) phyllite and argillaceous (shaley) phyllite. In the former, the grains are tightly packed with siliceous cement, and hence the rock is very hard. In the latter, due to the presence of the clayey matrix, the rock is soft though it has more sheets of mica. The rocks are derived from sandy and clayey shale by low-order metamorphism. Arenaceous phyllite can bear sufficient load and is safe for the structural foundation (Subinoy, 2013).

4) Slate

Slate is a fine-grained rock with a characteristic cleavage called "slaty cleavage," a new structure imposed by metamorphism. The rock can be split into big, smooth-surfaced sheets along the cleavage planes. The rock is generally grey, green, red, and black and formed from intense metamorphism of clastic sedimentary rocks like shale and claystone. Slate is a hard rock and is suited for the foundation. However, if it occurs on hill slopes, it can cause damage due to the breaking of large sheets of rock along the cleavage planes that slide down the hill. Large slabs of slates can be used for various purposes, such as the construction of walls and fencings.

5) Quartzite

Quartzite is white to light gray. This metamorphic rock consists mainly of recrystallized quartz derived from the metamorphism of sandstone. Quartz is the dominant content with iron oxide minerals present as an impurity. Some varieties contain more than 90% quartz with feldspar and micas in small amounts. With recrystallization, the boundaries of quartz become tight and interlocking assuming high strength. The rock is rugged and splits into sharp edges. It is suitable for the foundation and is also the most durable construction material.

6) Marble

Marble is the metamorphic equivalent of limestone. It is found in various colors, such as white, grey, and purple. It contains dominantly calcite and subordinately dolomite, iron oxide, graphite, and some quartz. Marble possesses good strength and durability and is resistant to meteoric weathering but slightly susceptible to chemical erosion.

7) Mylonite

Mylonite is a fine-grained rock with a flint-like appearance. The rock shows banding or streaking caused by extreme granulation and shearing of coarse-grained rocks. It is the product of dislocation metamorphism that disturbs the banding or other structure of original rocks, but the groundmass may contain lenses of rocks in elongated conditions. The rock should not be used in construction sites due to its inherent weakness caused by shearing.

2.3.5 The Rock Cycle

"Stable as a rock."this familiar expression implies that a rock is permanent, unchanging over time, but it isn't. In the time frame of Earth history, a span of over 4.57 billion years, atoms making up one rock type may be rearranged or moved elsewhere, eventually becoming part of another rock type. Later, the atoms may move again to form a third rock type, and so on. Geologists describe to the gradual transformation of Earth materials from one rock type to another as the rock cycle (Figure 2-25), one of many examples of cycles acting in or on the Earth (Marshak, 2013).

By following the arrows in Figure 2-25, you can see many paths around or through the rock cycle. For example, igneous rock may weather and erode to produce sediment, which lithifies to form sedimentary rock. The new sedimentary rock may become buried so deeply that it transforms into metamorphic rock, which could partially melt and produce magma. This magma later solidifies to form new igneous rock. We can symbolize this path as:

igneous → sedimentary → metamorphic → igneous.

But alternatively, the metamorphic rock could be uplifted and eroded to form new sediment, which is later buried and lithified to form new sedimentary rock without melting. This path takes a shortcut through the cycle that we can symbolize as:

igneous → sedimentary → metamorphic → sedimentary.

Likewise, the igneous rock could be metamorphosed straight without initial turning to sediment. This metamorphic rock could then be eroded to produce deposition that eventually becomes sedimentary rock, defining another shortcut path:

igneous → metamorphic → sedimentary.

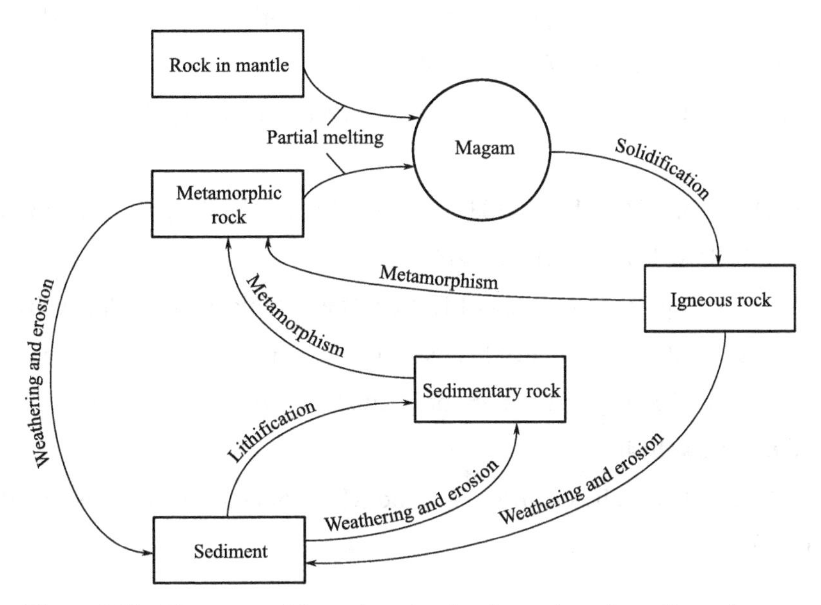

Figure 2-25 The stages of the rock cycle, showing various alternative pathways

Reference

[1] ATTEWELL P B, FARMER I W. Composition of Rocks[M/OL]// ATTEWELL P B, FARMER I W. Principles of Engineering Geology. Dordrecht: Springer, 1976[2021-12-07]. https://doi.org/10.1007/978-94-009-5707-7_1.

[2] BELL F G. Engineering geology[M/OL]. Amsterdam: Butterworth-Heinemann, 2007[2021-12-07]. https://www.elsevier.com/books/engineering-geology/bell/978-0-7506-8077-6.

[3] BLYTH F G H, FREITAS M H D. A geology for engineers[M].7th ed. Oxford: Elsevier Butterworth-Heinemann, 2005.

[4] CARLSON D H, PLUMMER C C, HAMMERSLEY L. Physical geology: earth revealed[M]. New York: McGraw-Hill, 2011.

[5] HALDAR S K, TISLJAR J. Introduction to Mineralogy and Petrology[M/OL]. Oxford: Elsevier, 2013[2021-12-07]. https://www.elsevier.com/books/introduction-to-mineralogy-and-petrology/haldar/978-0-12-408133-8.

[6] KLEIN C, PHILPOTTS T. Earth materials: introduction to mineralogy and petrology[M]. 2nd ed. Cambridge: Cambridge University Press, 2017.

[7] LUTGENS F K, TARBUCK E J, TASA D G. Essentials of geology[M/OL]. 13th ed. Hoboken, New Jersey: Pearson Education, 2018[2021-12-07]. https://www.pearson.com/us/higher-education/program/PGM1240907.html.

[8] MARSHAK S. Essentials of geology[M/OL]. New York: WW Norton, 2013[2021-12-07]. https://wwnorton.com/books/9780393667523.

[9] MCLEAN A C, GRIBBLE C D. Geology for civil engineers[M]. 2nd ed. London: CRC Press, 2015.

[10] PLUMMER C C, CARLSON D H, HAMMERSLEY L. Physical geology[M]. 15th ed. New York: McGraw-Hill Education, 2016.

[11] SUBINOY G. Engineering geology[M]. New Delhi, India: Oxford University Press, 2013.

[12] ZHANG S, LI Z, LUO K, et al. Discovery of carbon-based strongest and hardest amorphous material[J/OL]. arXiv preprint arXiv:2011.14819, 2020[2021-12-07]. https://doi.org/10.1093/nsr/nwab140.

[13] Zhu, J, Cui, G. Hydraulic engineering geology[M]. 5th ed. China Water and Power Press, 2017.(In Chinese)

Exercises

Choice Questions (Single Choice)

(1) An example of igneous rock is:

(a) marble (b) pegmatite

(c) phyllite (d) slate

(2) An example of metamorphic rock is:

(a) gabbro (b) arkose

(c) quartzite (d) diorite

(3) An example of sedimentary rock is:

(a) marble (b) sandstone

(c) quartzite (d) schist

(4) As for the crystal system, the crystallization of minerals can be grouped into:

(a)ten systems (b) six systems

(c) five systems (d) eight systems

Review Questions

(1) Name the three main rock types of the earth's crust. Discover the source of each of these rock types.

(2) Explain the three essential types of metamorphism and their effect.

(3) Describe briefly the following rock types: granite, dolerite, sandstone, limestone, conglomerate, marble, gneiss, schist, and phyllite.

(4) List and define the main physical properties used to recognize a mineral. Which minerals can react with acid to produce CO_2?

(5) How can you ascertain the hardness of a mineral? What is the Mohs hardness scale?

(6) Describe how a clastic sedimentary rock forms from its unweathered parent rock.

(7) How does dolomite differ from limestone, and how does dolomite form?

(8) How do metamorphic rocks differ from igneous and sedimentary rocks?

(9) Describe briefly the thoughts of the rock cycle.

3
Geological History

3.1 Introduction

The Earth's history is mainly the history of the development and evolution of the Earth's crust. As for the Earth's age, the vastness of geologic time is difficult for humans to conceive. It is rare for someone to live a hundred years, but a person would have to live 10000 times that long to witness geologic progress that takes a million years. Geologists working in the domain or with maps or illustrations in a workroom are involved with relative time— unraveling the series of geologic events. For example, a geologist looking at the photo of Sinian formation in Yichang can determine that the lower layers of the sedimentary rocks are older than the above layers (Figure 3-1). But this informs us nothing about how long ago any of the rocks formed. To decide how many years ago rocks formed, we need the specialized approaches of radioactive isotope dating. With the help of isotopic dating, we have been able to find out that the rocks in the lowermost portion of the Grand Canyon are well out of a billion years old. The Huangling massif is located to the northwest of Yichang city, and the oldest rocks are found in the Huangling massif with ages of 3 billion years.

This chapter explains applying several basic principles to decipher a sequence of events responsible for geologic features. These principles can be used in many aspects of geology— as, for example, in understanding geologic structures. Understanding the complex history of mountain belts also requires knowing the techniques for determining the relative ages of rocks. Choosing age relationships between geographically extensively separated rock units is indispensable for understanding the geologic history of a zone, a continent, or Earth. Substantiation of the plate-tectonics notion depends on the intercontinental connection of rock elements and geologic events, piecing together manifestation that the continents were once one great body. Extensive use of fossils led to the advancement of the classic geologic time scale.

Originally based on relative age relationships, the geologic time scale subdivisions have now been allocated numerical ages in thousands, millions, and billions of years through radioactive isotopic dating. We can regard the geologic time scale as a schedule to which events and rock units can be exposed.

Figure 3-1 Sinian formation scarp in Yichang

3.2 The present is the key to the past

3.2.1 Setting the stage for studying the past

Until recently, people in most cultures believed that geologic time began about the same time that human history began and that our planet has been virtually unchanged since its birth.

This view was challenged by James Hutton (1726-1797), a Scottish gentleman farmer and doctor. Hutton lived during the Age of Enlightenment, when, sparked by the discovery of physical laws by Sir Isaac Newton, many people sought natural, rather than supernatural, explanations for features of the world around them. While wandering in the highlands of Scotland, a region where rocks were well exposed, Hutton noted that many features (such as ripple marks and cross beds) found in sedimentary rocks resembled features that he could see forming today in modern depositional environments. These observations led Hutton to speculate that the formation of rocks resulted from the processes he could see occurring today. Hutton's idea came to be known as the principle of uniformitarianism. According to this principle, physical processes that operate in the modern world also performed in the past at roughly the same rates. These processes were responsible for forming geologic features preserved in outcrops. More concisely, the principle can be stated as: the present is the key to the past. Hutton deduced that the development of individual geologic features took a long time and that not all components formed simultaneously, so the Earth must have a history that includes a succession of slow geologic events. Since no one in recorded history has seen the entire process of sediment first turning into rock and then later rising into mountains, Hutton also realized that there must have been a long time before human history began.

3.2.2 Relative age versus numerical age

Like historians, geologists strive to establish both the sequence of events that produced an array of geologic features (such as rocks, structures, and landscapes) and, when possible, the date on which each event happened. We specify the age of one feature to another in a sequence as its relative age and the age of a attribute given in years as its numerical age (or, in older literature, its "absolute age"). Geologists have learned to determine the relative age long before being able to determine the numerical age. We, therefore, then look at the principles leading to the determination of relative age.

3.3 Geologic principles for defining relative age

Building from the work of Nicolas Steno (1638-1686), Hutton, and others, the British geologist Charles Lyell (1797-1875) laid out a set of valuable geologic principles in the first contemporary textbook of geology. Charles Lyell is known as the founder of modern geology. These principles, defined below, continue to provide the essential frame within which geologists study the record of Earth history and decide relative ages.

(1) The principle of uniformitarianism states that physical processes we observe operating today also operated in the past, at roughly comparable rates, so the present is the key to the past (Figure 3-2). It is possible to bond a particular geological period with the lithology, order of superposition, and fossil content concerned by in-depth study. From the present-day observation, we know that the various types of rocks with distinctive minerals and compositions are formed under different contexts (dry, humid, glacial, marine, fluvial, lacustrine, and terrestrial). It meant that various types of rocks had been formed in the past, too, depending on different environmental conditions that had dominated during distinct geological periods. This fact helps in connection and also in uncovering the part history of the earth.

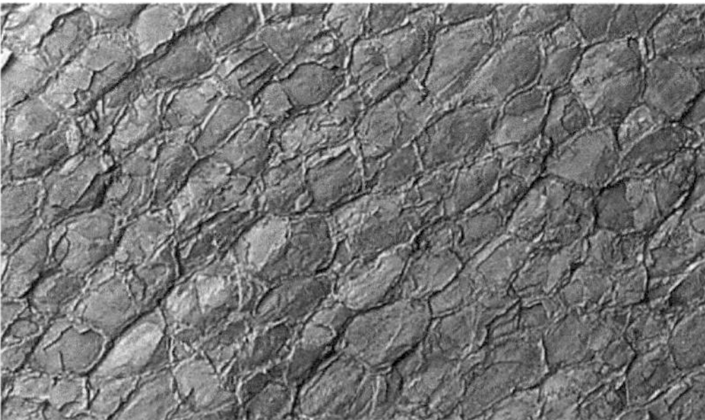

Figure 3-2 Uniformitarianism: The processes that formed cracks in the dried-up mud puddle on the left also formed the mudcracks preserved in the ancient, solid rock on the right. We can see these ancient mudcracks because erosion removed the adjacent bed

(2) The principle of original horizontality indicates that layers of sediment, when initial deposited, are fairly horizontal (Figure 3-3) because sediments accumulate on low relief surfaces (such as floodplains or the sea floor) in a gravitational domain. If the sediments were deposited on a steep slope, they would probably slip downslope before being concealed and lithified. With this principle in view, geologists infer that examples of folds and tilted beds stand for the effect of deformation after deposition.

(3) The principle of superposition indicates that in a succession of sedimentary rock layers, each layer must be younger than the one below. The layer of sediment cannot accumulate unless there is already a substrate on which it can assemble. Therefore, the layer at the bottom of a sequence is the oldest, and the layer at the head is the youngest (Figure 3-4).

(4) The principle of lateral continuity indicates that sediments generally collect in continuous sheets within a given area. If, for now, you discover a sedimentary layer cut by a gorge, you can infer that the layer once stretched the area that was later eroded by the river that formed the gorge (Figure 3-5).

Figure 3-3 Horizontal bedding in an outcrop of sandstone in Wisconsin

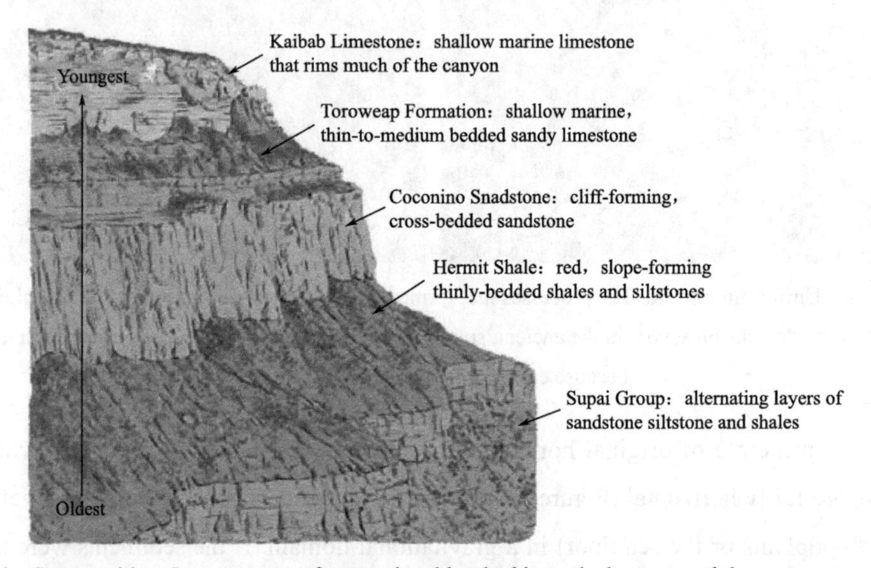

Figure 3-4 Superposition: In a sequence of strata, the oldest bed is on the bottom, and the youngest on top (After Lutgens et al., 2018)

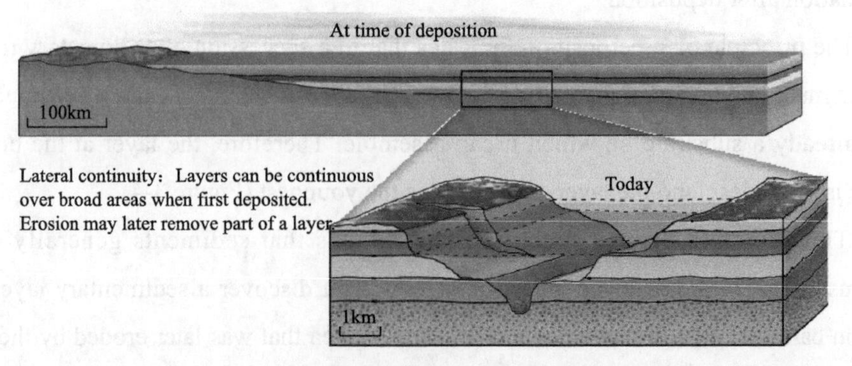

Figure 3-5 Lateral continuity: Layers can be continuous over broad areas when first deposited. Erosion may later remove part of a layer

(5) The principle of cross-cutting relations indicates that if one geologic object cuts across another, the object that has been cut is older. For example, if an igneous dike cuts across a sequence of sedimentary beds, the beds must be more senior than the dike (Figure 3-6). If a fault cuts across and displaces sedimentary layers, then the fault must be younger than the layers. But if a seam of sediment buries a fault, the deposition must be younger than the fault.

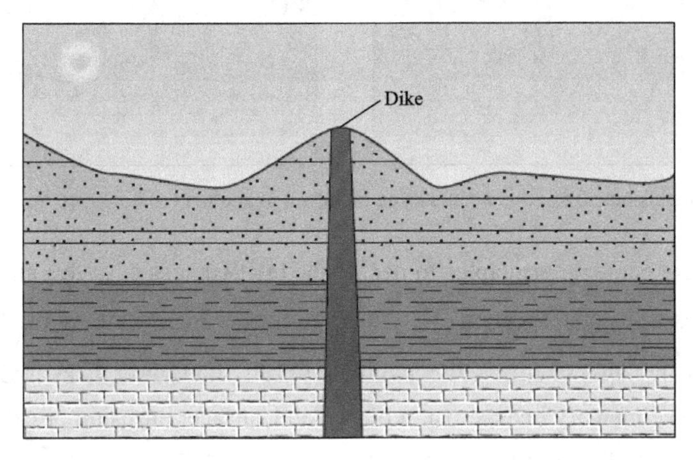

Figure 3-6 Cross-cutting relations: The dike cuts across the sedimentary beds, so the dike is younger

(6) The principle of baked contacts indicates that an igneous intrusion "bakes" (metamorphoses) encompassing rock, so the rock that has been baked must be older than the intrusion (Figure 3-7).

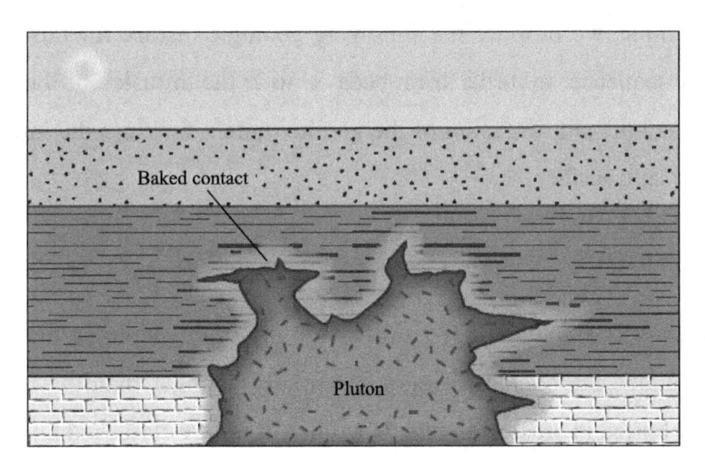

Figure 3-7 Baked contact: The pluton baked the adjacent rock, so the adjacent rock is older

(7) The principle of inclusions indicates that a rock containing an inclusion (a portion of another rock) must be younger than the inclusion. For example, a conglomerate comprised of basalt pebbles is younger than the basalt, and a sill including sandstone fragments must be

younger than the sandstone (Figure 3-8).

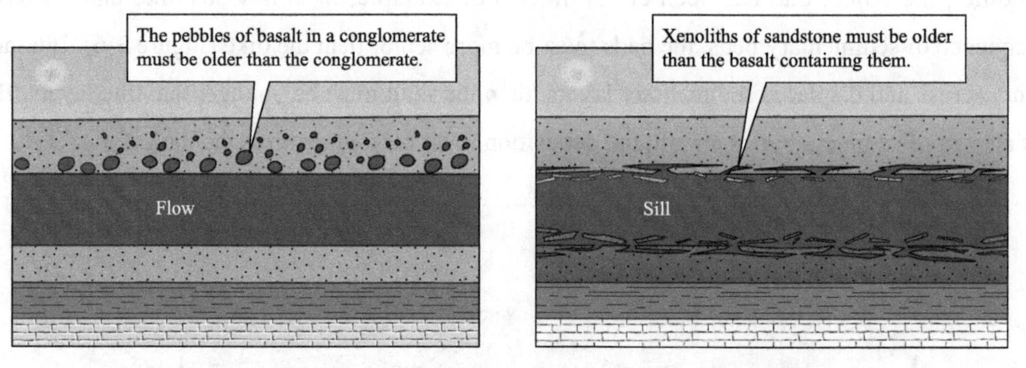

Figure 3-8 Inclusions: Rock that occurs as an inclusion in another rock must be the older of the two

Geologists use geologic principles to decide the relative ages of rocks, structures, and other geologic attributes of a given location. They then go further by explaining the formation of each element to be the outcome of a specific "geologic event". Examples of geologic events include sedimentation of sedimentary beds; erosion of the land surface; intrusion or extrusion of igneous rocks; deformation (folding and faulting); and a sequence of metamorphism. The progression of events in relative age that have generated the rock, structure, and landscape of a region is termed the region's geologic history.

We can apply these principles to determine the features'relative ages shown in Figure 3-9. In so doing, we develop a geologic history of the region, defining the relative ages of events there. For this example, we propose the following geologic history for this region: deposition of the sedimentary sequence in order from beds 1 to 7; the intrusion of the sill; folding of the sedimentary beds and the sill; intrusion of the granite pluton; faulting; the intrusion of the dike; erosion to form the land surface.

3.3.1 Unconformities

An unconformity is a contact that represents a gap in the geologic history, with the rock unit immediately above the contact being substantially younger than the rock beneath. Most unconformities are buried in erosion covers. Unconformities are classified into three types:

(1) disconformities (parallel unconformities).

(2) angular unconformities.

(3) nonconformities.

with each type having important implications for the geologic history of the area in which it occurs.

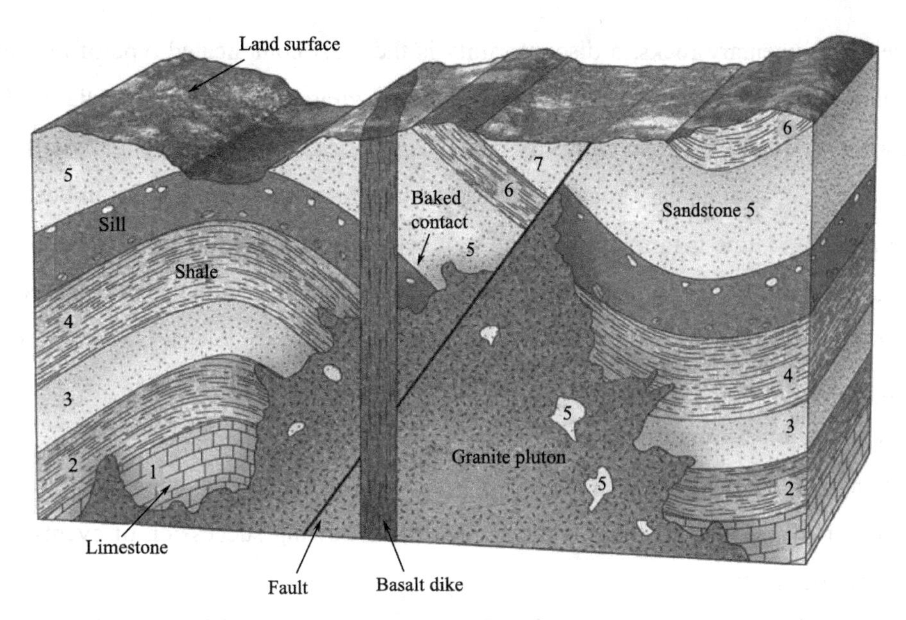

Figure 3-9 Geologic principles help us unravel the sequence of events leading to the development of the features shown above. Layers 1 to 7 were deposited first. Intrusion of the sill came next, followed by folding, intrusion of the granite pluton, faulting, intrusion of the dike, and erosion (Marshak, 2013)

1. Disconformities

In a disconformity, the contact representing the missing rock strata separates beds parallel to one another. Possibly what has happened is that older rocks were eroded parallel the horizontal bedding plane; additional deposition later buried the erosion top (Figure 3-10).

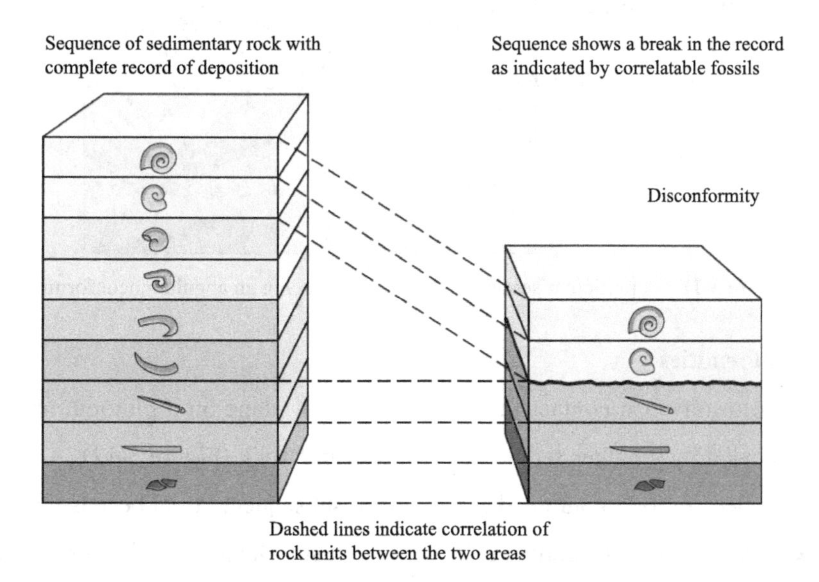

Figure 3-10 Schematic representation of a disconformity. The disconformity is in the block on the right

Because it often comes out to be just another sedimentary contact (or bedding plane) in a sequence of sedimentary rocks, a disconformity is the most complicated type of unconformity to detect in the field. Rarely, a revealing weathered zone is retained immediately below a disconformity. Usually, the disconformity can be discovered only by studying fossils from the beds in a series of sedimentary rocks. Suppose certain fossil beds are absent, indicating that a portion of geologic time is missing from the sedimentary record. In that case, it can be deduced that a disconformity is present in the sequence. Although it is most likely that some rock layers are missing because erosion followed sedimentation, in some cases, neither erosion nor deposition took place for a substantial amount of geologic time.

2. Angular unconformities

The angular unconformity is a contact in which younger strata cover an erosion surface on an inclined or folded layered rock. It means that the following succession of events, from the oldest to the youngest:

(1) deposition and solidification of sedimentary rock (or lithification of lava flows if the rock type is volcanic);

(2) upheaval accompanied by folding or tilting of the layers;

(3) erosion; and

(4) additional deposition (usually preceded by subsidence) on top of the erosion surface (Figure 3-11).

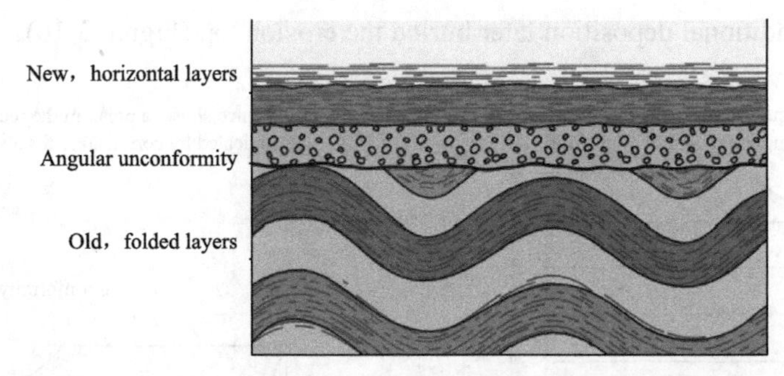

Figure 3-11 A particular sequence of events producing an angular unconformity

3. Nonconformities

The nonconformity is a contact in which an erosion plane on a plutonic or metamorphic rock has been coated by younger sedimentary or volcanic rock (Figure 3-12). A nonconformity usually indicates deep or long-continued erosion prior subsequent burial because metamorphic or plutonic rocks create in significant depths in Earth's crust. The geologic history implicated by a nonconformity is:

(1) Crystallization of metamorphic or igneous rock at depth.

(2) Erosion of at least several kilometers of the overlying rock (the significant amount of erosion further implies substabtial uplift of this portion of the Earth's crust).

(3) Deposition of new sediment, which eventually becomes sedimentary rock, on the ancient erosion plane.

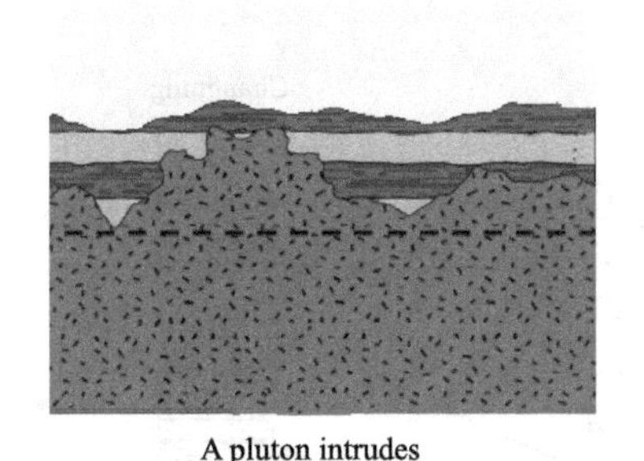

A pluton intrudes

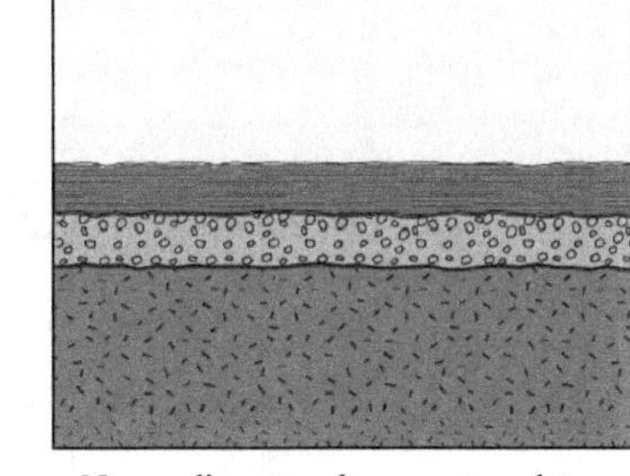

New sedimentary layers accumulate above the erosion surface

Figure 3-12 A particular sequence of events producing nonconformities

3.3.2 Correlation

We can summarize information about the sequence of sedimentary strata at a location by drawing a stratigraphic column. Typically, we draw columns to scale so that the relative thicknesses of layers portrayed on the column reflect the thicknesses of the layers in the outcrop. Then, we divide the sequence of strata represented in a column into stratigraphic formations, a sequence of beds of a specific rock type or group of rock types that can be traced over a reasonably broad region. Where did the conception of a stratigraphic formation come from? While excavating canals in England, William Smith (1769-1839) discovered that formations cropping out at one locality resembled formations cropping out at another, in that their beds looked similar and contained similar fossil assemblages (Marshak, 2013). In other words, Smith was able to define the stratigraphic relationship between the strata at one locality and the strata at another, a process now called correlation. In geology, correlation usually means determining the time equivalency of rock units.

1. Lithologic correlation

Under certain circumstances, two regions can be correlated, assuming similar rock types in two areas formed simultaneously. This method should be used with extreme caution, in particular when correlated rocks are common ones. Figure 3-13 shows the section of the Qingbaikou Group in the Neoproterozoic in Jixian and Changping. The Jixian section has relatively complete stratum development and a simple structure: rich in fossils and extensive in thickness. It is a typical section of the Meso-New Proterozoic in northern China. According to lithological characteristics, the Qingbaikou Group can be divided into the Xiamaling Formation dominated by shale, the Longshan Formation dominated by sandstone, and the Jingeryu Formation dominated by marl. The Longshan Formation is further divided into the lower sandstone section and the upper section. The shale and sandstone section. The Changping section is not far from Jixian. Compared with the Jixian section, the same division can be made.

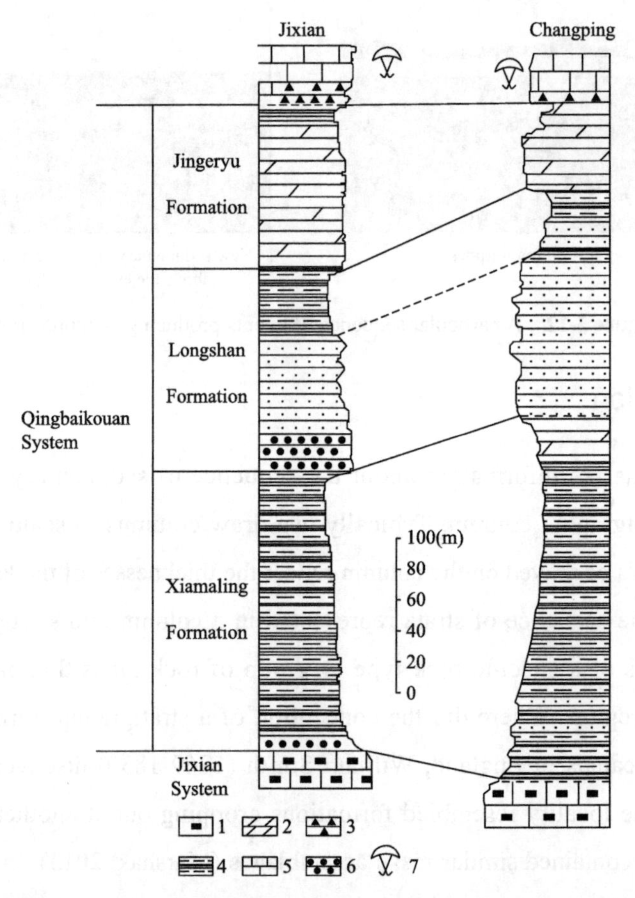

Figure 3-13 Comparison of the Division of Neoproterozoic in Jixian and Changping

1—siliceous limestone, dolomite; 2—marl; 3—breccia limestone; 4—shale; 5—sandstone; 6—breccia; 7—trilobite (Fu, 1994 ; Du & Tong, 2009)

Correlation by the similarity of rock types is more sound if a very uncommon succession of rocks is involved. If you find in one region a layer of gray shale on top of red sandstone that, in turn, overlies basalt of a former lava flow and then find the same series in another area, you probably would be correct in concluding that the two rows formed at basically the same time.

In some regions, a key bed, a distinguishing layer, can correlate rocks over great distances. A sample is a layer of volcanic ash generated from a massive eruption and distributed over a substantial portion of a continent.

2. Fossil correlation

Fossils are commonly found in sedimentary rocks, and their presence is essential for correlation. Sediments buried plants and animals that lived when the rock formed, and their fossil remains are preserved in sedimentary rock. Most of the fossil species found in rock formations are now extinct — 99.9% of all species that have ever lived are vanished.

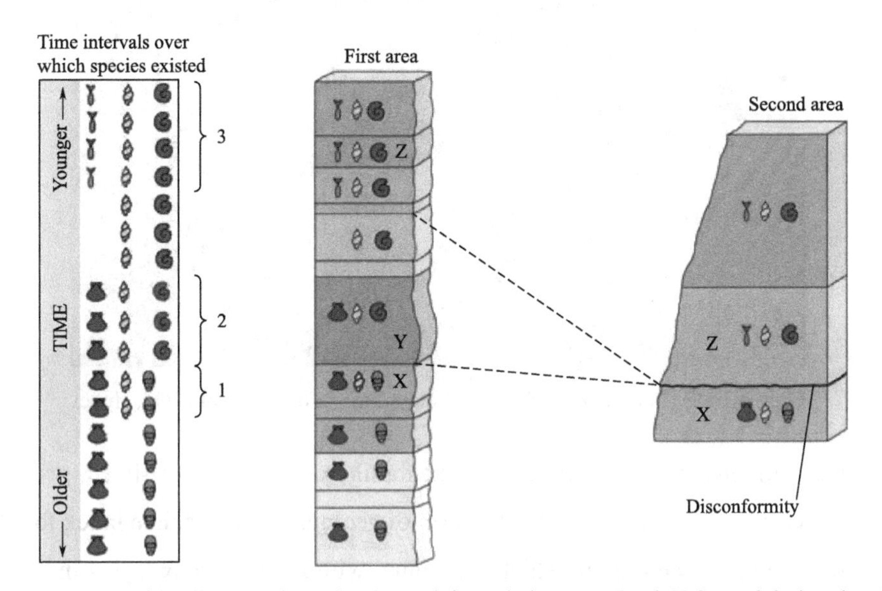

Figure 3-14 The use of fossil assemblages for determining relative ages. Rock X formed during time interval 1. Rock Y formed during time interval 2. Rock Z formed during time interval 3. In the second area, fossils of time interval 2 are missing. Therefore, the surface between X and Z is a disconformity (Carlson et al., 2011).

In a thick succession of sedimentary rock layers, the fossils nearer the bottom (that is, in the older rock) are more contrary to today's living body than are those near the top. At the end of the eighteenth century, naturalists accomplished that the fossil remains of creatures of a series of "former worlds" were maintained in the Earth's sedimentary rocks. In the early nineteenth century, an English investigator named William Smith realized that specific fossil species characterize various sedimentary layers and that fossil species follow one another through the layers in a predictable sequence. Smith's finding of this principle of animal succession allowed rock layers in separate places to be connected based on their fossils. We now understand that animal succession works because there is an evolutionary record of life on Earth. Species evolve, exist for a time, and go vanished. Because the identical species never develop twice (extinction has been forever), any period of time in Earth history can be determined by the species that lived in that period. Paleontologists, specialists in the research of fossils, have firmly and carefully over the years discovered many species of fossils and determined the time progression in which they existed. Consequently, sedimentary rock layers anywhere in the world can be allocated to their correct place in geologic history by selecting the fossils they comprise.

Figure 3-15 The trilobites from the Late Silurian Formation of Chibi, Hubei

Ideally, a geologist expects to find an index fossil from a very short-lived, geographically pervasive species known to exist in a specific term of geologic time. A single index fossil permits the geologist to compare the rock in which it is found with all other rock layers that include the fossil.

Many fossils are of little application in time determination because the species thrived during too extensive a segment of geologic time. Sharks, for example, have been in the oceans for a long time, so exploring an ordinary tooth of a shark in a rock is not very useful in definitive the rock's relative age.

A single fossil that is not an indicator fossil is not very valuable for determining the rock's age. However, finding several species of fossils in a layer of rock is generally more useful for dating rocks than a single fossil because the sediment must have been deposited when all the species represented existed. Figure 3-14 depicts five species of fossils, each of which lived over a long period. Where diverse combinations of these fossils are discovered in three rocks, the time of formation of each rock can be charged to a limited period (Carlson et al., 2011).

Some fossils are reserved in geographic appearance, representing organisms suited to particular environments. But many former organisms lived over most of the Earth, and fossil assemblages from these may be used worldwide. Fossils in the horizontal layers of the Chibi are comparable to the ones collected in the east of Yunan and many other places in the world (the trilobite in Figure 3-15). We can, hence, connect these rock elements and say they formed during the same general duration of geologic time.

3.4 Numerical age and geologic time

3.4.1 Numerical age

Counting the yearly growth rings in a tree trunk will indicate the age of a tree. Similarly, the sediment layers deposited each year in glacial lakes can be counted in determining how long these lakes have been measured and replicated in many different laboratories. Therefore, if we can decide the ratio of a specific radioactive element and its decay goods in a mineral, we can estimate how long ago that mineral crystallized. Determining the age of rock through its radioactive elements is known as isotopic dating (previously called radiometric dating). Geologists who specialize in this critical field are known as geochronologists.

Half-lives of isotopes, whether short-lived, such as used in medicine or long-lived, such as used in isotopic dating, have been discovered not to vary out statistical anticipations. The half-life of each radioisotope we utilize for dating rocks has not changed with physical conditions or chemical activeness, nor could the rates have been distinct in the past. It would break the principles of physics for decay rates to have been alternative in the past. Furthermore, when several isotopic dating schemes are carefully done on only an ancient igneous rock, the same age is gained within calculable borders of error. It affirms that the decay constants for each scheme are actually constant.

Comparing isotopic ages with relative age relationships approve the reliability of isotopic dating. For example, a dike that crosscuts rocks containing Cenozoic fossils gives us a relatively young isotopic age (less than 65 million years old). Conversely, a pluton truncated by overlying sedimentary rocks with the earliest Paleozoic fossil bears a fairly old age (greater than 544 million years). Many thousands of similar determinations have proven the reliability of radiometric dating.

3.4.2 Numerical age is preferred to absolute age

Geologists have explored the world for localities where they can understand cross-cutting relations between datable igneous rocks and sedimentary rocks or for beds of datable volcanic rocks interbedded with sedimentary rocks. By isotope dating of igneous rocks, they could provide numerical ages for the boundaries between all geological periods. For instance, work from around the world proves that the Cretaceous period commenced roughly 145 million years ago and terminated 65 million years ago.

The discovery of new data may cause the numbers defining the boundaries of periods to change, so the term numerical age is preferred to absolute age. Around 1995, new dates on rhyolite ash layers above and below the Cambrian-Precambrian boundary showed that this boundary occurred 542 million years ago, in contrast to previous, less definitive studies that had placed the border at 570 million years ago.

3.4.3 The standard geologic time scale

For the convenience of study and references and a relative comparison of the ages of a different sequence of rocks found in various places on the earth's surface, it is necessary to have a proper framework of geological time (Figure 3-16). This demand is met by the internationally accepted "geological time scale". It's like the calendar of a year. A year is divided into different months, each month into weeks, each week into days, each day into hours, etc. The geological time scale is similarly subdivided into smaller and smaller units to suit the stratigraphical study and correlation. Geologists have divided the total of geologic history into units of various lengths. Together, they create the geologic time scale of Earth history (Figure 3-17) (Cohen et al., 2013). The central units of the time scale were delineated during the nineteenth century, principally by researchers in Western Europe and Great Britain. Because radiometric dating was unreachable at that time, the full-time scale was established using relative dating means. It was only in the twentieth century that radiometric methods permitted numerical dates to be added. A Global Boundary Stratotype Section and Point (GSSP) is an globally agreed-upon reference point on a stratigraphic section that defines the lower boundary on the geologic time scale (Wikipedia,2021). The International Commission conducts the effort to determine GSSPs on Stratigraphy, a part of the International Union of Geological Sciences.

The geological time scale has been subdivided into 4 Eons recently. They are the Phanerozoic Eon, Proterozoic Eon, Achaean Eon, Hadean Eon.

Geologists can apply fossils in the rock to refer the age of the rock to the standard geologic time scale, an international relative time scale. Based on fossil accumulation, the geologic time scale splits geologic time. The geologic time scale has had enormous meaning as a unified concept in the physical and biological sciences. The working out of the evolving chronology by consecutive generations of geologists and other scientists has been an outstanding human accomplishment. The geologic time scale, representing a wide fossil record, consists of three eras divided into periods, which are, in turn, subdivided into epochs.

The duration of the period between any two such consecutive extraordinary events is referred to as an Era (erathem). Precambrian indicates the vast amount of time headed the

Paleozoic Era (which begins with the Cambrian Period). The Paleozoic Era (meaning “old life”) started with the appearance of complex life (trilobites, for example), as indicated by fossils. Rocks older than Paleozoic contain few fossils. It is because creatures with shells or other complex parts, easily preserved as fossils, did not evolve until the beginning of the Paleozoic.

The Mesozoic Era (meaning “middle life”) accompanied the Paleozoic. On land, dinosaurs converted the powerful animals of the Mesozoic. We are in the Holocene (or Recent) Epoch of the Quaternary Period of the Cenozoic Era (meaning “new life”). The Quaternary also covers the most modern ice ages, which were a portion of the Pleistocene Epoch.

Remarkably, the fossil record indicates mass extinctions, in which many species became defunct, occurred several times in the geologic past. The two most significant mass extinctions define the boundaries between the three eras.

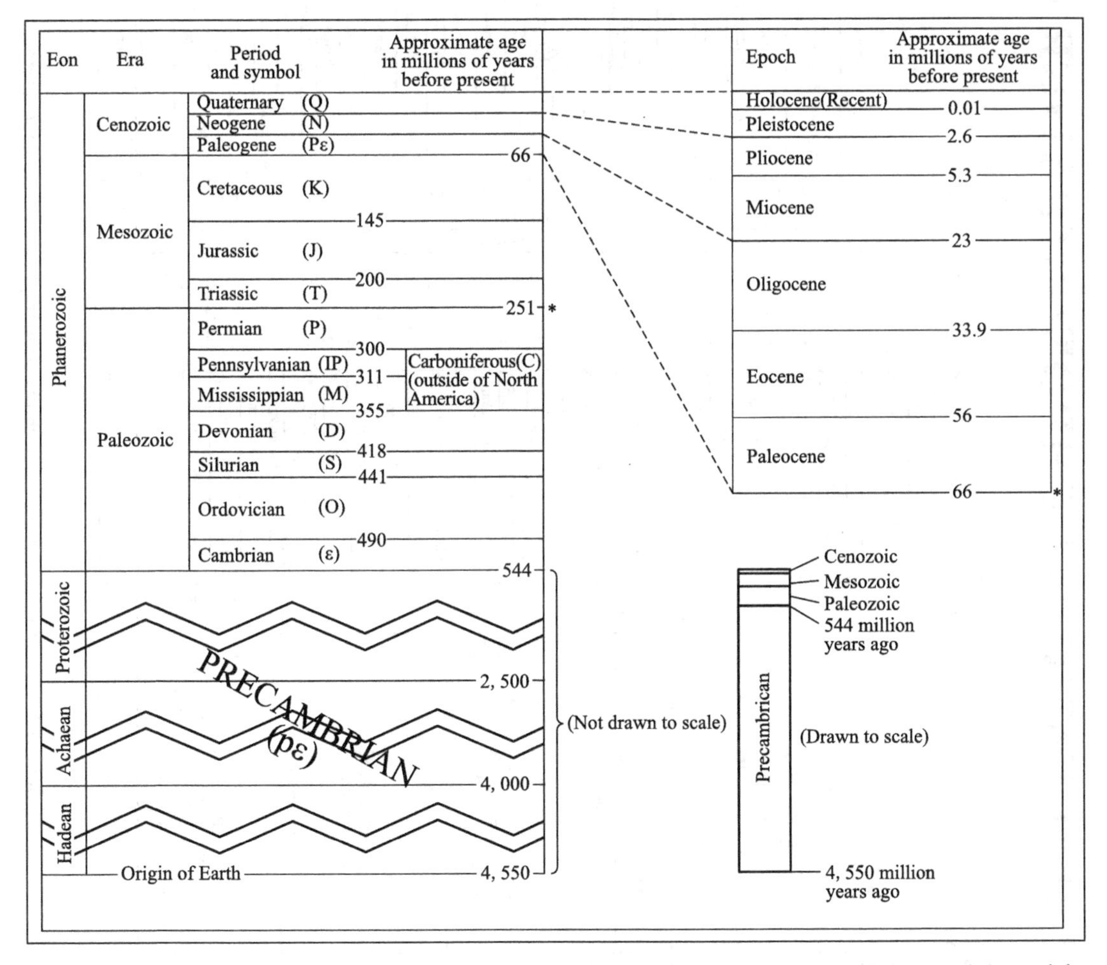

Figure 3-16 The simplified geologic time scale. The small diagram to the right shows the Precambrian and the three eras on the same scale. Note that the Precambrian accounts for almost 90% of geologic time

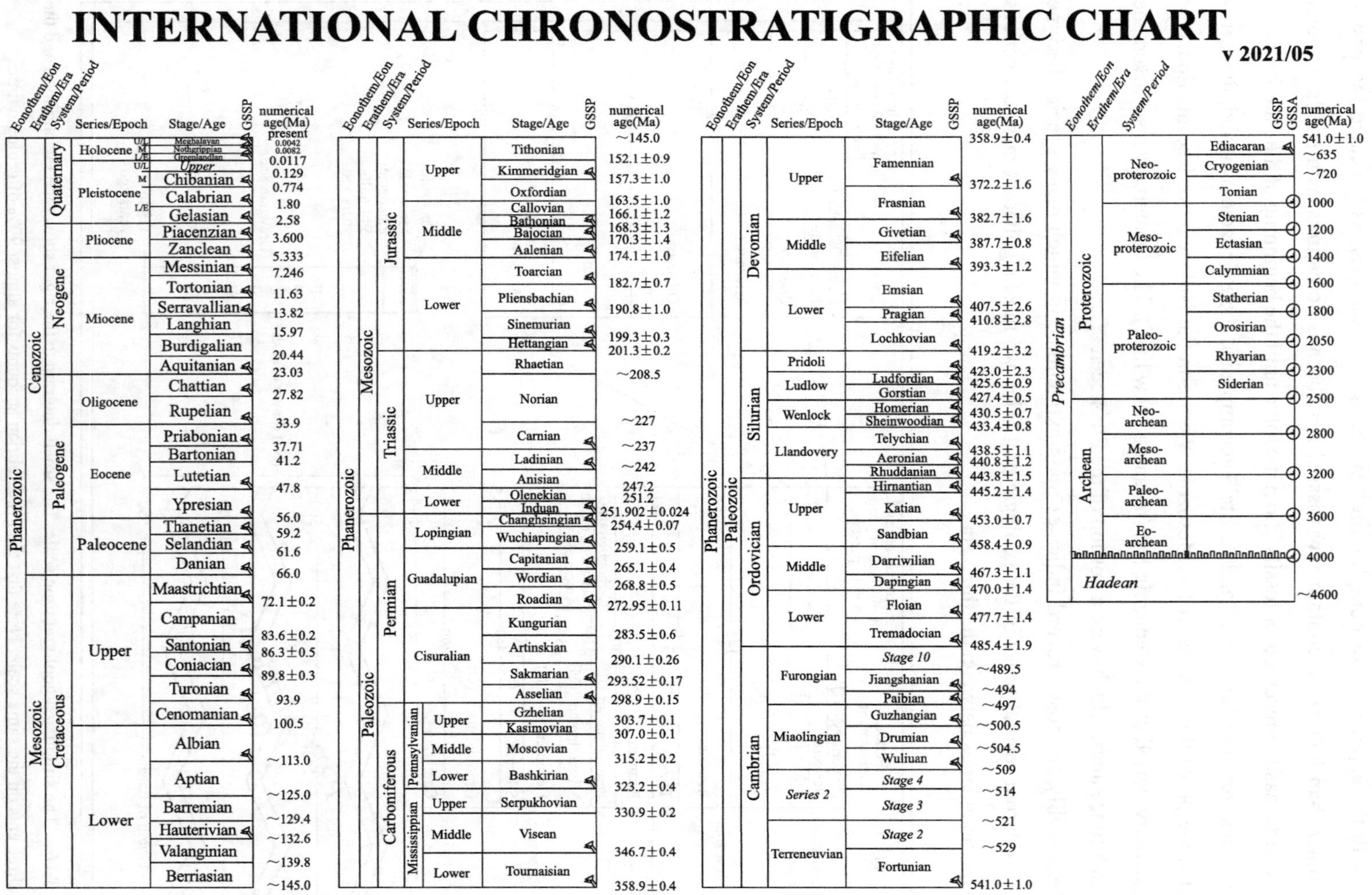

INTERNATIONAL CHRONOSTRATIGRAPHIC CHART

v 2021/05

Eonothem/Eon	Erathem/Era	System/Period	Series/Epoch	Stage/Age	numerical age(Ma)
					present
Phanerozoic	Cenozoic	Quaternary	Holocene U/L	Meghalayan	0.0042
			Holocene M	Northgrippian	0.0082
			Holocene L/E	Greenlandian	0.0117
			Pleistocene U/L	Upper	0.129
			Pleistocene M	Chibanian	0.774
			Pleistocene L/E	Calabrian	1.80
				Gelasian	2.58
		Neogene	Pliocene	Piacenzian	3.600
				Zanclean	5.333
			Miocene	Messinian	7.246
				Tortonian	11.63
				Serravallian	13.82
				Langhian	15.97
				Burdigalian	20.44
				Aquitanian	23.03
		Paleogene	Oligocene	Chattian	27.82
				Rupelian	33.9
			Eocene	Priabonian	37.71
				Bartonian	41.2
				Lutetian	47.8
				Ypresian	56.0
			Paleocene	Thanetian	59.2
				Selandian	61.6
				Danian	66.0
	Mesozoic	Cretaceous	Upper	Maastrichtian	72.1±0.2
				Campanian	83.6±0.2
				Santonian	86.3±0.5
				Coniacian	89.8±0.3
				Turonian	93.9
				Cenomanian	100.5
			Lower	Albian	~113.0
				Aptian	~125.0
				Barremian	~129.4
				Hauterivian	~132.6
				Valanginian	~139.8
				Berriasian	~145.0

Eonothem/Eon	Erathem/Era	System/Period	Series/Epoch	Stage/Age	numerical age(Ma)
					~145.0
Phanerozoic	Mesozoic	Jurassic	Upper	Tithonian	152.1±0.9
				Kimmeridgian	157.3±1.0
				Oxfordian	163.5±1.0
			Middle	Callovian	166.1±1.2
				Bathonian	168.3±1.3
				Bajocian	170.3±1.4
				Aalenian	174.1±1.0
			Lower	Toarcian	182.7±0.7
				Pliensbachian	190.8±1.0
				Sinemurian	199.3±0.3
				Hettangian	201.3±0.2
		Triassic	Upper	Rhaetian	~208.5
				Norian	~227
				Carnian	~237
			Middle	Ladinian	~242
				Anisian	247.2
			Lower	Olenekian	251.2
				Induan	251.902±0.024
	Paleozoic	Permian	Lopingian	Changhsingian	254.4±0.07
				Wuchiapingian	259.1±0.5
			Guadalupian	Capitanian	265.1±0.4
				Wordian	268.8±0.5
				Roadian	272.95±0.11
			Cisuralian	Kungurian	283.5±0.6
				Artinskian	290.1±0.26
				Sakmarian	293.52±0.17
				Asselian	298.9±0.15
		Carboniferous	Pennsylvanian Upper	Gzhelian	303.7±0.1
				Kasimovian	307.0±0.1
			Pennsylvanian Middle	Moscovian	315.2±0.2
			Pennsylvanian Lower	Bashkirian	323.2±0.4
			Mississippian Upper	Serpukhovian	330.9±0.2
			Mississippian Middle	Visean	346.7±0.4
			Mississippian Lower	Tournaisian	358.9±0.4

Eonothem/Eon	Erathem/Era	System/Period	Series/Epoch	Stage/Age	numerical age(Ma)
					358.9±0.4
Phanerozoic	Paleozoic	Devonian	Upper	Famennian	372.2±1.6
				Frasnian	382.7±1.6
			Middle	Givetian	387.7±0.8
				Eifelian	393.3±1.2
			Lower	Emsian	407.5±2.6
				Pragian	410.8±2.8
				Lochkovian	419.2±3.2
		Silurian	Pridoli		423.0±2.3
			Ludlow	Ludfordian	425.6±0.9
				Gorstian	427.4±0.5
			Wenlock	Homerian	430.5±0.7
				Sheinwoodian	433.4±0.8
			Llandovery	Telychian	438.5±1.1
				Aeronian	440.8±1.2
				Rhuddanian	443.8±1.5
		Ordovician	Upper	Hirnantian	445.2±1.4
				Katian	453.0±0.7
				Sandbian	458.4±0.9
			Middle	Darriwilian	467.3±1.1
				Dapingian	470.0±1.4
			Lower	Floian	477.7±1.4
				Tremadocian	485.4±1.9
		Cambrian	Furongian	*Stage 10*	~489.5
				Jiangshanian	~494
				Paibian	~497
			Miaolingian	Guzhangian	~500.5
				Drumian	~504.5
				Wuliuan	~509
			Series 2	*Stage 4*	~514
				Stage 3	~521
			Terreneuvian	*Stage 2*	~529
				Fortunian	541.0±1.0

Eonothem/Eon	Erathem/Era	System/Period		numerical age(Ma)
				541.0±1.0
Precambrian	Proterozoic	Neoproterozoic	Ediacaran	~635
			Cryogenian	~720
			Tonian	1000
		Mesoproterozoic	Stenian	1200
			Ectasian	1400
			Calymmian	1600
		Paleoproterozoic	Statherian	1800
			Orosirian	2050
			Rhyacian	2300
			Siderian	2500
	Archean	Neoarchean		2800
		Mesoarchean		3200
		Paleoarchean		3600
		Eoarchean		4000
	Hadean			~4600

Figure 3-17 The most recent complete chronostratigraphic chart from the international commission on stratigraphy

3.4.4 What is the age of the earth?

During the 18th and 19th centuries, before discovering isotopic dating, scientists developed a great variety of clever resolutions to the question, "How old is the Earth?"—all of which have since been proven wrong. William Kelvin(1824-1907), a 19th-century physicist renowned for his discoveries in thermodynamics, made the most influential scientific estimation of the Earth's age of his time. Kelvin measured how long it would take for the Earth to cool down from a temperature as hot as the Sun's and inferred it is about 20 million years old.

Kelvin's survey contrasted with those promoted by followers of Hutton, Charles Lyell(1797-1875), and Charles Darwin (1809-1882), who argued that the thoughts of uniformitarianism and evolution were right. The Earth must be much older. They claimed that physical processes that form the Earth and create its rocks and the process of natural choice that yields the variety of species all take a very long time. Geologists and physicists advanced to argue the age issuance for many years. The way to a resolution didn't appear until 1896 when Henri Becquerel(1852-1908) declared the invention of radioactivity. Geologists instantly understood that the Earth's center was producing heat from the decay of radioactive substances. This understanding uncovered one of the flaws in Kelvin's thought: Kelvin presumed no new heat was created after the Earth was first formed. Because radioactivity regularly produces new heat in the Earth, the planet has cooled down much more slowly than Kelvin had judged and could be much older. The invention of radioactivity invalidated Kelvin's view of the Earth's age and led to the evolution of isotopic dating.

Since the 1950s, geologists have searched the planet to recognize its oldest rocks. Rocks younger than 3.85Ga are pretty standard (Recall that "Ga" means "billion years ago."). Rock samples from several locations (China, Canada, Greenland, and USA) have yielded as old as 4.03Ga. It is reported that a zircon from the Paleozoic captured zircon with an age of 4079 ± 5Ma was obtained from the volcanic rocks in the western section of the North Qinling, China (Wang et al., 2007). Specific clastic grains of the mineral zircon have yielded dates of up to 4.40Ga, indicating that rock as old as 4.40Ga did once exist. Isotopic dating of Moon rocks bears dates of up to 4.50Ga, and dates on meteorites have admitted ages as old as 4.57Ga. Geologists consider 4.57Ga meteorites to be remnants of planetesimals like those from which the Earth first created. Thus, these dates are near to the age of the Earth's origin, for models of the Earth's formation pretend that all objects in the Solar System evolved at approximately the same time from the identical nebula.

3.4.5 Comprehending geologic time

The immensity of geological time (sometimes referred to as deep time) is hard to grasp. One way of visualizing deep time is to imagine driving from Jiuqunao to Lianziya, a distance of approximately 25km, where each kilometer represents 10 million years—this is a very, very slow trip. Travel highlights consistent with Earth's history are illustrated in Figure 3-18. Note that if you live to be 100 years old, your life is represented by less than the width of a curb on a pavement.

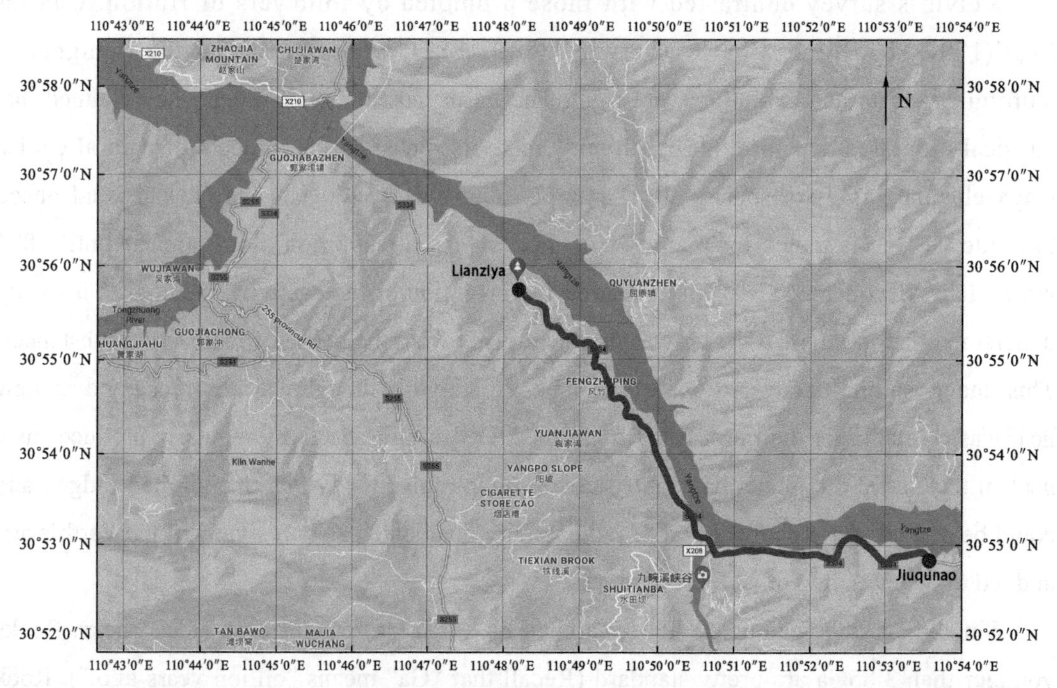

Figure 3-18 Going from Jiuqunao to Lainziya, a distance of approximately 25km, each kilometer represents 10 million years

Imagining our endurance as taking less than a frame of a movie can be very humbling. From the view of being attached in that one last frame, geologists would like to know what the whole film is like or, at least, get an outline of the most dramatic parts of the film.

Reference

[1] CARLSON D H, PLUMMER C C, CARLSON L. Physical geology: earth revealed [M]. New York City : McGraw-Hill, 2011.

[2] COHEN K M, FINNEY S C, GIBBARD P L, FAN J X. (2013; updated) The ICS

International Chronostratigraphic Chart[EB/OL]. Episodes 36: 199-204.[2021-05-07]. http://www.stratigraphy.org/ICSchart/ChronostratChart2021-05.pdf.

[3] DU Y S, TONG J N. An introduction to paleobiohistory[M].2nd ed. Wuhan : China University of Geosciences Press, 2009.

[4] WIKIPEDIA. Global Boundary Stratotype Section and Point[EB/OL].[2021-07-25]. https://en.wikipedia.org/wiki/Global_Boundary_Stratotype_Section_and_Point.

[5] WANG H L, CHEN L, SUN Y. Zircon xenoliths from The Ordovician volcanic rocks in the western Part of the North Qinling Mountains[J], China : Chinese Science Bulletin, 2020, 52(14) : 1685-1693.

[6] LUTGENS F K, TARBUCK E J, TASA D G. Essentials of geology (13th Edition)[EB/OL].[2020-08-21]. https://www.pearson.com/us/higher-education/program/PGM1240907.html.

[7] MARSHAK S. Essentials of geology[M]. Norton : WW Norton, 2013.

[8] FU Y Q. A concise course of Paleontological geohistory[M]. Beijing : Geological Publishing House, 1994.

Exercises

Choice Questions (Single Choice)

(1) "Within a series of undisturbed sedimentary rocks, the beds get younger going from bottom to top" is the principle of:

(a) original horizontality (b) superposition

(c) crosscutting (d) none of the preceding

(2) Eras are subdivided into:

(a) periods (b) eons

(c) ages (d) epochs

(3) Periods are subdivided into:

(a) eras (b) epochs

(c) ages (d) time zones

(4) Which is not a type of unconformity?

(a) disconformity (b) angular unconformity

(c) nonconformity (d) triconformity

(5) A geologist could use the principle of composition to determine the relative age of:

(a) fossils (b) metamorphism

(c) shale layers (d) xenoliths

(6) A contact among parallel sedimentary rock that records missing geologic time is:

(a) a disconformity (b) an angular unconformity

(c) a nonconformity (d) a sedimentary contact

Review Questions

(1) Explain the concept of uniformitarianism.

(2) Compare numerical age and relative age.

(3) Explain the principles that allow us to determine the relative ages of geologic events.

(4) How does the principle of fossil succession will enable us to determine the relative ages of strata?

(5) How does an unconformity develop?

(6) Can you make an explanation about the types of unconformity? You can add a simple graph to support your answer.

(7) How do geologists obtain an isotopic date? What are some of the pitfalls in getting a reliable one?

(8) What is the age of the oldest rocks on Earth? What is the age of the oldest rocks known?

4
Geologic Structures

4.1 Introduction

This chapter explains how rocks respond to tectonic forces caused by the movement of lithospheric plates and how geologists study the resulting geologic structures. Studying structural geology is very much like looking at the architecture of the crust and trying to relate how rocks that were once deposited underwater in horizontal layers are now bent (folded) and broken (faulted) many kilometers above sea level.

The geological structure is the foundation of geological research. To understand earthquakes, for instance, one must understand how faults are generated and behave. Appreciating how major mountain belts and the continents have evolved calls for a comprehension of faulting and folding. Understanding plate-tectonic theory as a whole also requires a knowledge of structural geology. The plate-tectonic theory was developed primarily to explain certain structural features. In areas of active tectonics, the location of geologic structures is essential in selecting safe sites for significant infrastructures such as schools, hospitals, dams, bridges, nuclear power facilities, etc. Also, understanding structural geology can help us appreciate the problem of finding more of Earth's dwindling natural resources better (Carlson et al., 2011 & Plummer et al., 2016).

4.2 Tectonic forces at work

4.2.1 Stress and Strain in the Earth's Lithosphere

Tectonic forces deform parts of the lithosphere, particularly along plate margins, and deformation may cause a change in the orientation, location, and shape of a rock body. In Figure 4-1, horizontal rock layers have been deformed into wavelike folds broken by faults. The layers have been distorted, probably by tectonic forces that pushed or compressed the layers together until they were shortened by buckling and cracking.

When studying deformed rocks, structural geologists typically refer to stress, a force per unit area. Where stress can be measured, it is expressed as the force per unit area at a particular point; however, it is difficult to measure stress in buried rocks. We can observe the effects of past stress (caused by tectonic forces and confining pressure from burial) when rock bodies are exposed after uplift and erosion. From our observations, we may be able to infer the principal directions of stress that prevailed. Also, we can observe the effect of forces on a rock that was stressed for exposed rocks. Strain is the change in shape or size (volume), or both, in response to applied stresses.

When stresses act parallel to a plane, shear stress along the plane is developed. It is much like putting a deck of cards and shearing the deck by moving your hands in opposite directions. Shear stress results in a shear strain parallel to the direction of the stresses. Shear stresses occur along actively moving faults.

Figure 4-1 Deformed sedimentary beds exposed in a parking. The sedimentary layers is contorted into folds and broken by smaller faults

4.2.2 Mechanical behavior of rock under pressure

Rocks behave as elastic, ductile, or brittle materials, depending on the amount and rate of stress applied, the type of rock, and the temperature and pressure under which the rock is strained.

The behavior is elastic if a deformed material recovers its original shape after the stress is reduced or removed. For example, if tensional stress is applied to a rubber band, it will stretch as long as the stress is involved. Still, once the stress is removed, the rubber band returns (or recovers) to its original shape (if its behavior is linear elastic). The plasticine will behave elasticity if molded into a ball and bounced. Most rocks at very low stresses act elastically. However, once the stress applied exceeds the elastic limit (Figure 4-2), the rock will deform permanently, just as the rubber band will break if stretched too far.

A rock that behaves in a ductile or plastic manner will bend while under stress and does not return to its original shape after the stress is removed. The plasticine acts as a ductile material unless the rate of strain is rapid. Rocks exposed to elevated pressure and temperature during regional metamorphism also behave in a ductile manner and develop a planar texture, or foliation, due to the alignment of minerals. As shown in Figure 4-2, material behaving in a ductile manner does not require much of an increase in stress to continue to strain (relatively flat curve). Ductile behavior results in permanently deformed rocks mainly by folding or bending of rock layers. A rock exhibiting brittle behavior will fracture at stresses higher than its elastic limit or once the stresses are greater than the strength of the rock. Rocks typically exhibit brittle behavior at or near Earth's surface, where temperatures and pressure are low. Under these conditions, rocks favor breaking rather than bending. Faults and joints are examples of structures that form by the brittle behavior of the crust.

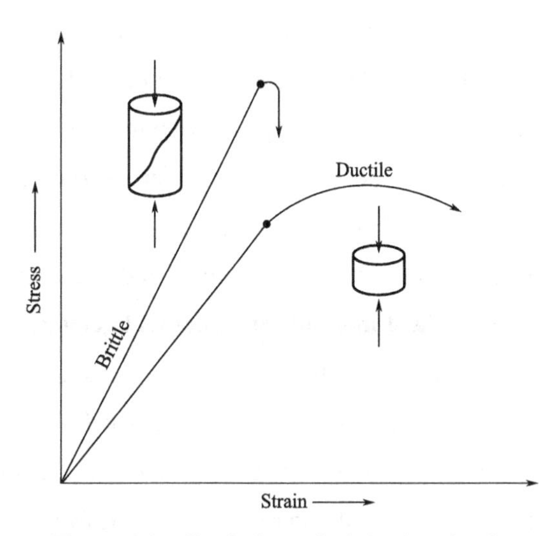

Figure 4-2 Graph shows the behavior of rocks with increasing stress and strain. Elastic behavior occurs along the straight-line portions of the graph. At stresses greater than the elastic limit, the rock will deform as a ductile material or break, as shown in the deformed rock specimens

A sedimentary rock exposed to the Earth's surface is brittle and will fracture if you hammer it. How then do sedimentary rocks, such as those shown in Figure 4-1, become bent (or deformed in a ductile way)? The answer is that stress increased very slowly, or the rock was deformed under considerable confining pressure (buried under more rock) and higher temperatures. However, there are some fractures (faults) disrupting the bent layers in Figure 4-1. It tells us that although the rock was ductile in the beginning, the amount of stress increased or the strain rate increased, and the rock fractured eventually.

4.3 Structures as a record of the geologic past

Some geologic activities that give us clues to the past have been described in earlier chapters. Batholiths, stocks, dikes, and sills, for example, are keys to past igneous activity. This chapter is mainly concerned with types of structures that can provide a record of crustal deformation that is no longer active. Ancient structures that are now visible at the Earth's surface were once buried and are exposed through erosion. The study of geological structures is of more than academic interest. The oil and mining industries, for instance, employ geologists to research and map geological structures associated with oil and metal ore deposits. Understanding and mapping geologic structures are also crucial for evaluating problems related to engineering decisions and seismic risks, such as determining the most appropriate sites for building dams, large bridges, or nuclear reactors, and even houses, schools, and hospitals.

1. Field methods of structural geology

In an ideal situation, a geologist studying structures would fly over an area and see the local and regional patterns of bedrock from above. Sometimes this is possible, but very often, soil and vegetation conceal the bedrock. Therefore, geologists ordinarily use observations from several individual outcrops (geological formation exposed on the surface) to determine geologic structures'patterns (Figure 4-3). The characteristics of rock at each outcrop in an area are plotted on a map using appropriate symbols. With the data collected, a geologist can make inferences about those parts of the site that cannot be observed.

A geologic map, which uses standardized symbols and patterns to represent rock types and geologic structures, is typically produced from the field map for a given area. A map of the classification and distribution of rock units, the occurrence of structural features (folds, faults, joints, etc.), ore deposits, and so forth are plotted. Sometimes special features, such as deposits by former glaciers, are included, but these may be shown separately on a different geologic map. The content of the geologic map will be introduced in Chapter 5.

2. Strike and Dip

According to the principle of original horizontality, sedimentary rocks and some lava flows and ash falls are deposited as horizontal beds or strata. Where these originally horizontal rocks are found tilted indicates that tilting must have occurred after deposition and lithification (Figure 4-4). Geologists make an effort to measure and record geological structures because they are essential to understanding the geological history of an area. One of the key features to measure is the orientation or attitude (Earle, 2015). When studying a geological map of an area, the extent and direction of the inclination should be reported. By convention, this is determined by plotting

Figure 4-3 Geology students mapping the fractured rocks

the relationship between a surface of an inclined bed and an imaginary horizontal plane. You can understand the relationship by carefully examining Figure 4-5, which represents exposed sedimentary layers along a lake (the lake surface provides a convenient horizontal plan for this discussion).

The strike is the compass direction of a line formed by the intersection of an inclined plane with a horizontal plane. In this example, an inclined plane is a bedding plane. You can see from Figure 4-5 that the beds are striking from north to south. Usually, only the northerly direction (of the strike line) is given, so we say that beds strike north a certain number of degrees east or west (such as N50° E).

Observe that the dip angle is measured downward from the horizontal plane to the bedding plane (an inclined plane). The dip angle (30° in the figure) is measured within a vertical plane perpendicular to both the bedding and horizontal planes.

The direction of dip is the compass direction in which the angle of dip is measured. If you could roll a ball down a bedding surface, the compass direction in which the ball rolled would be the direction of dip. The dip direction is the azimuth of the dip direction as projected to the horizontal (Wikipedia,2021). The dip angle is always measured at a right angle to the strike—perpendicular to the strike line, as shown in Figure 4-5. Because the beds could dip away from the strike line in either of two possible directions, the general direction of dip is also specified, in this example, it is west.

Figure 4-4 Tilted sedimentary beds at a quarry in Chibi City. The direction of dip is toward the left(Photo by Tao Wang)

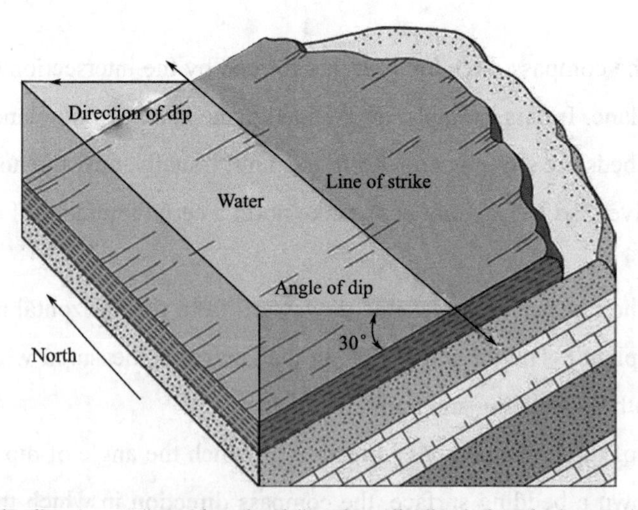

Figure 4-5 Strike, the direction of dip, and angle of dip. The strike line is found where an inclined bed intersects a horizontal plane (as shown here by the water). The dip direction is always perpendicular to the strike, and in the direction the bed slopes. The dip angle is the vertical angle of the inclined bed as measured from the horizontal

A specially designed instrument called the geological compass or a Brunton pocket transit (after the inventor) is used by geologists for measuring the strike and dip (Figure 4-6). The pocket transit contains a compass, a level, and a device for measuring angles of inclination. Besides recording strike and dip measurements in a field notebook, a geologist who is mapping an area draws strike and dip symbols on the field map for each outcrop with dipping or tilted beds. On the map, the intersection of the two lines at the center of each strike and dip symbol represents the location of the outcrop where the strike and dip of the bedrock were measured. The long line of the symbol is aligned with the compass direction of the strike. The small tick, which is always drawn perpendicular to the strike line, is put on one side or the other, depending on the two directions the beds' dip. The angle of dip is given as a number next to the appropriate symbol on the map. Thus, ^{30}Y indicates that the bed is dipping 30° from the horizontal toward the northwest and the strike is northeast (assuming that the top of the page is north). The orientation of the bed would be written N45° E, 30° NW.

Figure 4-6 Geology student measuring the strike and dip of an inclined limestone bed in the Three Gorges Region (Photo by Tao Wang)

Beds with vertical dip require a unique symbol because they dip neither to the left nor the right of the direction of the strike. The symbol used is X , which indicates that the beds are striking northeast and that they are vertical (N30° E, 90°).

4.4 Folds

Folds are bends or wavelike features in the layered rock. Folded rock can be compared to several layers of rugs or blankets that have been pushed into a series of arches and troughs. Folds in rock often can be seen in road cuts or other exposures (Figure 4-7). When the arches and troughs of folds are concealed (or when they exist on a grand scale), geologists can still determine the presence of folds by noticing repeated reversals in the direction of dip taken on outcrops in the field shown on a geologic map.

The fact that the rock is folded or bent shows that it behaved as a ductile material. Yet the rock exposed in outcrops is generally brittle and shatters when struck with a hammer. The rock is not metamorphosed (most metamorphic rock is intensely folded because it is ductile under the high pressure and temperature environment of deep burial and tectonic stresses). Perhaps folding took place when the rock was buried at a moderate depth where higher temperature and confining pressure favor ductile behavior. Alternatively, folding could have taken place close to the surface under a very low rate of strain.

Figure 4-7 Folded sedimentary rock layers exposed at Jiuwan Stream, Yichang

1. Geometry of Folds

Determining the geometry or shape of folds can have important economic implications because many oil and gas deposits and some metallic mineral deposits are located in the folded rocks. The geometry of folds is also essential in unraveling how a rock was strained and how it might be related to the movement of tectonic plates. Folds are usually associated with the shortening of rock layers along convergent plate boundaries, but are also commonly formed where the rock has been sheared along a fault. Because folds are wavelike forms, two basic fold geometries are —anticlines and synclines (Figure 4-8).

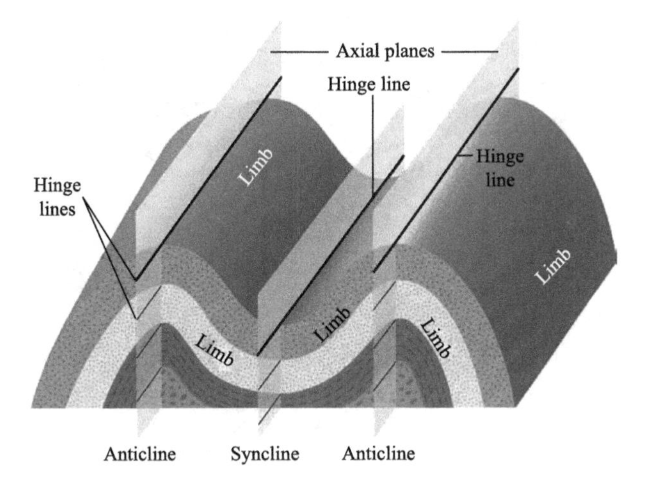

Figure 4-8 Diagrammatic sketch of two anticlines and a syncline illustrating the axial planes, hinge lines and fold limbs

An anticlinal is an arcing fold with the oldest rocks in the centre of the fold. Usually the rock layers dip away from the hinge line (or axis) of the folds. The counterpart of an anticline is a syncline, a fold shaped like a trough with the youngest rocks in the center of the fold. The layered rock usually dips toward the syncline's hinge line. In the series of folds shown in Figure 4-8, two anticlines are separated by a syncline. Each anticline and adjacent syncline share a limb. Note the hinge lines on the crests of the two anticlines and bottom of the syncline. Similar hinge lines could be located in the hinge areas at the contacts between any two adjacent folded layers. For each anticline and the syncline, the hinge lines are contained within the shaded vertical planes. Each of these planes is an axial plane, an imagined plane containing all the hinge lines of a fold. The axial plane divides the fold into its two limbs.

It is important to remember that anticlines are not necessarily related to ridges nor synclines to valleys because valleys and ridges are nearly always erosional features. In an area eroded to a plain, the presence of underlying anticlines and synclines is determined by the direction of dipping beds in exposed bedrock, as shown in Figure 4-9.

Figure 4-9 also illustrates how determining the relative ages of the rock beds can tell us whether a structure is an anticline or a syncline. Observe that the oldest exposed rocks are along the hinge line of the anticline. The lower layers in the original flat-lying sedimentary or volcanic rock were moved upward and are now at the core of the anticline. Initially, the youngest rocks in the upper layers were folded downward and are now exposed along the synclinal hinge line.

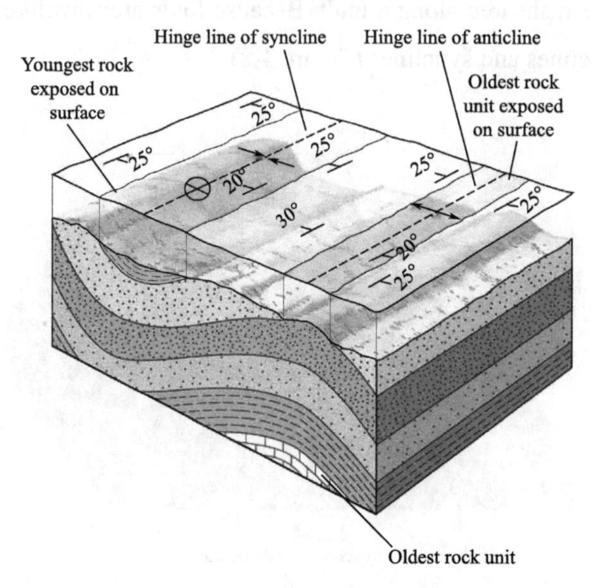

Figure 4-9 By measuring the strike and dip of exposed sedimentary beds in the field and plotting them on a geologic map (top surface), geologists can interpret the geometry of the geologic structure below the ground surface(Modified from Plummer et al.,2016)

2. Plunging Fold

The examples presented so far are horizontally hinged folds, and these are the most easily visualized. However, in nature, anticlines and synclines are apt to be plunging folds—that is, folds in which the hinge lines are not horizontal. On a surface leveled by erosion, the patterns of exposed strata (beds) resemble vs. or horseshoes (Figure 4-10 and Figure 4-11) rather than the parallel, striped patterns of layers in non-plunging folds. However, plunging anticlines and synclines are distinguished from one another in the same way as are non-plunging folds by directions of dip or by relative ages of beds.

A plunging syncline contains the youngest rocks in its center, and the V points in the direction opposite of the plunge. Conversely, a plunging anticline includes the oldest rocks in its core, and the V points in the same direction as the plunge of the fold.

3. Structural Domes and Structural Basins

A dome is a structure in which the oldest rocks are found in the middle, and the beds dip away from a core point. In cross-section, a dome looks like an anticline and is frequently called a doubly plunging anticline. In a basin, the beds dip toward a central point, and the youngest

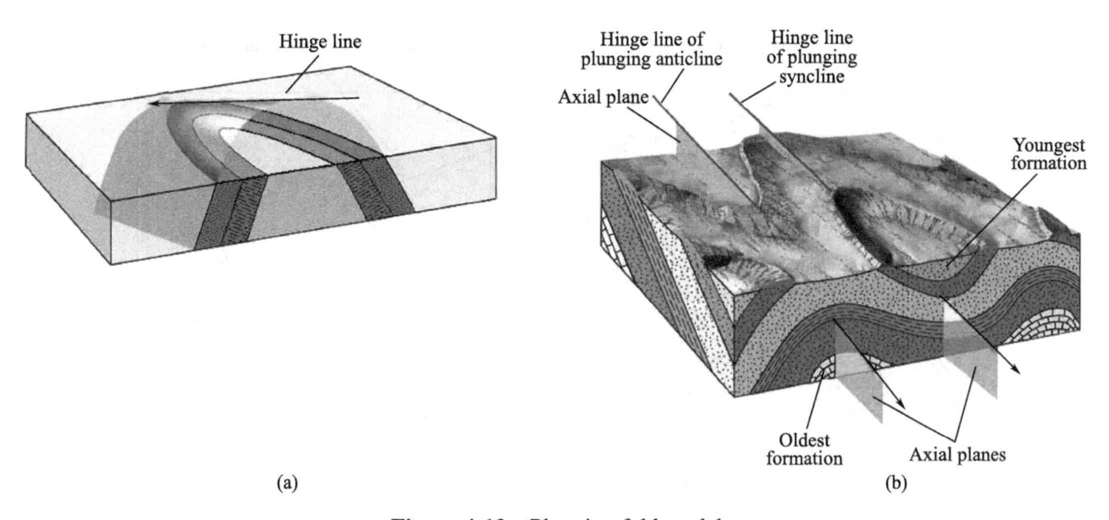

Figure 4-10 Plunging fold model

(a) Plunging fold that is cut by a horizontal plane has a V-shaped pattern; (b) Plunging anticline on left and right and plunging syncline in center. The hinge lines plunge toward front of block diagram and lie within the axial planes of the folds

rocks are exposed in the center of the structure (Figure 4-12); in cross-section, it is comparable to a syncline. A structural basin is like a set of fitted bowls, and if the set of bowls is turned upside down, it is comparable to a structural dome.

The study of Huangling dome and Zigui basin in the Three Gorges area is hot in structure geology (Figure 4-13). Domes and basins tend to be featured on a marvelous scale (some are more than a hundred kilometers), formed by uplift somewhat greater (for domes) or less (for basins) than that of the rest of a region. Domes of similar size are found in other parts of the Middle West of the U.S., and smaller domes are located in the Rocky Mountains. Domes and anticlines are usually essential to the world' s petroleum resources.

Figure 4-11 Rock layers dip away from the center of a plunging anticline exposed at Sheep Mountain in Wyoming. Anticline plunges toward the bottom of the photo (After Calson et al., 2011. Photo by Michael Collier)

4. Further Description of Folds

Folds occur in numerous varieties and sizes. Some are studied under a microscope, while others may have adjacent hinges ten kilometers apart. Figure 4-14 shows several of the more common types of folds based on appearance. Open folds (Figure 4-14a) have limbs that dip gently, and the angle between the limbs is large. All other factors

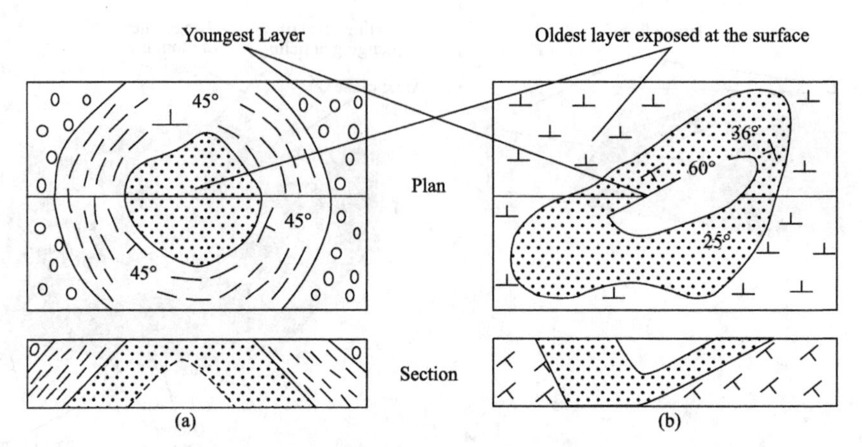

Figure 4-12 Special cases of folds

(a) Structural dome; (b) Structural basin

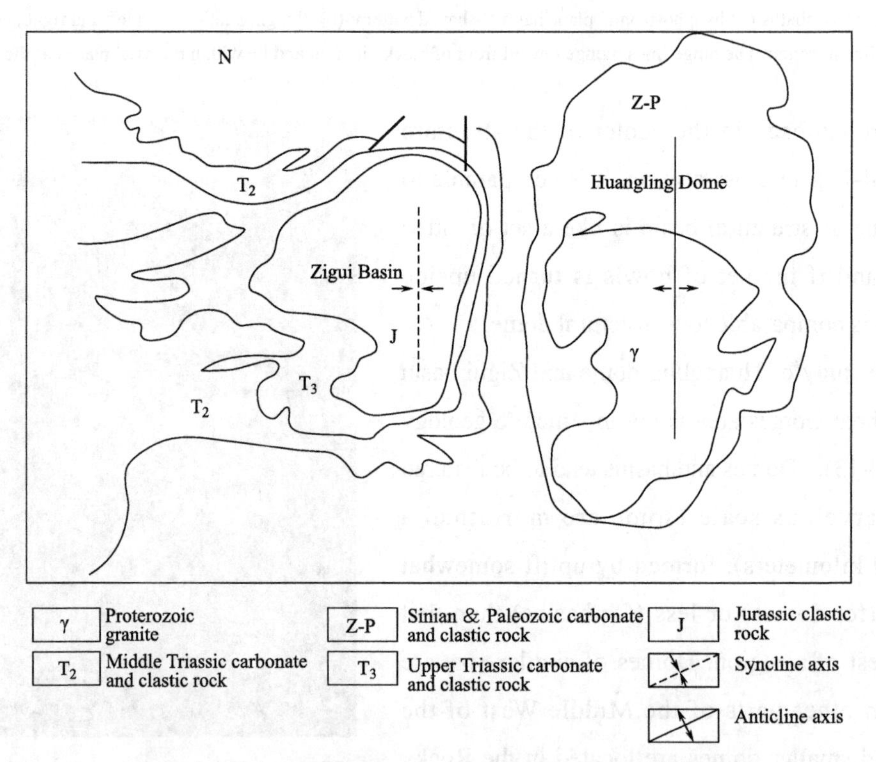

Figure 4-13 Huangling Dome and Zigui Basin in Zigui, Yichang

are equal. The more open the fold, the less tense it has been as a result of shortening. By contrast, if the angle between the limbs of the fold is slight, then the fold is tight. An isoclinal fold, where the limbs are almost parallel to each other, involves an even more significant shortening or shear deformation (Figure 4-14b).

If the axial plane is tilted to such a degree that the fold limbs dip in the same orientation, the fold can be classified as an overturned fold (Figure 4-14c). Studying an outcrop where only the overturned limb of a fold is exposed, you would likely infer that the youngest bed is at the top. However, the principles of superposition cannot be applied to determine the top and bottom for overturning beds. You should either see the rest of the fold or find primary sedimentary structures inside the beds, such as mud fissures that indicate the top of origin or upwards.

Recumbent folds (Figure 4-14d) are overturned so that the limbs are essentially horizontal. Recumbent folds are found in the cores of mountain ranges such as the Canadian Rockies, Alps, and the Himalayas and record extreme shortening and shearing of the crust typically associated with plate convergence.

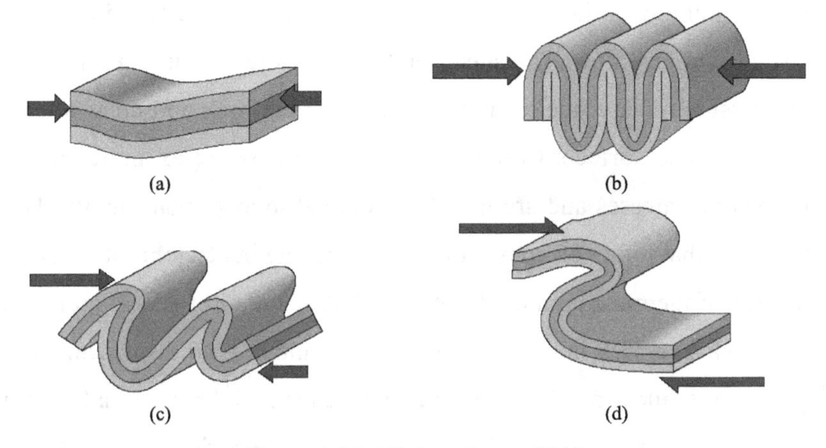

Figure 4-14 Various types of folds

(a) Open folds; (b) Isoclinal folds; (c) Overturned anticline; (d) Recumbent folds

Folds are also classified according to the attitude of their axes and their appearance in cross-sections perpendicular to the fold trend(Encyclopaedia Britannica, 2018). A symmetrical fold(Figure 4-15a) is one in which the axial plane is vertical(also called upright fold). An asymmetrical fold(Figure 4-15b) is one in which the axial plane is inclined. An overturned fold, or overfold(Figure 4-15c), has the axial plane inclined to such an extent that the strata on one limb are overturned. A recumbent fold(Figure 4-15d) has an essentially horizontal axial plane.

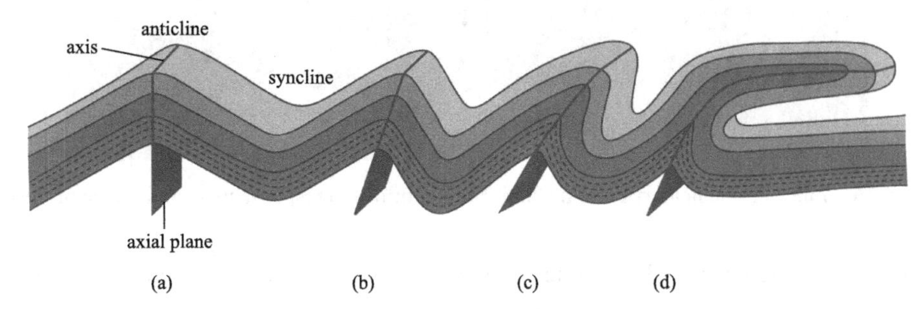

Figure 4-15 Folds are classified according to the attitudes of their axes and limbs

(a) symmetrical fold; (b)asymmetrical fold; (c)overfold; (d)recumbent fold

4.5 Fractures in rock

If a rock is brittle, it will rupture. Commonly, there is some movement or displacement. If essentially no shear displacement occurs, a fracture or crack in the bedrock is called a joint. If the rock on either side of a fracture moves parallel to the fracture plane, the fracture is a fault. Most rocks at or near the surface are brittle, so nearly all exposed bedrock is jointed to some extent.

4.5.1 Joints

The columnar jointing is the form of a hexagonal column resulting from the contraction of a cooling, solidified lava flow. The sheet jointing is a type of jointing due to expansion, is caused by the pressure release due to removal of overlying rock and creates tensional stress perpendicular to the land surface. Columnar and sheet joints are examples of fractures that form from nontectonic stresses and are therefore referred to as primary joints. This chapter is concerned with joints that form not from cooling or unloading but tectonic stresses.

Joints are one of the most commonly observed structures in rocks (Figure 4-16). A joint is a fracture or crack in a rock body along which substantively no displacement has resulted. Joints usually form at shallow depths in the crust where the rock breaks in a brittle way and are pulled apart slightly by tensional stresses caused by bending or regional uplift. A joint set can be defined where joints are oriented approximately parallel to one another.

Figure 4-16 Joints in sedimentary rock in Chibi, Hubei, formed in response to tectonic force of the region

Geologists sometimes find valuable ore deposits by exploring the orientation of joints. For example, hydrothermal solutions may transfer upward through a set of joints and deposit quartz and economically essential minerals such as gold, silver, copper, and zinc in the cracks (Figure 4-17).

Accurate information about joints is also crucial in the planning and constructing large engineering projects, particularly hydropower dams and tunnels. The possibility of dam failure or tunnel rupture may make that site too dangerous if the bedrock at a proposed location is intensely jointed. The movement of polluted groundwater from unlined tailings and abandoned mines may also be controlled by joints, which results in complex and costly cleanups.

Figure 4-17 Fractures in altered granitic rock are filled with quartz, copper, and iron sulfides (chalcopyrite and pyrite). Silver Bell mine, Arizona (After Carlson et al.,2010)

4.5.2 Faults

Faults are fractures in the bedrock along which moving has taken place, and the displacement may range from a few centimeters to thousands of kilometers. For plenty geologists, an active fault is considered as one movement that has taken place during the last 11000 years. Most faults, however, are no longer active (Plummer et al., 2016).

The fault is exposed in an outcrop where the nature of past movement ordinarily can be determined. The geologist explores dislocated beds or other rock features that might show how much displacement has occurred and the relative direction of motion. In some faults, the contact between the two displaced sides is very narrow. In others, the rock has been broken or ground to a fractured or pulverized mass sandwiched between the displaced sides (Figure 4-18).

1. Classification of faults

Geologists describe fault movement in terms of the direction of slippage: dip-slip, strike-slip, or oblique-slip (Figure 4-19). In a dip-slip fault, movement is parallel to the dip of the fault surface, and a strike-slip fault indicates horizontal motion parallel to the strike of the fault

Figure 4-18 Fault in Three Gorges Zone is marked by a 0.5m wide zone of broken rocks that offset rock layers

surface. An oblique-slip fault has both strike-slip and dip-slip components.

1) Dip-slip Faults

The movement is up or down along with the dip of the inclined fault surface in a dip-slip fault. The hanging wall is the side of the fault above the inclined fault surface, whereas the side below is called the footwall (Figure 4-20).

These terms came from miners who extracted along the fault pursuing veins of mineralized rock (ore). As they mined, their feet were on the lower football block, and they could hang their light on the upper surface or hanging-wall block (Figure 4-20).

Normal and reverse faults, the most frequent types of dip-slip faults, are distinguished from each other based on the relative displacement of the footwall block and the hanging-wall block. The hanging-wall block moved down relative to the footwall block in a normal fault (Figure 4-21 and Figure 4-22). The relative movement is illustrated on a geological map by a pair of arrows because geodetic measurement of faults indicates that both blocks move during the slip. A normal fault induces extension or lengthening of the crust. The hanging-wall block drives downward along the fault to make up for the pulling apart of the rocks. There is an extension of the crust. Sometimes a block bounded by two normal faults will drop down, creating a graben, as shown in Figure 4-21(c). Note that "graben" is the German expression for "ditch." Rifts are grabens engage with various plate boundaries, either along mid-oceanic ridges or on continents. The Rhine Valley in Germany and the Red Sea are cases of grabens.

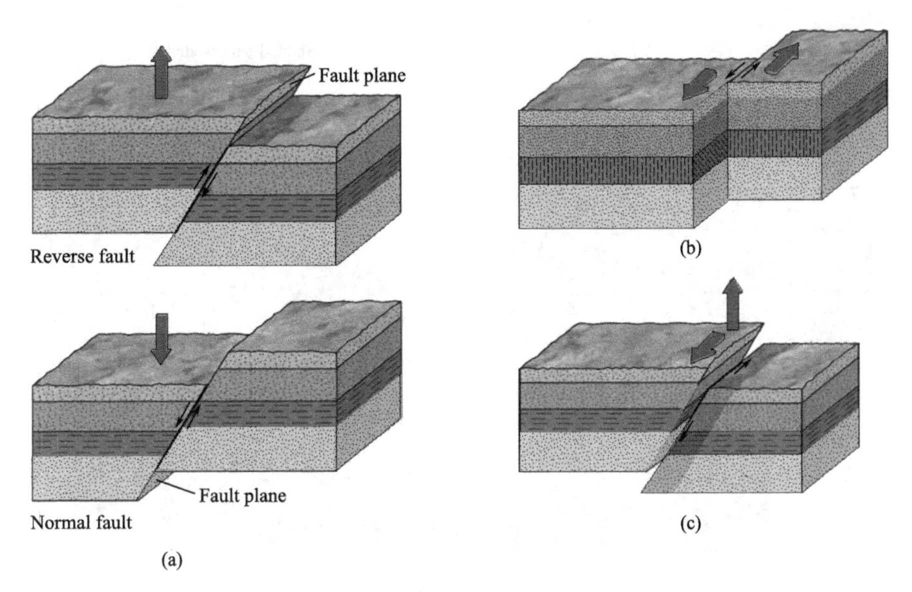

Figure 4-19 Three types of faults illustrated by displaced blocks

(a) Dip-slip movement; (b) Strike-slip movement; (c) Oblique-slip movement. Black arrows show dip-slip and strike-slip components of movement

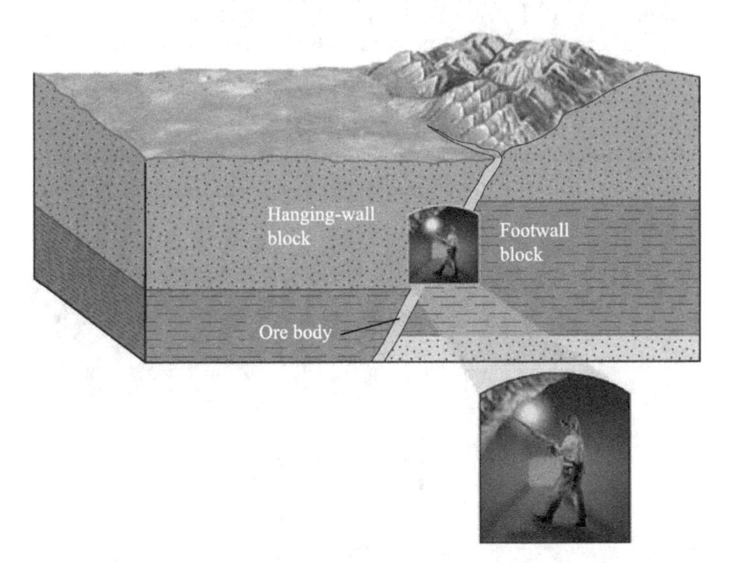

Figure 4-20 Relationship between the hanging-wall block and footwall block of a fault. The upper surface where a miner can hang a lantern is the hanging wall, and the lower surface below the fault is the footwall

It becomes a fault-block mountain range if a block bounded by normal faults is uplifted adequately (It is also called a horst, the opposite of a graben). The Teton mountains and Sierra Nevada mountains are remarkable examples of the fault-block mountain. Nevada's Basin and Range province and portions of adjoining states are also described by numerous mountain ranges (horsts) divided from connecting valleys by normal faults. Normal fault planes typically dip at steep angles (60°) at shallow depths but may become curved or even horizontal at depth.

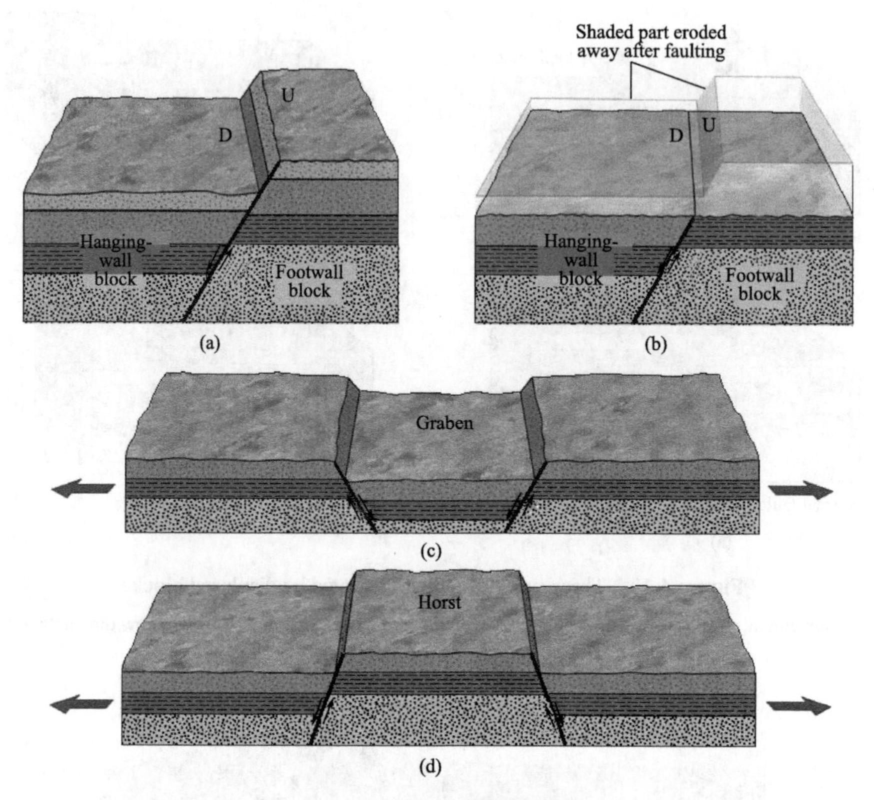

Figure 4-21 Normal faults

(a) Illustration shows the fault before erosion and the three-dimensional relationships of the fault; (b) The same area after erosion; (c) Graben; (d) Horst. Arrows indicate horizontal spreading of the crust

Figure 4-22 Normal faults with prominent horst block offset volcanic ash layers in southern Oregon (modified from Carlson et al., 2011)

In a reverse fault, the hanging-wall block has moved up regarding the footwall block (Figure 4-23 and Figure 4-24). As shown in Figure 4-23, horizontal compressive stresses generate reverse faults. Reverse faults shorten the crust.

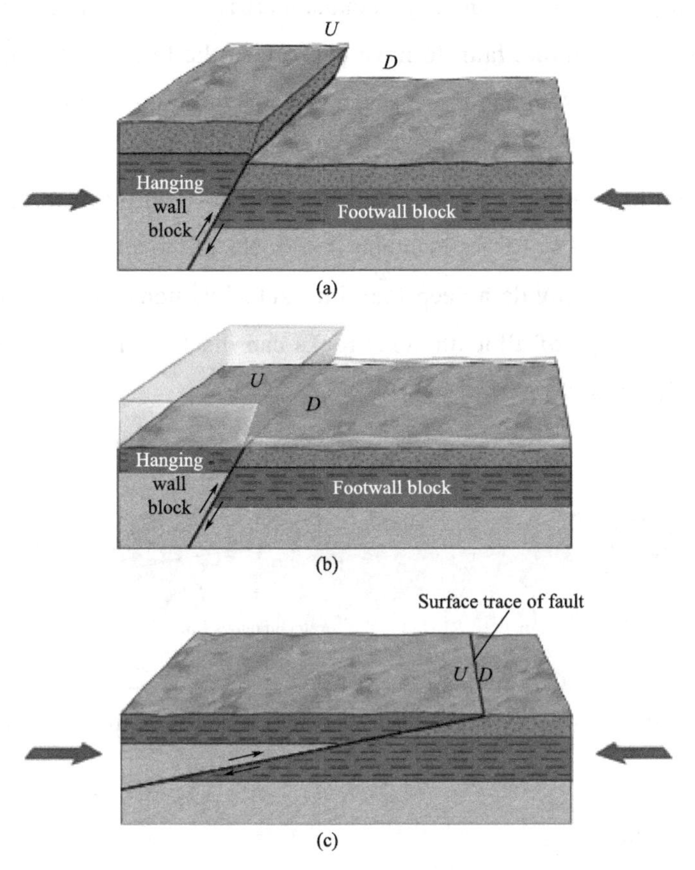

Figure 4-23 Reverse faults

(a)The fault is unaffected by erosion. Arrows indicate shortening direction;(b) The diagram shows the area after erosion; (c) Thrust fault has a lower dip angle and accommodates more shortening by stacking rock layers on top of one another

Figure 4-24 The reverse fault offsets volcanic ash beds, southern Oregon. Hanging wall has moved up relative to the footwall. The fault has been eroded and covered by younger sediments (After Carlson et al., 2011)

A thrust fault is a reverse fault in which the dip of the fault plane is at a low angle (< 30°) or even horizontal (Figure 4-23c). In some mountain areas, it is not rare for the upper plate (or hanging-wall block) of a thrust fault to have overridden the lower plate (footwall block) for several tens of kilometers. Regional thrust sheets, such as those shown in the Canadian Rockies, are considered allochthonous. Allochthonous rocks rest in thrust contact on autochthonous rocks, which retain their original location because they have not been thrust. Regional thrust faults serve to separate rocks of the allochthon from rocks of the autochthon. Windows through allochthonous cover can provide a deep look into autochthonous rocks, which are otherwise concealed. Isolated klippe of allochthonous rocks can disclose the former extensiveness of overthrust strata (Figure 4-25).

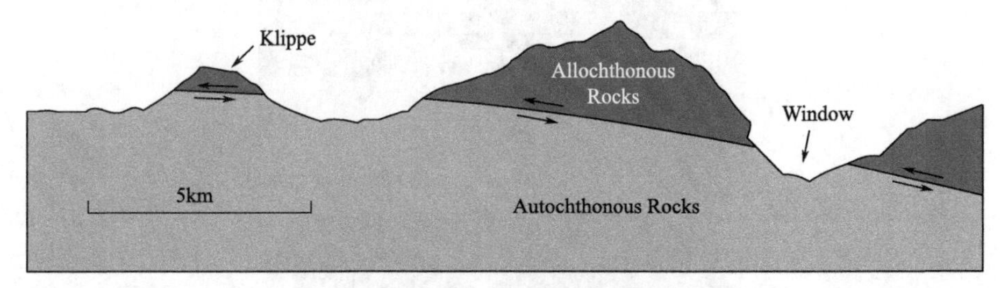

Figure 4-25 Cross-sectional view of allochthonous and autochthonous rocks, as well as klippe and window(After Davis et al., 2012)

2) Strike-slip Faults

A fault where the movement is predominantly horizontal and therefore parallel to the fault's strike is called a strike-slip fault. The displacement along a strike-slip fault is either left-lateral or right-lateral and can be determined by looking across the fault. For instance, if a recent fault displaced a stream (Figure 4-26a), a person walking along it would stop where the fault truncates it. If the person looks across the fault and sees the stream displaced to the right, it is a right-lateral fault. In a left-lateral fault, a stream or other displaced feature would appear to the left across the fault. Again, we cannot tell which side moved, so pairs of arrows are used to indicate relative movement.

Significant strike-slip faults, such as the Tan-Lu fault, typically define a zone of faulting that is more than 2400km in China and cuts through different tectonic units in eastern China. The Tan-Lu Fault Zone is the central fault zone in a series of NE-trending giant fault systems on the East Asian continent. It extends for more than 2400km in China and cuts through different tectonic units in eastern China (Shen et al., 2000). It has a magnificent scale and complex structure. It is the junction zone of the differential movement of the crustal fault blocks. It was formed in the Mesoproterozoic. The southern section of the Tan-Lu fault belt (south of Tancheng) was formed at the end of the Triassic when it was a strike-slip fault east of the Qinling-Dabie collision zone between the Yangtze plate and the Sino-Korean plate.

The surface trace of an strike-slip fault is usually defined by a remarkable linear valley that has been more easily decreased where the rock has been ground up along the fault during movement. The linear valley may involve lakes or sag ponds where the impervious fault rock causes groundwater to surface. The trace of the fault may also be pronounced by offset surface exhibits, such as streams, fences, and roads or by distinct rock units.

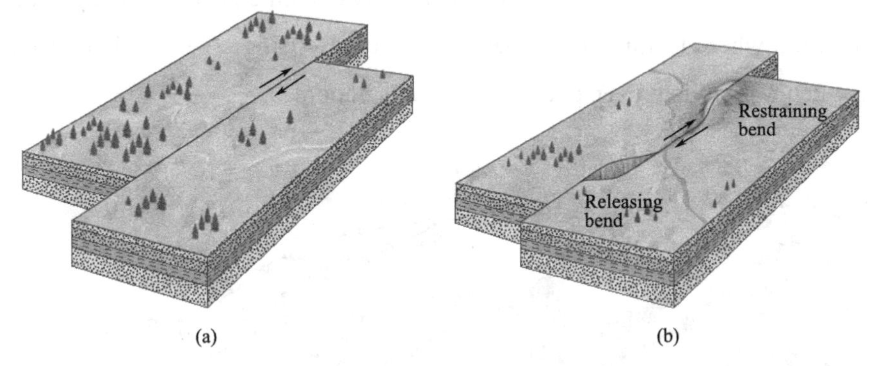

Figure 4-26 Strike-slip faults

(a) Right-lateral strike-slip fault offsets a river channel. Looking across the fault, you would need to walk to the right to find the continuation of the river;(b) Strike-slip movement along curved faults produces gaps or basins at releasing bends where the lithosphere is pulled apart or shortening and hills where it is pushed together at restraining bends(Plummer, et al., 2016)

Strike-slip faults that have suffered a large amount of offset do not remain straight for long distances, and they may either bend or step over to another fault that is matched. Depending on the direction of the bend, the stratigraphy is either pulled apart (releasing bend) or pushed together (restraining bend) (Figure 4-26b). Normal faults form dealing with the pulling apart at the releasing bends and folds, and thrust faults form at the restraining bends to integrate the pushing of the stratigraphy. Death Valley is an excellent example of a deep graben formed along a releasing bend in the lately formed plate boundary along the eastern side of the Sierra Nevada Mountains in California.

2. Faults identification

When conducting geological surveys in the field, due to the influence of various factors such as weathering and denudation, valley cutting, and Quaternary loose rock and soil cover, the existence of faults and the types of faults are often not directly observed or difficult to distinguish. Therefore, it is necessary to judge based on some phenomena formed by faults in topography, structures marks, stratum distribution, and other aspects(Xu, 1989; Li & Liu,2010;

Hu, 2015).

1) Geomorphological marks

(1) Fault cliff and fault triangular facets

Both sides of a vast fault often cause sudden changes in the landform. For example, a mountainous area suddenly becomes a plain along the fault line, and at the same time, the ridge is cut into a cliff (that is, a fault scarp) at the cross-cutting fault. The cliffs are often triangular, called the fault triangular facets (Figure 4-27). In addition, fault fracture zones are susceptible to weathering and denudation in mountainous areas, and valleys are often formed along faults. Geologists often call them "Ten ditches, Nine faults." Although this is not wholly real, it is more common in areas where tectonic movements are strongly rising. When a fault crosses a river valley, it is often washed into deep pools at the river's bottom.

Figure 4-27 Fault triangular facets

(2) A break in the ridge

When some mountain ranges have transverse or oblique faults in the direction of extension, the ridges that make up the mountain range will stagger each other, called the staggered ridge. The staggered ridge is often caused by the relative displacement of the two blocks of the fault, and the contact zone between the plain and the mountain that crosses the strike of the mountain is often a more significant fault (Figure 4-28).

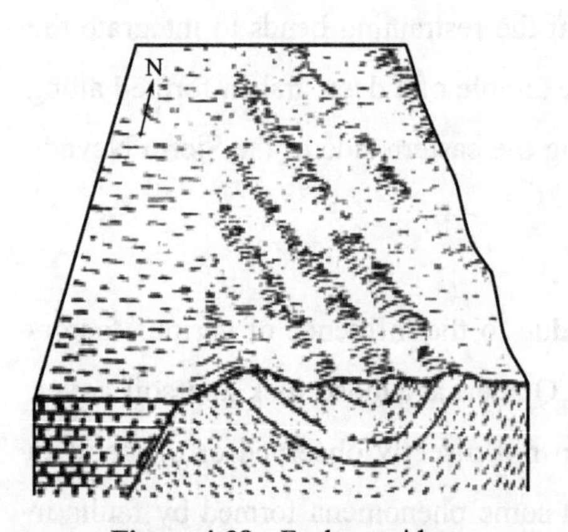

Figure 4-28 Fault-cut ridge

(3) Beaded lakes and depressions

Faulted depressions caused by fault activities often form beaded lakes and valleys. For example, the spring water of Yuquan Mountain in Beijing rises along the fault line, and the fault along the northern foot of the Qinling Mountains in Shaanxi has a series of famous hot springs exposed in Lintong Huaqing Lake, Huxian County, and

Meixian County. Along the Xiaojiang fault zone in Yunnan, there are a series of lake basins, such as Caohai lake, Songming Lake, Yangzonghai Lake, Dianchi Lake, and Fuxian Lake, which are distributed in a beaded pattern from north to south.

(4) Linear springs

Fault fracture zones are often good channels for groundwater. When a fault cuts off the groundwater aquifer, the groundwater will seep out of the ground along the fault zone to form spring water. The spring water outcrops are linearly distributed along the fault line, resembling a string of beads. The linear hot springs are mostly related to modern active faults.

(5) Abrupt changes in the mountains and plains

The contact zone between the plain and the mountain that crosses the mountain strike is often an enormous fault. For example, a reverse fault along the mountain strike develops between the Longmenshan orogenic belt and the Sichuan Basin (Figure 4-29).

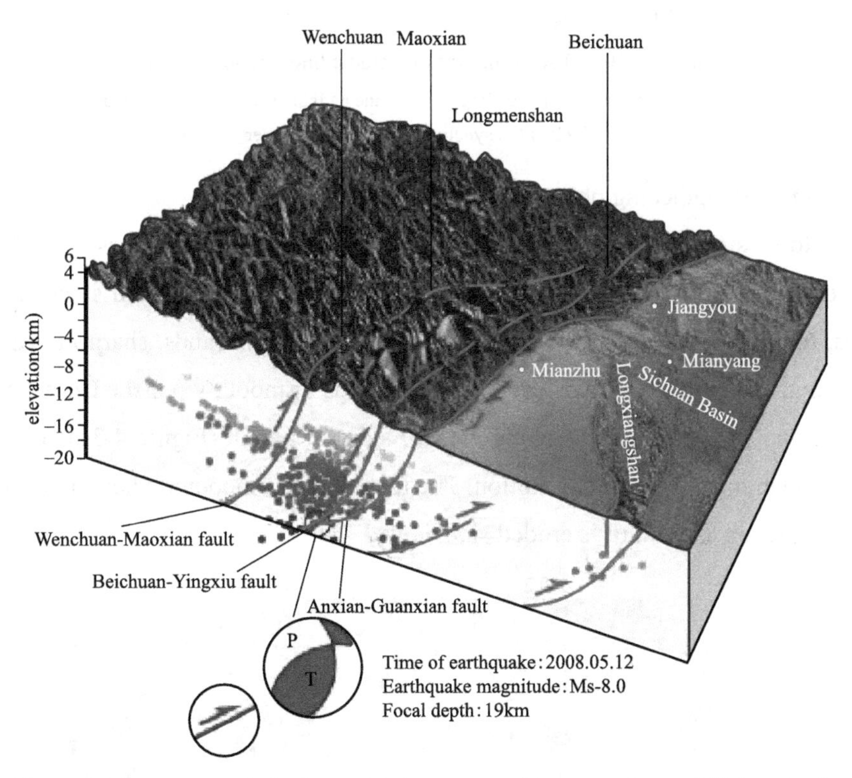

Figure 4-29 a reverse fault along with the contact of Longmenshan and Sichuan basin

2) Structures marks

Fault activities always form or leave many structural phenomena, which are also vital signs for judging faults.

(1) Discontinuities of tectonic lines

Any linear or planar geological body, such as stratum and mineral layer, extends along its direction. If these linear or planar geological bodies are suddenly interrupted, staggered, and no longer continue on the plane or the section, it indicates that there is a fault. The interruption of the structural line caused by the fault can show the discontinuity of the structure from the view of the plane or section. To determine the fault's existence and the mismatch distance, the corresponding part of the misalignment should be ascertained as much as possible (Figure 4-30).

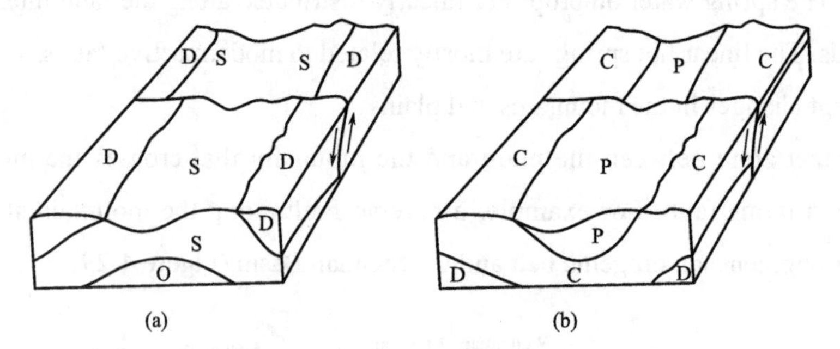

Figure 4-30 Discontinuities of tectonic lines caused by faults

(a) When the fault transects the fold core, the descending block of the anticlinal core becomes narrower; (b) the descending block of the synclinal core becomes wider

(2) Tectonic strengthening phenomenon

The tectonic strengthening caused by fault activity is an essential basis for the possible existence of faults. Structural strengthening includes sudden changes and steepening of rock formations, the sudden appearance of joint and cleavage narrow bands, sharp increase of small folds, crushing, and various slickensides. A slickenside is a smoothly polished surface caused by frictional movement between rocks along the two sides of a fault (Figure 4-31). This surface is generally striated in the direction of motion. The texture feels smoother when the hand is moved in the same direction as the fault's eroded side moved.

Figure 4-31 Slickenlines on the fault surface

In addition, the fault lens-like is also a phenomenon in which faults cause structural strengthening. The rocks in the fault zone or on both sides of the fault plane are broken into lenticular breccia blocks of different sizes. Generally, tens of centimeters to one or two meters in length, structural lenses are often produced in groups, belts, or superimposed. The lens-like structure is generally formed by cutting the rock into rhombic blocks by two sets of conjugate shear joints produced by extrusion. The edges and corners of the rhombuses are ground off (Figure 4-32), including the plane of the long axis and the middle axis of the lens (That is, the largest flat surface), or parallel to the fault plane, or intersect with the fault plane at a slight angle.

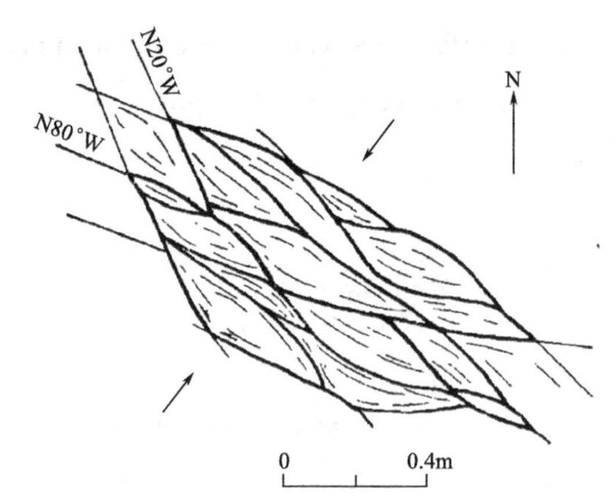

Figure 4-32 Tectonic lens-like in the Xiaojiuhua compressional fault zone, Nanjing

Sometimes, a series of complex and closed isocline folds composed of small folds are seen in the fault zone or on both sides of the fault. The crumpled belt is generally produced in a weaker layer. The axis of small folds is sometimes inclined to one side, occasionally steep, but the overall appearance of the axis is often oblique to the fault plane, and the acute angle intersecting with the fault plane generally indicates the direction of movement of the opposite disk.

(3) Complex crumples on either side of the fault

In or on both sides of the fault zone, a series of complex and closed secondary folds can sometimes be seen forming a crumple zone which is generally produced in weaker rock formations and is controlled by fault properties and lithology (Figure 4-33).

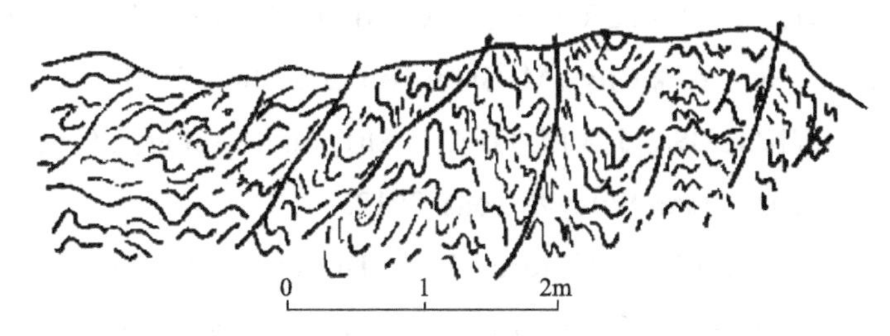

Figure 4-33 The crumple phenomenon near the fault zone

3) Stratigraphic marks

When the fault's strike is roughly parallel to the strike of the rock formation, the fault causes a block to rise or fall. After the ground is flattened by denudation, observe along the slope of the surface, and you will see repeated occurrences of the same strata or missing phenomena that should have appeared (Figure 4-34).

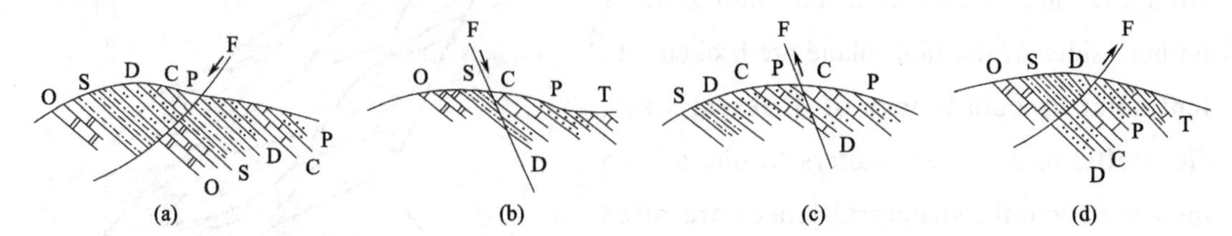

Figure 4-34 The repeated and missing strata with strike-slip faults

4) Magmatic activity and mineralization

Large faults, especially those with deep cuts, are often channels and storage sites for magma and hydrothermal fluid migration. If the rock mass, mineralized zone, or hydrothermal alteration zone is distributed intermittently along a line (belt), it often indicates significant faults or fault zones.

5) Sudden changes in lithofacies and thickness

If an area's sedimentary facies and thickness change drastically along a line (belt), it may result from fault activity. There are two situations in which faults cause sharp changes in lithofacies and thickness. One is the activity of synsedimentary faults that control sedimentary basins and sedimentation, which causes significant changes in the sedimentary environment along the fault on the two sides of the fault, thus causing differences in lithofacies and thickness. In another case, the movement of faults over long distances brings disparate lithofacies zones far apart into direct contact.

Reference

[1] CARLSON D H, PLUMMER C C, HAMMERSLEY L. Physical geology: earth revealed[M]. 9th ed. New York City : McGraw-Hill Education, 2011.

[2] DAVIS G H, REYNOLDS S J, KLUTH C F. Structural geology of rocks and regions (3rd edition)[M]. New Jersey : John Wiley & Sons, 2012.

[3] EARLE S. Physical Geology. Victoria, B.C.: BCcampus[EB/OL].[2021-09-01] https://opentextbc.ca/geology/chapter/12-4-measuring-geological-structures/.

[4] HU M, ZHOU X. Tectonic geology[M]. Beijing : Petroleum Industry Press, 2015.

[5] Li Z, Liu S. Tectonic geology[M]. Beijing : Geological publishing house, 2010.

[6] PLUMMER C C, CARLSON D H, HAMMERSLEY L.Physical geology[M]. 15th ed. New York City : McGraw-Hill Education, 2016.

[7] SHEN Z, ZHAO C, YIN A, LI Y, JACKSON D D, FANG P, DONG D. Contemporary crustal deformation in east Asia constrained by Global Positioning System measurements[J]. Journal of Geophysical Research: Solid Earth, 2000, 105(B3) : 5721-5734. https://doi:10.1029/1999jb900391.

[8] SONG H, ZHANG C, WANG G. Structural Geology[M]. Beijing : Geological Publishing House, 2013.

[9] FOLDS. Encyclopaedia Britannica[EB/OL]. [2018-05-09]. https://www.britannica.com/science/fold#/media/1/211999/5532.

[10] WIKIPEDIA. Strike and dip[EB/OL]. [2021-09-28]. https://en.wikipedia.org/wiki/Strike_and_dip.

[11] Xu K, ZHU Z. Tectonic geology[M]. 2nd ed. Beijing : Geological Publishing House, 1989.

Exercises

Choice Questions(Single Choice)

(1) The direction of a line generated by the intersection of an inclined plane with a horizontal plane is called:

(a) strike (b) direction of dip

(c) angle of dip (d) axis

(2) Folds in a rock show that the rock performed in a:

(a) ductile (b) elastic

(c) brittle (d) all of the preceding

(3) An anticline is:

(a) a fold formed similar to an arch with the youngest rocks revealed in the center of the fold

(b) a trough-shaped fold with the oldest rocks revealed in the center of the fold

(c) a fold formed similar to an arch with the oldest rocks revealed in the center of the fold

(d) a trough-shaped fold with the youngest rocks revealed in the center of the fold

(4) A structure in which the layers dip away from a central point and the oldest rocks are exposed in the center is called a:

(a) basin (b) syncline

(c) dome (d) monocline

(5) Which is not a type of fold?

(a) open (b) isoclinal

(c) overturned (d) recumbent

(e) thrust

(6) Fractures in bedrock along which movement has taken place are called:

(a) joints (b) faults

(c) cracks (d) folds

(7) In a normal fault, the hanging-wall block has moved________relative to the footwall block.

(a) upward (b) downward

(c) sideways (d) rotationally

(8) Normal faults occur where:

(a) there is horizontal shortening

(b) there is horizontal extension

(c) the hanging wall moves up

(d) the footwall moves down

Review Questions

(1) Most anticlines have both limbs dipping away from their hinge lines. For which kind of fold is this not the case?

(2) What is the distinction between a joint and a fault?

(3) If you locate a dip-slip fault while fieldwork, what kind of testimony would you look for to determine whether the fault is normal or reverse?

(4) What factors control whether a rock performs as a brittle material or a ductile material?

(5) What is the difference between the strike, the direction of dip, and the angle of dip?

(6) How does a structural dome vary from a plunging anticline?

5
Geological Maps

5.1 Introduction

A map is an image or representation of the Earth's surface, showing how places on Earth are linked together by distance, direction, and size. Maps are a way to show part of the Earth's surface on a piece of flat paper that can be carried easily. A map is not a photo of the Earth's surface. It can show many things that an image cannot show, and therefore a map seems different in many ways from a photo of the Earth's surface. Maps have been used for centuries. People who create maps as their profession are called cartographers. To make a geological map, you need a topographic base map to plot your geological observations in the field (John & Lisle, 2004). But, what is a Topographic Map? The distinctive feature of a topographic map is that contour lines show the shape of the Earth's surface. Contours are imaginary lines that combine points of equal elevation on the land's surface above or below a mentioning surface, such as mean sea level. Contours make it possible to measure the height of mountains, depths of the ocean bottom, and steepness of slopes (Figure 5-1).

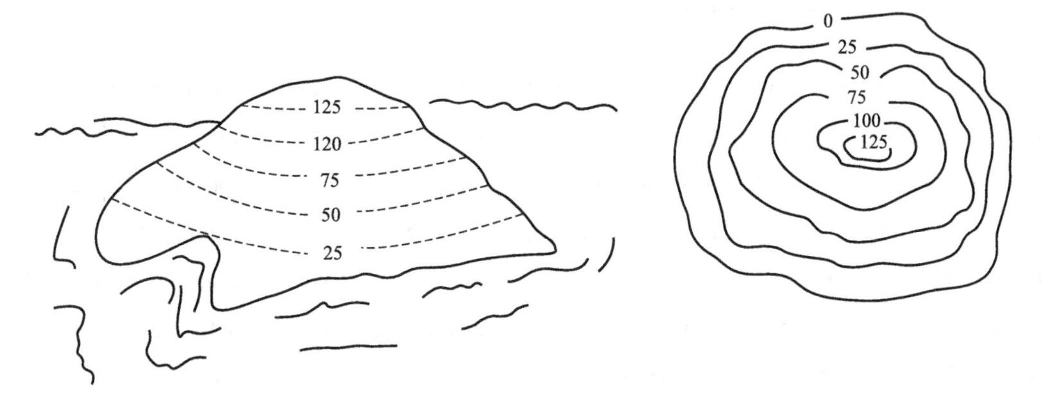

Figure 5-1 Topographic maps showing a three dimensional parts of earth (left) in two dimensions (right) by using contour lines

In a perfect situation, a geologist studying structures would fly over a domain and see the local and regional models of bedrock from above. Sometimes this is probable, but soil and vegetation usually cover the bedrock. Therefore, geologists ordinarily use observations from some individual outcrops (exposures of bedrock at the surface) to determine geologic structures' patterns. The characteristics of rock at each outcrop in an area are plotted on a map using appropriate symbols. With the data collected, a geologist can make inferences about those parts of the site that cannot be observed. A geologic map, which uses standardized symbols and patterns to interpret rock types and geologic structures, is typically created from the field map for a given region. The type and distribution of rock units are plotted on such a map, the occurrence of structural features (folds, faults, joints, etc.), ore deposits, and so forth. Sometimes surficial features, such as deposits by former glaciers, are included, but these may be shown separately on a different geologic map.

Geological map reflects various geological phenomena and conditions, and it is compiled from geological survey data and is one of the main results of geological survey. The planning and design of engineering projects are all based on geological maps. Therefore, it is critical to study how to read and analyze geological maps.

5.2 Types of geological maps

Geological maps can fall into four main groups. These are geological reconnaissance maps, regional geological maps, detailed geological maps, and specialized maps. Small-scale maps covering extensive regions are usually compiled from information selected from one or more of these groups.

5.2.1 Geological reconnaissance maps

Reconnaissance maps are established to clear the geology of a zone as quickly as possible. They are generally produced on a scale of 1 : 250000 or more minor, sometimes much smaller. Some reconnaissance maps are made by photogeology, that is, by interpreting geology from aerial photographs, with only a minimum of work done on the ground to identify rock types and dubious structural features, such as lineaments. Reconnaissance maps were drawn up by plotting the main geological features that operated a light aircraft or helicopter with brief visits to the ground. Airborne methods are beneficial in areas with short exposure seasons.

Reconnaissance may have given the outline of rock distribution and general structure, and the geology must be studied in more detail. Most maps are commonly at a scale of 1 : 50000 or

1 : 25000, although some resulting maps will probably be published in 1 : 100000.

5.2.2 Regional geological maps

Regional geological maps should be plotted on a reliable basement. An accurate geological map will lose much of its point if superimposed on an inadequate topographic base. Regional geological mapping done on the ground may be supported by systematic photogeology. It should be emphasized that photogeological evidence is not inferior to information obtained on the ground, although it may differ in character. Some geological features seen on aerial photographs cannot even be detected on the ground, while others can even be more conveniently followed on photos than in surface exposures. Geological mapping should incorporate any techniques that can help plot the geology, depending on the available survey budget, including geophysics, pitting, augering, drilling, and even satellite images where available.

5.2.3 Detailed geological maps

Scales for detailed geological maps may be anything from 1 : 10000 and larger. Such maps are made to investigate specific problems that have arisen during smaller-scale mapping, or from discoveries made during mineral exploration, or perhaps for the preliminary investigation of a dam site or other engineering projects.

5.2.4 Specialised thematic maps

There are several specialized maps. These include large-scale maps of small areas made to record specific geological features in great detail. Some are for research, others for economic interest, such as open-pit mine plans at scales from 1 : 1000 to 1 : 2500; underground geological mine plans at 1 : 500 or larger; and engineering site investigations at similar scales. There are many other types of maps with geological affiliations too. These include geophysical and geochemical maps, foliation and joint maps, and sampling plans. Most are superimposed over an outline of the geology or drawn on transparencies to be superimposed on geological maps to study their relationship with the solid geology (Lisle, 2004).

For instance, a hydrogeological map is a map reflecting the hydrogeological data of a region. It can be divided into a water-bearing map of formation, groundwater chemical composition map, phreatic contour map, comprehensive hydrogeological map, etc. An engineering geological map is compiled for engineering. It reflects the engineering geology conditions of the mapping area and gives a thorough evaluation of the natural conditions of the building. It integrates the results obtained by various engineering geological survey methods: surveying, exploration, test, etc., and is compiled through analysis and synthesis.

5.3 Common symbols used on geologic maps

To create, manage and disseminate digital earth science information, some widely recognized standards are essential. In the past, many organizational units (e.g., projects or programs) have necessarily developed their standard practices for creating and managing digital map information. Because the resources needed to develop widely-accepted standards are difficult for a single organizational unit to justify, many of these standard practices developed in an ad hoc fashion, with scant input from other groups. Two notable mechanisms now exist in the United States to promote the coordination of widely-accepted standards: (1) the National Geologic Mapping Act of 1992 stipulates that necessary standards be developed by USGS and the Association of American State Geologists (AASG) to support the National Geologic Map Database (NGMDB); (2) the Federal Geographic Data Committee (FGDC), through its Geologic Data Subcommittee, is responsible for the development of standards to support geologic data management at the Federal level (USGS,2021).

Tables 5-1 to 5-3 provide references for the digital cartographic representation of geologic map features. The tables below are intended to support the users of geologic map information by providing line symbols, point symbols, and colors and patterns that can be used to portray the various features on geologic maps (USGSFGDC,2017; AQSIQ & SAC, 2015;CGS,2019).

Table 5-1 Bedding and contacts

DESCRIPTION	SYMBOL	
	USGS	Standards in China
Horizontal bedding		
Inclined bedding—Showing strike and dip	40	30
Vertical bedding—Showing strike		
Overturned bedding—Showing strike and dip	65	25
Unconformable contact—Identity and existence certain, location accurate		

Table 5-2 Symbols used for geological structures

DESCRIPTION	SYMBOL	
	USGS	Standards in China
Anticline—Identity and existence certain, location accurate		
Syncline—Identity and existence certain, location accurate		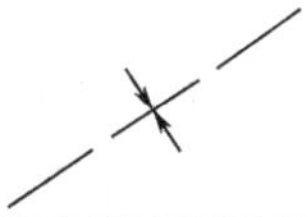
Small, minor dome		
Small, minor basin		
Normal fault—Identity and existence certain, location accurate. Ball and bar on downthrown block		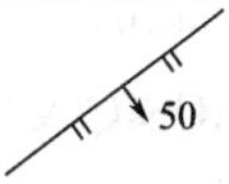 50
Reverse fault—Identity and existence certain, location accurate. Rectangles on upthrown block		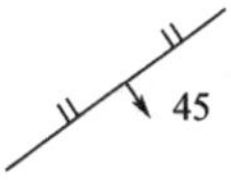 45
Strike-slip fault, right-lateral offset—Identity and existence certain, location accurate		
Strike-slip fault, left-lateral offset—Identity and existence certain, location accurate		
Thrust fault—Identity and existence certain, location accurate. Sawteeth on upper (tectonically higher) plate		

Table 5-3 Topographic and hydrographic features

DESCRIPTION	SYMBOL	
	USGS	Standards in China
Index topographic contour (1st option)	 300	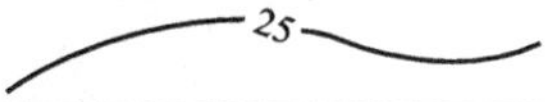 25
Index topographic depression contour		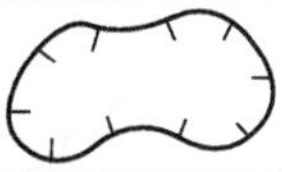
Water well, as shown on topographic maps or on general-purpose or smaller scale maps	 Well	
Spring, as shown on topographic maps or on general-purpose or smaller scale maps	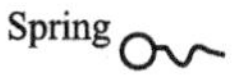 Spring	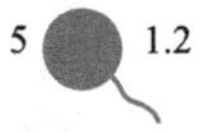 5 1.2
Geyser, fumarole, mud pot, or thermal spring, as shown on topographic maps or on general-purpose or smaller scale maps	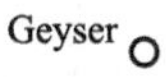 Geyser	
Tailings pond	 Tailings Pond	Nil

In China, with the deepening of geological survey and research, the rapid development of database technology, geological information application technology, and information construction, the amount of information reflected and carried by geological maps is increasing. The "Geological Symbols used for Regional Geological Maps" (BQTS,1989) issued and implemented in 1989 cannot fully meet the needs of regional geological map and spatial database construction at the present stage. Therefore, China Geological Survey initiated a project to revise the GB/T 958-1989 standard in 2009, and the revision was completed in May 2013. The newly revised "Geological Legends used for Regional Geological Maps" fully considers the present situation and demand of compilation, mapping, publication of regional geological maps, and the construction of spatial geological map database, and maximizes the number of legends. The promulgation and implementation of this standard have played a positive role in promoting the standardization of the compilation, drawing, printing, and publishing of geological maps in China (AQSIQ & SAC, 2015).

5.4 Uniformly dipping beds

If wandering at Wuhan University, you're going to be impressed by the layers of rock on Luojiashan. The layered structure results from the deposition of sediments in sheets or beds, which have a sizeable area compared to their thickness. When more beds of sediments are laid down on top, the structures resemble a pile of pages in a book (Figure 5-2). This stratified structure is known as bedding.

Figure 5-2 Dipping beds in Wuhan Unviersity

In some regions, the sediments exposed on the earth's surface still show their unmodified sedimentary structure, and the bedding is generally horizontal. In other parts of the world, especially those in ancient mountain belts, the layering structure is controlled by the buckling of the strata into corrugations or folds. The slope of the bedding varies from place to place.

5.4.1 Outcrop patterns of uniformly dipping beds

The geological map in Figure 5-3(a) illustrates the geographic distribution of two rock formations. The line on the map separating the formations has an irregular shape even though the contact between the formations is a flat surface (Figure 5-3b).

To understand the shapes described by the boundaries of formations on geological maps, it is important to realize that they represent a line (horizontal, plunging or curved) produced by the intersection in three dimensions of two surfaces (Figure 5-3b, d). One of those surfaces is the "geological surface"; in this case is the surface of contact between the two formations. The other is the "topographic surface"– the land. The topographic surface is not plane, but has exhibits such as hills, valleys and ridges. As shown in the block diagram in Figure 5-3(b), it is these irregularities or topographic features that generate the sinuous pattern of geological contacts that we observe on the maps. If the ground surface was plane (Figure 5-3d), the contacts would be straight lines on the map (Figure 5-3c).The extent to which topography influences the form of contacts depends also on the angle of dip of the beds. Where beds dip at a gentle angle, valleys and ridges will produce pronounced "meanders" (Figure 5-4a, b). Where beds dip steeply, the course of the contact is straighter on the map (Figure 5-4c, d, e, f). When contacts are vertical, their course on the map will be a straight line following the direction of the strike of the contact.

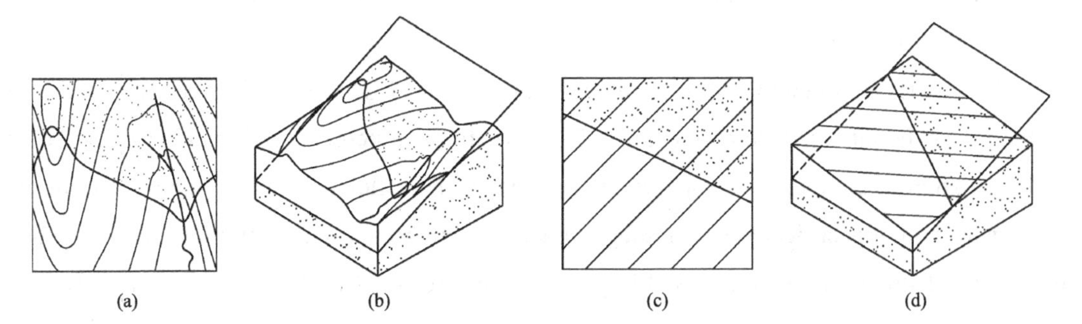

Figure 5-3 The concept of outcrop of a geological contact

5.4.2 V-shaped outcrop patterns

A dipping surface that crops out on a ridge or in a valley will result in a V-shaped

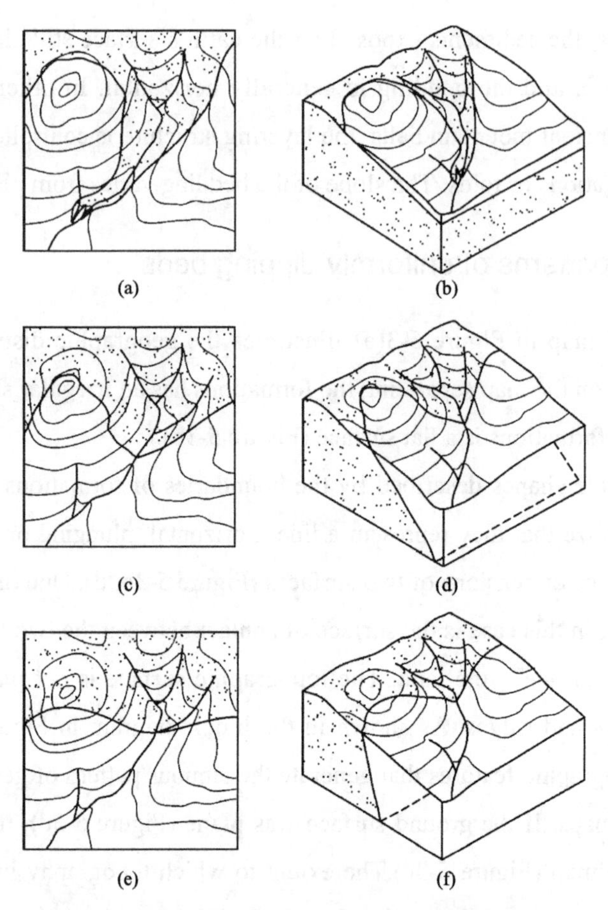

Figure 5-4 The effect of the angle of dip on the sinuosity of a contact's outcrop (from Lisle Richard, 2004)

outcrop. How the configuration of outcrops varies is dependent on the dip of the geological surface relative to the topography. In the case of valleys, the outcrop patterns vee upstream or downstream (Figure 5-5).

The law for deciding the dip from the type of vee (the "V shape rule") is easily memorized if one views the intermediate case (Figure 5-5a) where the outcrop vees in neither direction. It is the situation where the dip is equal to the gradient of the valley bottom. As soon as we tilt the beds away from this critical position, they will start to exhibit a V-shape. If we visualize the bed to be rotated slightly upstream, it will begin to vee upstream, at first veeing more sharply than the topographic contours defining the valley (Figure 5-5b).

The bed can be dipped still further upstream until it converts horizontal. Horizontal beds always generate outcrop patterns that parallel the topographic contours, and hence, the beds still vee upstream (Figure 5-5c). If the bed is tilted further upstream, it starts to dip upstream, and we retain a V-shaped outcrop, but now the vee is blunter than the vee exhibited by the topographic contours (Figure 5-5d).Downstream-pointing vees are generated when the beds dip downstream more steeply than the valley inclination (Figure 5-5e). Finally, vertical beds have straight outcrop sequences and do not vee (Figure 5-5f).

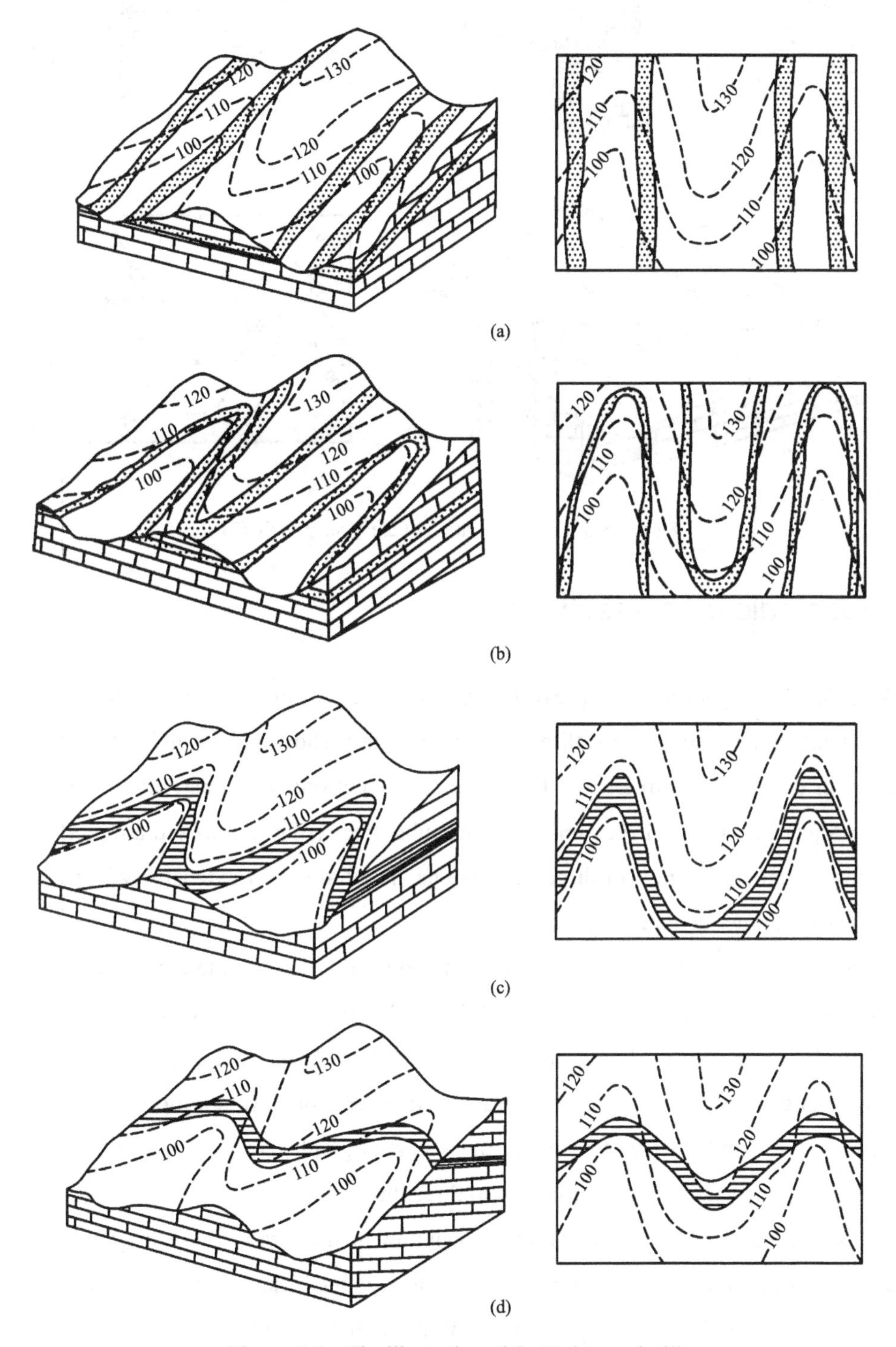

Figure 5-5 The illustration of the V-shape rule (1)

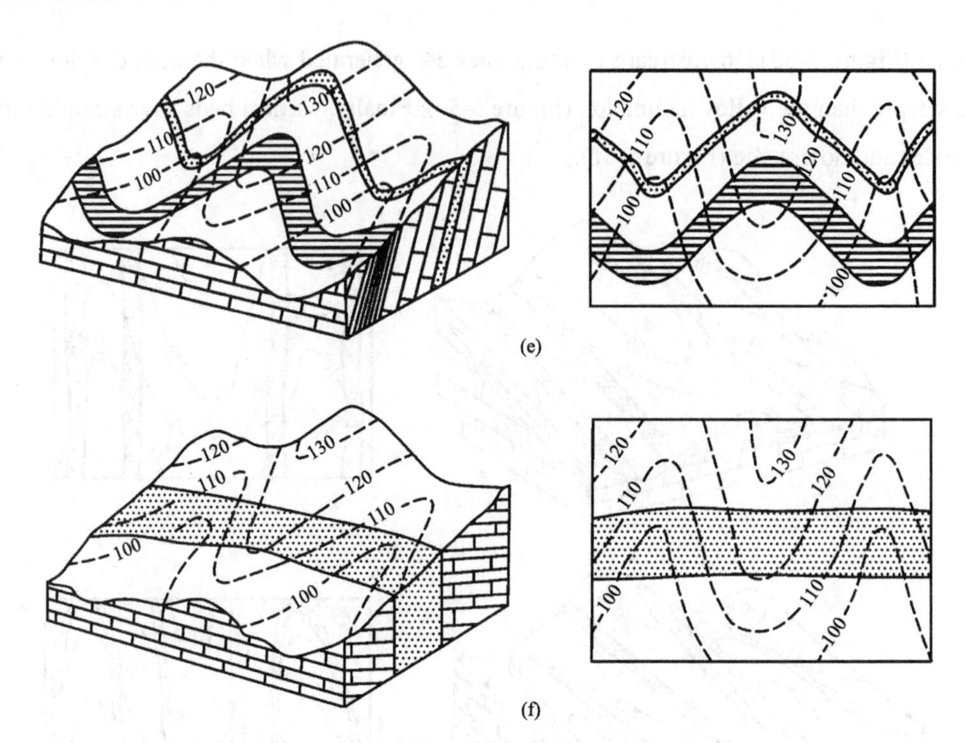

Figure 5-5 The illustration of the V-shape rule (2)

5.5 Geologic cross-sections

A geological cross-section represents a vertical section through part of the Earth. It is a bit like a watermelon cup or the wall of a quarry in that it shows the attitude of rock units and structures in the vertical dimension. Geologic cross-sections are constructed from geologic maps by projecting the dip of rock units into the subsurface (Figure 5-6) and are pretty valuable for helping visualize geology in three dimensions. They are used extensively in professional publications (Plummer et al., 2016).

No geological map can be regarded as complete until at least one cross-section has been drawn to show the geology at depth (John & Lisle, 2004). Cross-sections explain the structure of a region far more clearly than a planimetric map. They may be drawn as adjuncts to your fair copy map and simplified as text illustrations in your report. In addition to cross-sections, columnar sections can be drawn to show changes in stratigraphy from place to place, or "fence" or "panel" diagrams to show their variations in three dimensions.

Although much of this variety of drawing can now be done on a computer, you should learn the basics of this type of illustration by drawing them yourself, and you must also bear in your brain that you may not have a computer with you in the field. Drawn cross-sections are so often faced in professional life that a process is given below:

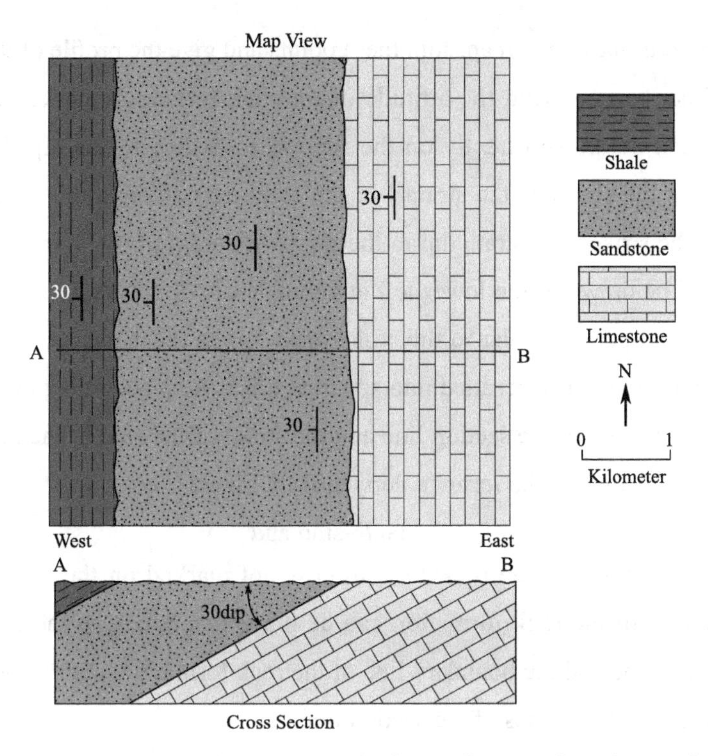

Figure 5-6 A geologic map and cross section of an area with three sedimentary formations. Beds strike north and dip 30° to the west. The geologic cross-section (vertical cut) is constructed between points A and B on the map

(1) Draw the line of section (A–A) on the face of the map, marking each end of the line with a short cross-line. Fasten the map to a drawing board with the section-line parallel to the bottom edge of the board.

(2) Draw a baseline on the tracing paper parallel to the section line on the map. Then draw a series of parallel lines at the chosen contour interval above it. Tape down a plastic ruler or steel straight-edge to not move, well below and parallel to the baseline. By sliding a triangle ruler along the straight edge, drop a perpendicular down to the appropriate elevation on the section paper from every point where the

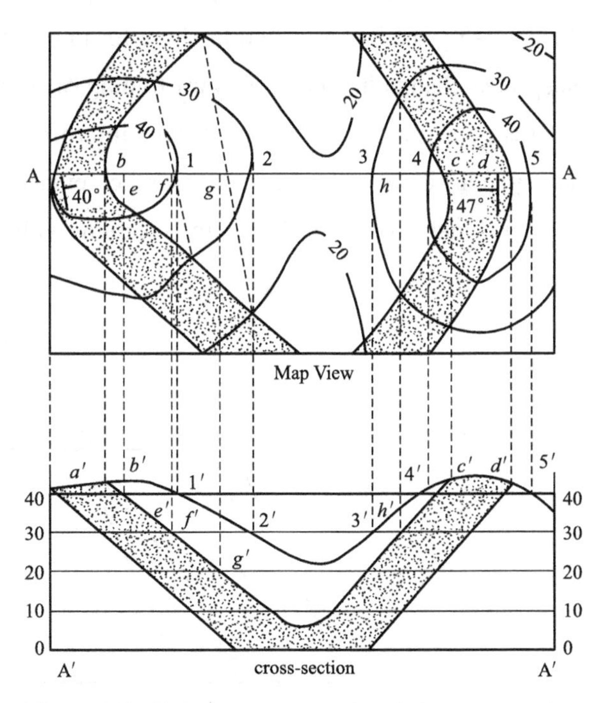

Figure 5-7 Drawing a cross-section, 1~5 represent points of intersection

section cuts a contour line on the map. Join these points and give the profile of the topography.

(3) Sometimes, we can find the attitude of the rock formation marked on the map. The strata boundary can be drawn directly on the section according to the dip if the section line is perpendicular to the strike of the strata. In the right part of Figure 5-7, the section line is perpendicular to the strike. The strata dip to the west with a dip of 47° . The strata boundary in the section should be drawn to the lower left at an angle of 47° to the horizontal line. Suppose the section line is not perpendicular to the strike of the strata. In that case, the dip angle of the strata in the section should be converted into apparent dip angle β according to the true dip angle α of and the angle θ between the section line and the strike of the strata. The apparent dip angle can be calculated according to the formula derived from Figure 5-8.

$$\tan\beta=\sin\theta\tan\alpha$$

(4) When the attitude of the rock formation is not marked on the map, the strike lines of different altitudes of the rock formation can be drawn according to the intersection of the topographic contour line and the boundary line of the rock formation, that is, the connection line of the same altitudes on both sides of the section line.

(5) Still using the triangular rule, drop a perpendicular wherever the section line crosses a geological contact on the map and lightly mark the position on the profile of the topography. Lightly sketch in the structure by extending "dip lines" and drawing contacts parallel to them. Then modify the interpretation to allow for thickening and thinning of beds and for any further suspected change in straightforward folding or tilting. Do not interpret geology to improbable depths beneath the surface. Test your interpretation by continuing the structure above the topographic surface. You have just as much evidence there as for your understanding below the surface. Finally, you can fill the cross-sections with legends.

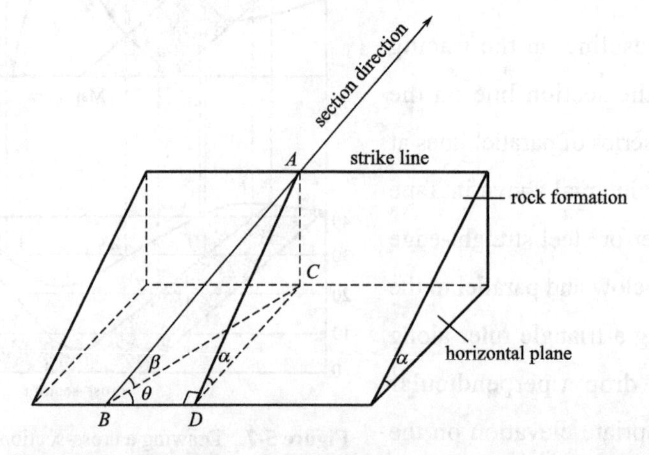

Figure 5-8 Conversion of true inclination and apparent inclination

5.6 Reading of geological map in the Heishan Town

According to the geological map of the Heishan Town (Figure 5-9 and Figure 5-10), the geological conditions of the site are analyzed as follows:

l. Scale and Topography

The scale of the geological map is 1∶10000; that is, 1cm represents 100m of field distance. The northwest part of this area is the highest, with an elevation of about 570m; the southeast is lower, about 100m; the relative elevation difference is about 470m. There is a hill in the east with an elevation of more than 300m. There are two northwest valleys along the terrain.

2. Formation lithology

The exposed strata in this area range from old to new. Paleozoic era includes limestone of the lower Devonian system(D_1), shale of the middle Devonian system(D_2), quartz sandstone of the upper Devonian system (D_3), shale interbedded with coal of the lower Carboniferous system (C_1), limestone of the middle Carboniferous system(C_2). Mesozoic era includes:

(1) Shale of the lower Triassic system(T_1)

(2) Limestone of the middle Triassic system(T_2)

(3) Marl of the upper Triassic system(T_3)

(4) Calcareous sandstone of Cretaceous system(K)

Cenozoic era includes sand and shale interbedded of Tertiary system(R). The Paleozoic strata have a large distribution area, and the Mesozoic and Cenozoic strata are exposed in the north and northwest.

In addition to sedimentary rock formations, granite veins (γ) are also invaded and exposed in the northeast. The intrusion in the strata before Triassic is the product of the Hercynian movement.

3. Geological structures

(1) Attitudes of formations. R is a horizontal formation; T and K are monoclinic beds with an orientation of 330°∠28°, and the D and C formations generally extend east-west or northeast.

(2) Folds. The Paleozoic strata formed three folds from D_1 to C_2 from north to south: anticline, syncline, and anticline in sequence. The fold axis is NE75°~80°.

① Northeast anticline: the older stratum of the core of the anticline is D_1, the north limb is D_2, and the orientation is 345°∠36°; the south limb is D_2, D_3, C_1, and C_2 from old to new, and the formation orientation is 165°∠36°; It is upright (symmetric) folds because the rock attitudes on the two limbs are symmetrical.

② Central syncline: the younger stratum in the core of the syncline is C_2, and the north limb is the south limb of the anticline as mentioned above; the outcropping strata in the south limb are C_1,D_3,D_2, and D_1. Their orientation is 345°∠ 56° ~58° . The two limbs have different inclination angles, so it is inclined synclines.

③ Southern anticline: the core is D_1, the two limbs are symmetrically distributed with D_2, D_3, and C_1, inclined anticlines.

These three folds occurred after the middle Carboniferous period (C_2) and before the early Triassic period(T_1) because the strata from D_1 to C_2 all undergo fold changes. The strata after T_1 are not affected by this fold. However, the formations of T_1-T_3 and K are monoclinic, and their orientations are different from the D and C formations. It may be a limb of another syncline or anticline, formed by another tectonic movement, which occurred after K and before R.

(3) Faults. There are two large faults F_1 and F_2 in this region. The formations are not continuous along the strike direction, the fault strike is 345° , and the dip angle of the fault surface is steep (F_1: 75 °∠ 65 ° ; F_2: 255 °∠ 65 °). Both faults are normal faults with crosscutting the synclinal axis and anticlinal axis. The outcrop width change of the C_2 formation at the syncline core on both sides of the fault shows that the formation between F_1 and F_2 moves down relatively, so the combined model of the F_1 and F_2 is a graben.

In addition, there are two faults: F_3 and F_4. The strike of the F_3 is 300° , and the strike of the F_4 is 30° , which are relatively small-scale strike-slip faults. The fault also formed after the middle Carboniferous period (C_2) and before the early Triassic period (T_1) because the fault did not dislocate the formation after T_1.

Analyzed from the time and space of the distribution of folds and faults in this area, they are in the same tectonic stress field and are formed by two tectonic movements before and after. Compressive stress mainly comes from the near north-west-north direction, so the fold axis is east-north-east. The two faults of F_1 and F_2 are normal faults formed by tensile stress, so the fault direction is roughly parallel to the direction of compressive stress, while F_3 and F_4 are strike-slip faults formed by shear stress.

4. Contact relationship

Between the Cretaceous (K) and the underlying upper Triassic (T_3), the Jurassic (J) is missing, but the orientation is roughly parallel, so it is a disconformity(parallel) contact. T_3, T_2, and T_1 are in conformity contacts.

The lower Triassic system (T_1) is in direct contact with the underlying Carboniferous system (C_1, C_2) and Devonian system (D_1, D_2, D_3) strata. They are angular unconformity contact because the Permian system (P) and upper Carboniferous system (C_3) are missing. All layers from C_2 to D_1 are in conformity contact.

The granite vein (γ) cuts through the Devonian system (D_1, D_2, D_3) and Lower Carboniferous system (C_1)strata and intrudes into it, so it is intrusive contact. Because it does not cut through the overlying lower Triassic system(T_1) strata, therefore, γ and T_1 are in depositional contact. It shows that the granite veins (γ) were formed after the early Carboniferous system (C_1) and before the early Triassic period (T_1), but they are relatively small in scale and present in the form of a vertical stock distributed from north-west-north to south-east-south.

5. A brief history of geological development

In the historical process of geological development, from the Devonian period to the middle Carboniferous period in this area, the crust was slowly rising and falling. The amplitude was minimal, and it has always been deposited. After the middle Carboniferous period, with the influence of the Hercynian movement, the earth's crust changed drastically, the rock layers were folded and fractured, and accompanied by magma intrusion, the area rose to land during the Permian period and suffered weathering and erosion. In the early Triassic period, it subsided into the ocean again and re-accepted marine deposits. In the late Triassic period, a large crust area rose gently and continuously to become land. During the Jurassic period, the crust was

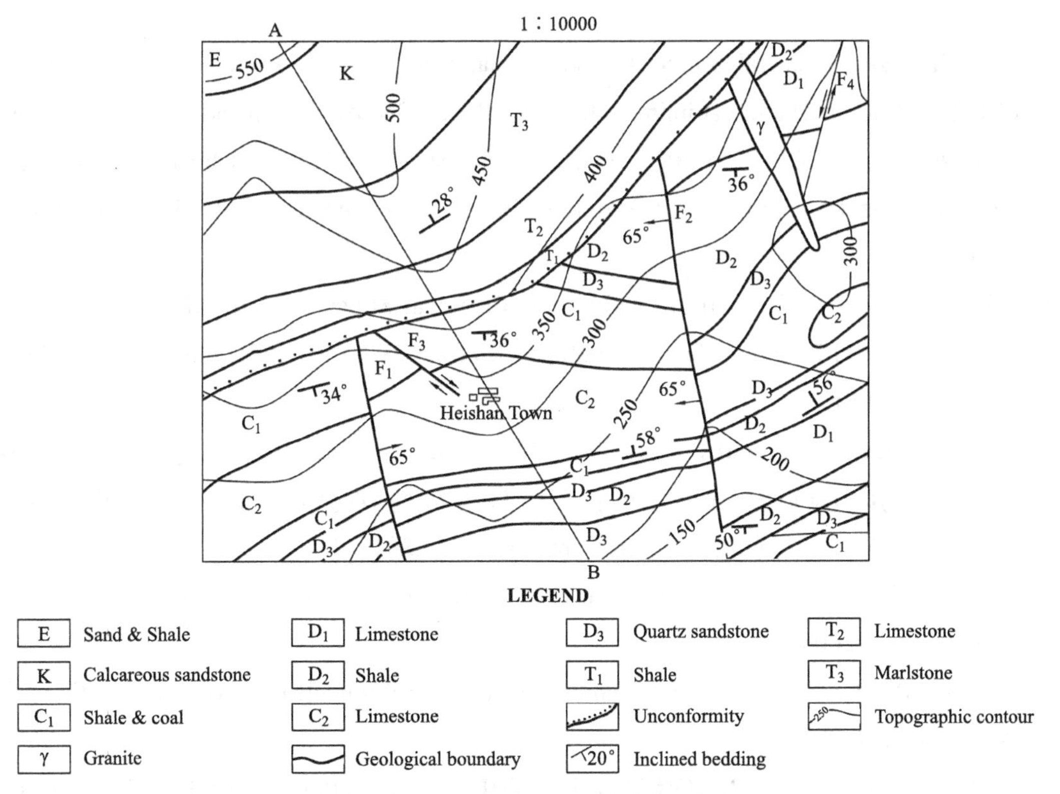

Figure 5-9 Geologic map of Heishan Town

weathered, eroded, and didn't accept deposition. Until the Cretaceous period, it slowly declined again in the shallow sea sedimentary environment. In the late Cretaceous period, it was affected by the Yanshan Movement again, so Triassic and Cretaceous system strata may form gentle folds. There were no drastic structural changes in the Cenozoic period, so the Paleogene system strata were horizontal attitudes.

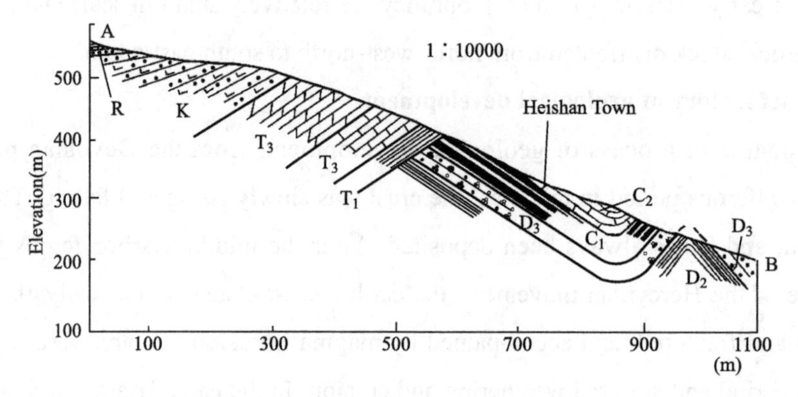

Figure 5-10 Cross-section of Heishan Town

Reference

[1] AQSIQ, SAC (General Administration of Quality Supervision, Inspection and Quarantine of P.R.C., Standardization Administration of P.R.C.). Geological Legend used for Regional Geological Maps : GB/T958-2015[S/OL]. Beijing: Standards Press of China, 2015: 2. http://c.gb688.cn/bzgk/gb/showGb?type=online&hcno=2C6718A842C6D8CE39587B60410C3BD8.

[2] BQTS (The State Bureau of Quality and Technical Supervision of P.R.C.). Geological Symbols used for Regional Geological Maps: GB/T958-1989[S/OL]. Beijing: Standards Press of China, 1989: 1.

[3] CGS(China Geological Survey, Ministry of Natural Resources). Specification for compiling hydrogeological survey maps. Part 1: Hydrogeological mapping (1∶50000) : DZ/T0329-2019[S/OL]. Beijing: Standards Press of China, 2019: 1.

[4] JOHN W, BARNES, LISLE R J. Basic Geological maping[M]. 4th ed. New Jersey : John Wiley & Sons, 2004.

[5] LISLE R J. Geological structures and maps: a practical guide[M]. Revise Edition Oxford : Pergamon Press, 2004.

[6] PLUMMER C C, CARLSON D H, HAMMERSLEY L. Physical geology[M]. 15th ed.

New York City : McGraw-Hill Education, 2016.
[7] U.S. Geological Survey. National Geologic Map Database-Standards and guidelines[S/OL]. [2021-11-02]. https://ngmdb.usgs.gov/Info/standards/.
[8] U.S. Geological Survey for the Federal Geographic Data Committee. FGDC Digital Cartographic Standard for Geologic Map Symbolization[S/OL].[2017-05-29].https://ngmdb.usgs.gov/fgdc_gds/geolsymstd/fgdc-geolsym-all.pdf.

Exercises

Choice Questions

(1) Topographic maps do not show:

(a) elevation (b) water

(c) bathymetry (d) cities

(2) A cartographer:

(a) draws pictures of an area (b) interprets maps

(c) arranges layers of maps (d) creates maps

Review Questions

(1) What is a map?

(2) Type of maps. Show examples.

(3) How are geological conditions (attitude, folds, faults, contact relationships, etc.) represented in geological maps?

(4) How to draw a section from a geological plan?

(5) When a fault crosscuts the synclinal axis or anticlinal axis in a geologic map, how to judge the type of the fault?

6 Groundwater

6.1 Introduction

Many communities obtain the water from rivers, lakes, or reservoirs, sometimes using aqueducts or canals to bring water from distant surface sources. Another source of water lies directly beneath most towns. This resource is groundwater, the water that lies beneath the ground surface, filling the pore space between grains in bodies of sediment and clastic sedimentary rock and filling cracks and crevices in all types of rock. Groundwater is a significant economic resource, particularly in the dry western areas of China, where surface water is scarce. Many towns and farms pump significant quantities of groundwater from drilled wells. Even cities (Wuhan, Nanjing, etc.) next to large rivers may pump their water from the ground because groundwater is commonly less contaminated and more economical to use than surface water.

In the construction of water conservancy and hydropower projects, groundwater often causes serious impairs to the construction, regular operation, and safety of buildings. The common reservoir, seepage in the dam area, volume expansion of some rocks when encountering water, dissolved caves, and sudden water gusher in the excavation of underground caverns are all related to groundwater activities. Obviously, on the one hand, groundwater is a valuable natural resource that should be fully and reasonably utilized; On the other hand, it is a disadvantageous factor in production and construction, which must be effectively prevented and controlled. This chapter mainly introduces some basic knowledge about groundwater, especially the void properties of rocks and the essential characteristics of several common types of groundwater.

6.2 Voids and water in rock and soil

There are various kinds of voids in rocks, and it is sometimes helpful to be aware of them. If the voids were created at the same time as the rock, they are referred to as primary openings.

The pores in sand and gravel, as well as other unconsolidated deposits, are primary openings. Lava tubes and other basalt openings are also primary openings. If the voids were formed after the rock was formed, they are referred to as secondary openings. The fractures in granite and consolidated sedimentary rocks are secondary openings. Voids in limestone, formed as groundwater slowly dissolves the rock, are an essential secondary opening (Heath, 1987).

6.2.1 Aquifers of natural formations

Aquifers are permeable geological formations that can store and transmit water. There is a wide range of natural formations with very different capacities for storing and transmitting water; in hydrogeological terms, these formations can be divided into four main groups (Table 6-1):

(1) Aquifers: can store and transmit water (gravels, sands, limestone materials, etc.); these are formations with a high drainage capacity where wells and boreholes can be drilled to satisfy human needs for water supplies, agriculture, livestock, or industry.

(2) Aquitards: can store vast quantities of water but transmit it with difficulty; these are often classified as semi-permeable formations (silts, silty or clayey sands) with medium to low drainage capacity; they cannot be used to produce flow rates for water supply requirements, but they play a vital natural role as water transmitting elements capable of vertical recharge over large surface areas.

(3) Aquicludes: can collect large quantities of water but cannot transfer it and drain with great pain; the water occupies the formation pores and cannot be released (clays, plastic clays, or clay silts); in classical hydrogeology, they are classified as impermeable, but in engineering geology, this concept is less precise, as even minimal drainage can cause problems in some projects.

(4) Aquifuges: formations that cannot store or transmit water. Among them are hard rock materials, such as granite and gneiss, and even compacted and non-karstified limestone. These substances are impermeable unless there are fractures and discontinuities which allow flow.

Table 6-1 Hydrogeological behaviour of geological formations

	Storage capacity	Drainage capacity	Transmission capacity	Characteristic capacity
Aquifer	High	High	High	Gravel, sand, limestone
Aquitard	High	Medium/Low	Low	Silt, silty sand, clayey sand
Aquiclude	High	Very low	Nil	Clays
Aquifuge	Nil	Nil	Nil	Granite, gneiss, marble

6.2.2 Types of aquifers

An aquifer is a mass of saturated rock or sediment through which water can flow quickly. Aquifers are both highly permeable and saturated with water. A well must be penetrated into an aquifer to reach a sufficient supply of water (Figure 6-1). Suitable aquifers include sandstone, conglomerate, well-jointed limestone, bodies of sand and gravel, and some fragmental or fractured volcanic rocks such as columnar basalt. These positive geologic substances are sought in "prospecting" for groundwater or looking for suitable sites to drill water wells (Carlson et al., 2011). There are three types of aquifer materials in terms of their texture, shown in Figure 6-2.

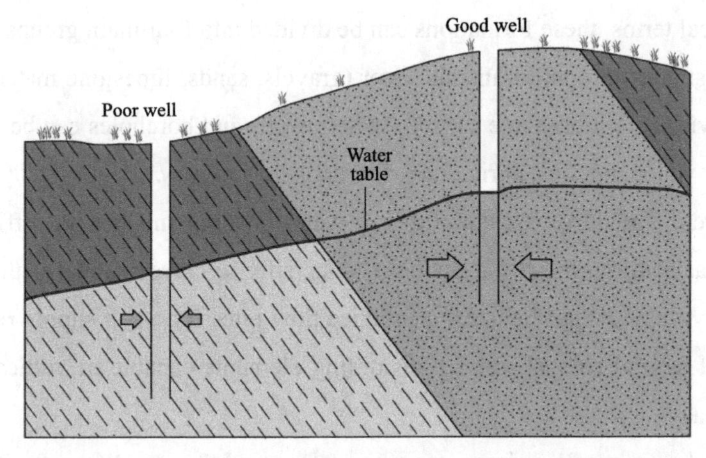

Figure 6-1 A well must be installed in an aquifer to obtain water. The saturated part of the highly porous sandstone is an aquifer, but the less permeable shale is not. Although the shale is saturated, it will not readily transmit water

The aquifers with permeability resulting from intergranular porosity are called porous aquifers (porous medium), which include gravels, sands, and in general, all detritus materials of at least sand grain size. The texture of the medium is made up of grains that allow water to be stored and circulate through the intergranular spaces. These pores may be filled with very fine granular material, reducing the capacity of the medium for water storage and transportation, or they may be filled with clay materials, reducing these characteristics to almost nothing. Sometimes, the grains themselves are made up of porous material, which provides even better properties for water storage. Because of their genesis, granular media are usually very homogeneous on reduced scales.

The second type comprises those where the permeability is due to discontinuities, caused either mechanically (fissured medium) or by dissolution (karst medium), which forms karstic and fissured aquifers and includes limestone, dolomite, granite, basaltic formations, etc., with the first two of these the most important. Karstification is a dissolution process caused by the action of water in carbonated, previously fissured formations; karstic aquifers are not very homogeneous on a small scale but can be considered more homogeneous if the working scale is large enough.

The third type comprises aquifers, where permeability is due to a combination of the causes outlined above and sometimes called "double aquifers," making them karstic, fissured, and porous (de Vallejo & Ferrer, 2011).

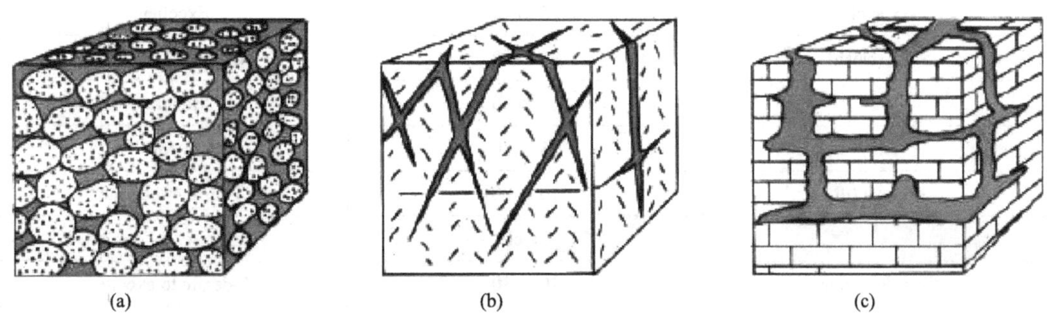

Figure 6-2 Types of aquifer by texture (Heath,1987)

(a) Porous medium;(b) Fissured medium; (c) Karst medium

6.2.3 Porosity and permeability

Porosity, the percentage of rock or sediment that consists of voids or openings, is a measurement of a rock's ability to hold water. Most rocks can hold some water. Some sedimentary rocks, such as sandstone, conglomerate, and many limestones, tend to have high porosity and, therefore, contain considerable water. A deposit of loose sand may have a porosity of 30% to 50%, but this may be reduced to 10% to 20% by compaction and cementation as the sand lithifies (Table 6-2). A sandstone in which pores are nearly filled with cement and fine-grained matrix material may have a porosity of 5% or less. Crystalline rocks, such as granite, schist, and some limestones, do not have pores but may hold some water in fractures and other openings.

Although most rocks can contain some water, they vary significantly in their ability to enable water to pass through them. Permeability refers to the potential of a rock to transmit fluid such as water or petroleum through pores and fractures. In other words, permeability estimates the relative ease of water flow and indicates how openings in a rock interconnect. The distinction between porosity and permeability is essential. A rock that holds a lot of water is called porous; a rock that allows water to flow easily through it is defined as permeable. Most sandstones and conglomerates are both porous and permeable, and an impermeable rock does not allow water to flow through it easily. Unfractured granite and schist are impermeable. Shale can have substantial porosity, but it has low permeability because its pores are too small to permit easy passage of water.

Table 6-2 Porosity and Permeability of Sediments and Rocks

	Porosity (%)	Permeability
	Sediment	
Gravel	25 to 40	Excellent
Sand (clean)	30 to 50	Good to excellent
Silt	35 to 50	Moderate
Clay	35 to 80	Poor
Glacial till	10 to 20	Poor to moderate
	Rock	
Conglomerate	10 to 30	Moderate to excellent
Sandstone		
Well-sorted, little cement	20 to 30	Good to very good
Average	10 to 20	Moderate to good
Poorly sorted, well-cemented	0 to 10	Poor to moderate
Shale	0 to 30	Very poor to poor
Limestone, dolomite	0 to 20	Poor to good
Cavernous limestone	up to 50	Excellent
Crystalline rock		
Unfractured	0 to 5	Very poor
Fractured	5 to 10	Poor
Volcanic rocks	0 to 50	Poor to excellent

6.3 The occurence of ground water

The occurrence (burial condition) of groundwater refers to the position of water-bearing strata in the geological section and the restriction of the aquiclude layer (aquitard layer). Groundwater can be divided into perched water, phreatic water, and confined water. Depending on the water-bearing medium (void) type, groundwater can be divided into pore water, fissured water, and karst water. The conditions of groundwater burial determine the relationship between water and the surrounding environment. The influence of natural factors varies according to burial conditions, and the conditions of recharge, runoff, and discharge conditions of groundwater are also different. Ground water problems encountered in industrial and agricultural production and engineering construction, the calculation methods, and the exploitation or pollution prevention measures are also different.

6.3.1 Water table

Responding to the gravity, water percolates down into the ground through the soil, cracks, and pores in the rock. The groundwater flow rate tends to decrease with depth because sedimentary rock pores tend to be closed by increasing amounts of cement and the weight of the overlying rock. Moreover, sedimentary rock overlying igneous and metamorphic crystalline basement rock, which usually has very low porosity (Plummer et al., 2016).

The subsurface zone in which all rock openings are filled with water is called the saturated zone (Figure 6-3). If a well were drilled downward into this zone, ground water would fill the lower part of the well. The water level inside the well marks the upper surface of the saturated zone; this surface is the water table.

The subsurface region in which all rock openings are packed with water is called the saturated zone (Figure 6-3). If a well were drilled down into this place, groundwater would fill the lower part of the well. The water level inside the well indicates the upper surface of the saturated zone; this surface is called the water table.

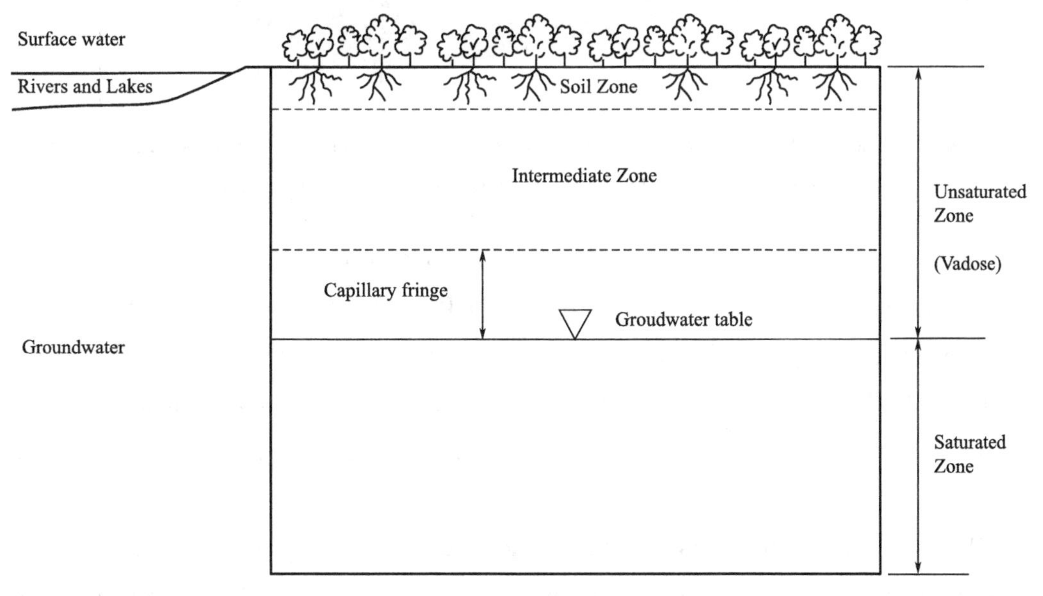

Figure 6-3 The water table marks the top of the saturated zone where water fills the rock pore space. Above the water table is the unsaturated zone in which rock openings typically contain both air and water

Most rivers and lakes are across the saturated zone. Rivers and lakes involve low places on the land surface, and groundwater flows into the saturated area into these surface sinks. The water level at the surface of lakes or rivers usually agrees with the water table. Groundwater also flows into pits and quarries cut below the water table.

Above the water table is a region where not all the sediment or rock cracks are filled with water. It is mentioned as the unsaturated zone. Within the unsaturated zone, surface tension causes water to be held above the water table. The capillary fringe is a transitional zone with higher moisture content at the base of the unsaturated zone just above the water table. Some of the water in the capillary fringe has been drawn or wicked upward from the water table (like water rising a paper towel if the corner is dipped in water) (Carlson et al., 2011). The capillary fringe is typically less than 1m thick but maybe much thicker in fine-grained sediments and thinner in coarse-grained sediments such as sand and gravel.

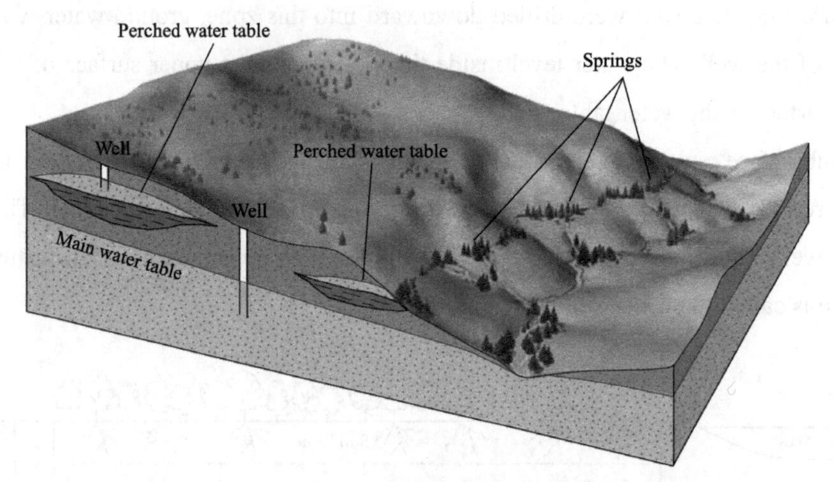

Figure 6-4 Perched water tables above lenses of less permeable shale within a large mass of sandstone. The less permeable shale impedes the downward percolation of water

The perched water table is the top of a groundwater body separated from the main water table with an unsaturated area beneath it (Figure 6-4). It can develop as groundwater accumulates over a low permeability shale lens in more permeable rock, sandstone. If the perched water table crosses the land surface, a line of springs can form along with the upper contact of the shale lens. Water perched above a shale lens can provide an insufficient water supply to a well and is unreliable over the long term.

The water table slope strongly influences groundwater velocity, and the steeper the slope of the water table, the faster groundwater moves. Topography essentially controls water-table gradients—the water table is roughly parallel to the land surface (particularly in humid regions). Even in highly permeable rock, groundwater will not move if the water table is flat.

6.3.2 Classification of ground water based on buried conditions

The perched water exists in the aeration zone, while the phreatic and confined water belong to the saturated zone, our main research object. These three different types of groundwater can occur in the loose porous medium, fissure medium, and karst medium of the rock.

1. Perched water

The perched water refers to the gravity water on the local aquiclude or aquitard in the vadose zone. It is formed by the accumulation of atmospheric precipitation, surface water, etc., blocked locally during the infiltration process. This kind of local aquiclude may be composed of lenses such as clay and loam in the area of loose sediments. The fissure medium of bedrock may be caused by the lack of cracks or the filling of the cracks. In the karst medium, it may result from poor development or the existence of non-soluble rock lenses due to differential dissolution.

The perched water has the following characteristics: its surface forms its top interface, and the surface is subject to atmospheric pressure but not hydrostatic pressure. It is a free surface that can rise and fall freely. Meteoric precipitation is the primary supply source of the perched water, so its recharge area is consistent with its distribution area. In some cases, it is possible to obtain infiltration recharge from nearby surface water. The perched water is discharged vertically through evaporation and slow infiltration through the aquiclude floor below it.

Meanwhile, under the action of gravity, it is discharged laterally at the edge of the floor (Figure 6-5). Because the distribution of the perched water is limited and the volume is small, it can only be used as a small temporary water supply source. It is also easy to deal with during excavation.

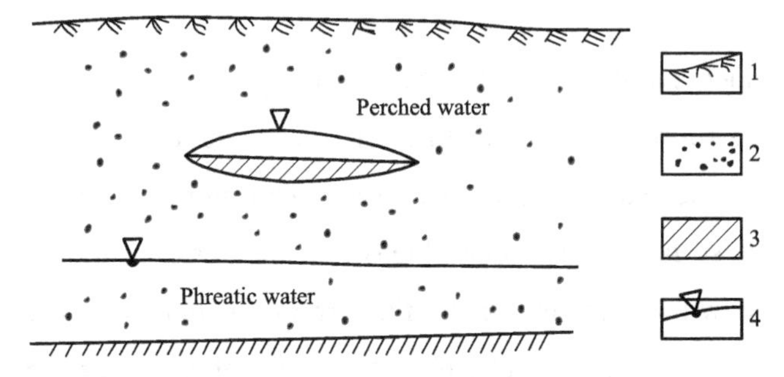

Figure 6-5 Schematic diagram of the perched water

1—ground surface; 2—aquifer; 3—aquiclude; 4—underground water table

2. Phreatic water

Occurs on the surface of a stable aquifer above, with a free surface of the aquifer gravity water is called phreatic water. The aquifer is called a phreatic aquifer, and the bottom surface of the aquifer is called a water-proof floor. The phreatic water surface can also be called the water table. The water table and water-proof floor constitute the top and bottom boundary of the phreatic aquifer. The distance between the water table and the ground is called the buried depth of the phreatic, and the distance between the water table and the water-proof floor is called the thickness of the phreatic aquifer. The height of the water table is called the water level (Figure 6-6).

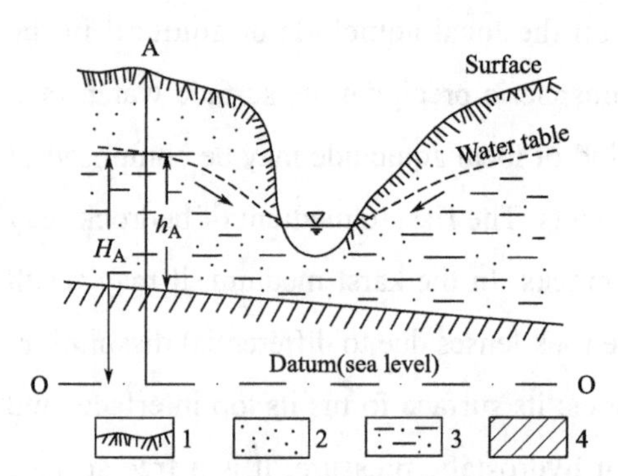

Figure 6-6 Schematic diagram of phreatic water
1—Topsoil; 2—Unsaturated zone; 3—Aquifer; 4—Aquiclude; H_A—Phreatic water level at point A; h_A—Thickness of the phreatic water at point A.

Phreatic water is closely related to the atmosphere and surface water through the zone of aeration. In its distribution range, it directly receives the infiltration recharge of atmospheric precipitation, surface water, and irrigation water leakage through the zone of vadose. The recharge area is generally consistent with the distribution area. The water table, buried depth, water quantity and water quality of the phreatic water are controlled and influenced strongly by meteorological and hydrological factors, which change with time and show significant seasonal variations. In the rainy season, phreatic water gets abundant recharge. The storage increases, the thickness increases, the water table rises, the buried depth decreases, and the salt content in the water is diluted due to the addition of freshwater. In the dry season, the replenishment is small, and the storage capacity is consumed due to the constant discharge of water. The thickness of the aquifer is thinner, the water table is decreased, the buried depth is increased, and the salt content in water is also increased.

Topographical fluctuations control the shape of the phreatic surface (water table). Usually, the undulations of the surface are similar to the terrain undulations, but they are gentler. The phreatic surface has a steep slope and a large buried depth in mountainous areas where cutting is strong. The phreatic surface is often buried tens of meters or even more than 100m below the surface. In plain areas with weak cutting and flat terrain, the phreatic surface fluctuates slowly, and the depth is only a few meters. The phreatic water table is close to the ground surface and even forms a swamp in low-lying terrain. In addition, phreatic water is susceptible to contamination due to its shallow burial and direct connection to the vadose zone.

3. Confined water

The gravity water filled in the aquifer between two stable aquicludes (aquitard) is called confined water, and this aquifer is called a confined aquifer. The bottom interface of the upper impervious layer and the lower impervious layer are respectively called the water-proof roof and the water-proof floor, which constitute the top and bottom interfaces of the confined aquifer. The vertical distance between the top and bottom interfaces of the aquifer is the thickness of the confined aquifer.

The various water levels that may be encountered in boreholes are illustrated in Figure 6-7. Borehole A is unlined and penetrates an impersistent zone of saturation supported on an impermeable zone of limited extent: "a" is a perched water table. Boreholes B_1 to B_4, which are cased only in their upper portion, enter the main zone of saturation and "b" is the main water table. Thus a water table is the level of water encountered in a borehole that is either unlined, partially lined, or lined with perforated casing. Borehole C penetrates the confined aquifer where groundwater pressure is sufficient to support a column of water in the hole. Unlike the other bores, this hole is lined with casing to its base so that the column of water in it balances the pressure of water in the ground at c': the level c is called a piezometric (or pressure) level for the confined aquifer.

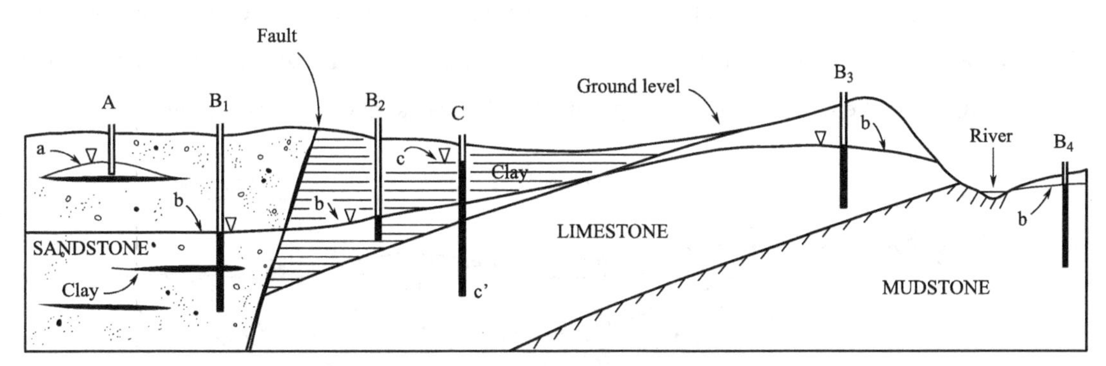

Figure 6-7 The main sandstone aquifer containing clay lenses is faulted against a limestone aquifer confined by clay and underlain by mudstone. Note: water tables can exist in aquicludes, as shown in the clay and mudstone (modified from Blyth & de Freitas, 2005)

When the borehole drills through the water-proof roof of the confined aquifer during drilling, groundwater is seen. At this time, the elevation of the water table is called the initial water level. After that, the water surface continued to rise, and it stabilized and no longer increased, reaching a certain height. At this time, the elevation of the water surface is called the stable water level, that is, the piezometric level of the confined aquifer at this point. The surface formed by the piezometric level of each point in a confined aquifer is the piezometric water table of the aquifer.

The vertical distance between the surface of the waterproof roof and the piezometric level at a certain point is called the pressure head (h in Figure 6-8). The value of the confined water head represents the hydrostatic pressure of the confined water acting on the waterproof roof at this point. When the piezometric level is higher than the ground, the pressure head of the confined water is called a positive head, and vice versa, it is called a negative head. When the aquifer is exposed, water can spray out of the surface when drilling in areas with positive water heads, usually called the artesian water. Wells that expose artesian water are called artesian wells. Drilling in areas with negative water heads, after the aquifer is exposed, the water table of the confined water is higher than the top interface of the aquifer but lower than the ground (Figure 6-8).

Bores drilled into an aquifer that has its piezometric level above ground-level will overflow and are termed artesian bores after the French province of Artois, where overflowing wells were first recorded. The London Basin was formerly an artesian basin because water levels in the Chalk outcrop north and south of London were higher than the ground level in central London, where wells drilled through the confining London Clay and into the Chalk would overflow. Pumping has since reduced the piezometric level to below ground level. In Queensland, Australia, the Great Artesian Basin is a famous example of such a structure, extending $1.56 \times 10^6 km^2$ west of its outcrop along the Great Divide and supplying water, via deep wells, to the arid country of central Queensland (Blyth & de Freitas, 2005). Piezometric levels above the aquifer supporting them, but do not reach ground level, are described as sub-artesian.

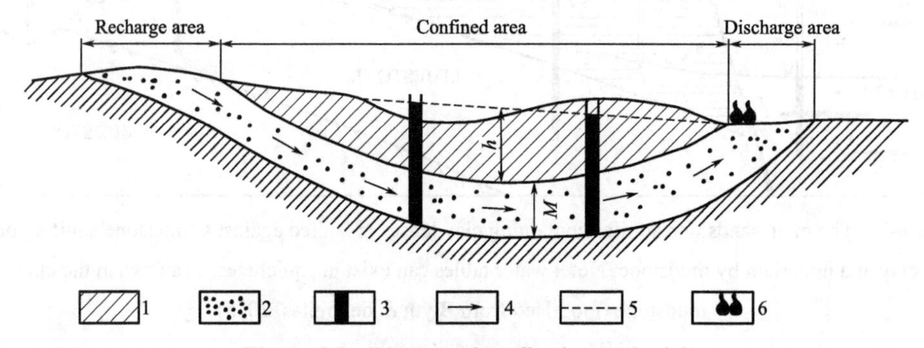

Figure 6-8 Diagram of confined water burial

1—aquiclude; 2—confined aquifer; 3—well; 4—ground water flow direction; 5—pressure water table; 6—A rising spring; h—pressure head

Due to different burial conditions, confined water is significantly different from phreatic and perched water. Its main features are as follows:

(1) The top surface of the confined aquifer is subjected to piezometric pressure, and the confined water is filled between two aquiclude layers.

(2) The location of the recharge area is higher so that the groundwater has higher potential energy. As a result of the transmission of hydrostatic pressure, the top surface of the confined

aquifer in other areas bears the atmospheric pressure and the pressure of the overlying strata and the hydrostatic pressure.

(3) The piezometric water level of a confined aquifer is an imaginary surface above its top interface. The pressurized water flows from the high point of the piezometric level to the low end of the piezometric level. When the amount of water in the aquifer changes, the piezometric level will also rise and fall, but the top interface of the aquifer and the thickness of the aquifer do not vary significantly.

(4) Due to the barrier of the upper impermeable layer, the confined aquifer is less closely related to the atmosphere and surface water than phreatic water. The distribution area of confined water is usually larger than its recharge area, and confined water resources are not readily replenished and restored as phreatic water. However, confined aquifers generally have an extensive distribution range and often have good multi-year adjustment capabilities.

(5) The natural dynamics, such as the level and volume of the confined water, are generally relatively stable. Confined water is usually not easily polluted, but once it is contaminated, purification is complicated. Therefore, when using confined water as a water supply source, the problem of water quality protection cannot be taken carelessly.

6.3.3 Wells

A well is a deep, cylindrical hole dug or drilled into the ground to penetrate an aquifer within the saturated zone (Figure 6-1). Typically, water that flows into the well from the saturated rock must be lifted or pumped to the ground. As Figure 6-9 displays, a well dug in a valley usually has to go down a shorter range to reach water than a well drilled on a hilltop. During dry seasons, the water table declines as water flows from the saturated zone to springs and rivers. Wells are not deep enough to intersect the lowered water table go dry, but the rising water table

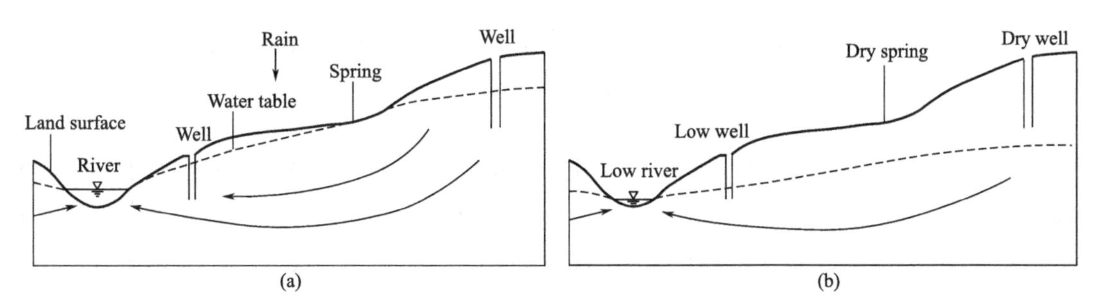

Figure 6-9 The water table in an unconfined aquifer increases during wet seasons and decreases during dry seasons as water flows out of the saturated zone into rivers

(a) Rainy season: water table and rivers are high; springs and wells flow readily; (b) Dry season: water table and rivers are low; some springs and wells dry up

during the next rainy season typically returns water to the wells. The addition of new water to the saturated zone is called recharge.

When water is elevated from a well, the water table is usually drawn down around the well into a depression mold like an inverted cone known as a cone of depression (Figure 6-10). This limited lowering of the water table, called a "drawdown," tends to alter the direction of groundwater flow by varying the slope of the water table. In slightingly used wells that are not pumped, drawdown does not happen, and a cone of depression does not appear. In a simple rural well with a container lowered at the end of a rope, the water cannot be extracted fast enough to reduce the water table significantly.

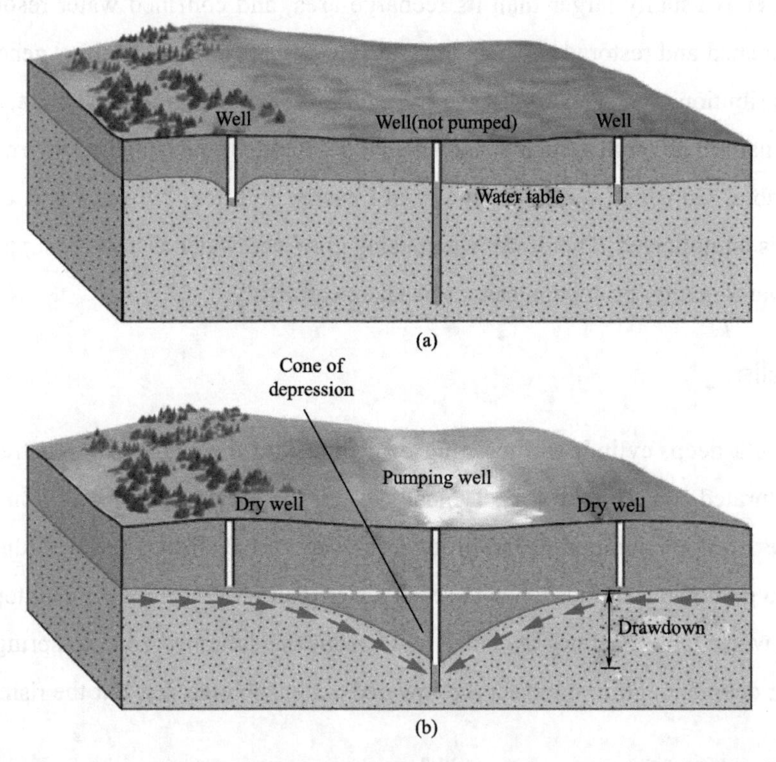

Figure 6-10 Pumping well lowers the water table into a cone of depression. If well is heavily pumped, surrounding shallow wells may go dry

In unconfined aquifers, the water rises in shallow wells to the level of the water table. In confined aquifers, the water is under pressure and grows in wells to a level above the top of the aquifer (Figure 6-10b). Such a well is termed an artesian well, and confined aquifers are also called artesian aquifers.

In some artesian wells, the water increases above the land surface, producing a flowing well that squirts continuously into the air unless capped (Figure 6-11). Flowing wells used to occur in

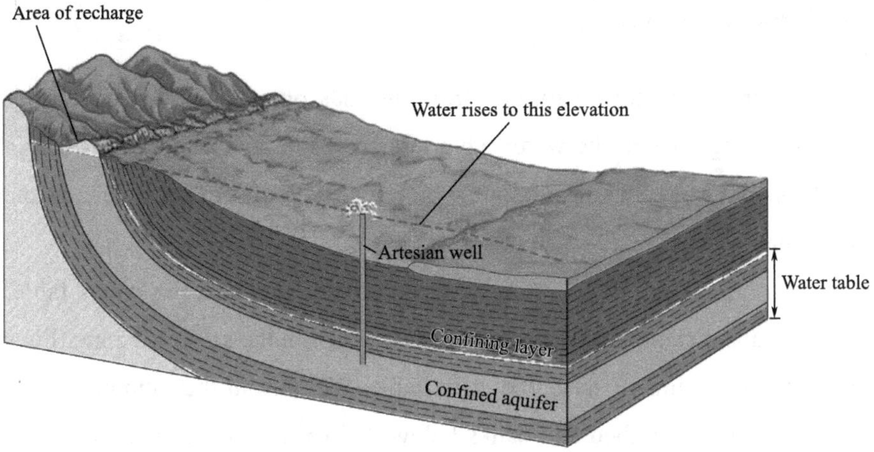

Area of recharge
Water rises to this elevation
Artesian well
Water table
Confining layer
Confined aquifer

Figure 6-11 The Dakota Sandstone in South Dakota is a particular kind of confined aquifer because it is dipped and exposed to the surface by erosion. Water in most wells was raised above the land surface when the aquifer was first tapped (Plummer et al., 2016)

South Dakota when the extensive Dakota Sandstone aquifer was first tapped (Figure 6-12). Still, continued use has lowered the water pressure surface below the ground surface in most parts of the state, and water still rises above the aquifer but does not reach the land surface.

Figure 6-12 Artesian well spouts water above land surface in South Dakota, early 1900s. Heavy use of this aquifer has reduced water pressure so much that spouts do not occur today (USGS & Plummer et al., 2016)

6.3.4 Springs and streams

Spring is where water flows naturally from the rock onto the land surface (Figure 6-13). Some springs discharge where the water table intersects the land surface, but they also occur where water flows out from caverns or along with fractures, faults, or rock contacts that come to the surface (Figure 6-14).

Climate determines the connection between streamflow and the water table. In rainy regions, most streams obtain streams; they receive water from the saturated zone (Figure 6-15a). The surface of these streams matches the water table. Water from the saturated area flows into the stream through the streambed and banks below the water table. Because of the addition of groundwater, the discharge of these streams increases downstream. Ponds, lakes, and swamps are found where the water table intersects the land surface over a broad area.

Figure 6-13 A large spring(Mizhong) flowing from limestone in Zigui,Hubei

In dry climates, rivers lose streams, losing water to the saturated zone (Figure 6-15b). The water filtering into the ground beneath a losing stream induces the water table to slope away from the stream. In arid climates, such as in a desert, a losing stream may be disconnected from the underlying saturated zone, and a groundwater mound continues beneath the stream even if the streambed is dry (Figure 6-15c).

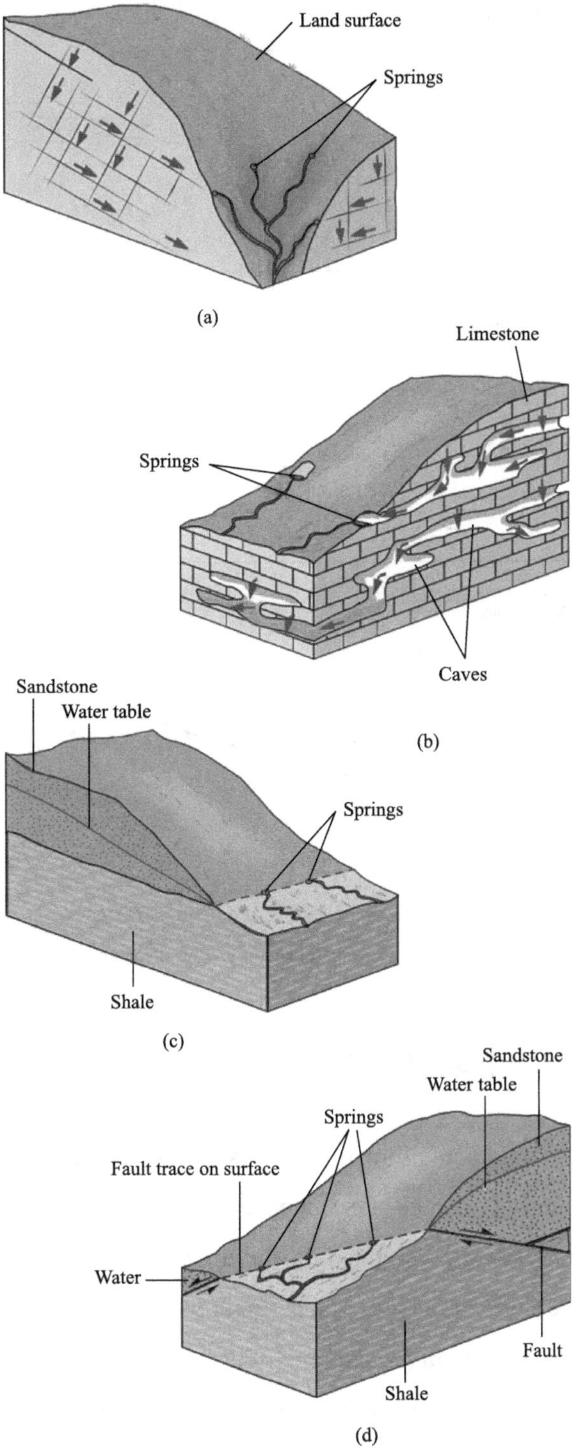

Figure 6-14 Springs can form in many ways

(a) Water moves along fractures in crystalline rock and forms springs where the fractures intersect the land surface; (b) Water enters caves along joints in limestone and exits as springs at the mouths of caves; (c) Springs form between a permeable rock such as sandstone and an underlying less permeable rock such as shale; (d) Springs can create faults when the permeable rock has been moved against the less permeable rock. Arrows show relative motion along the fault (Carlson et al., 2010)

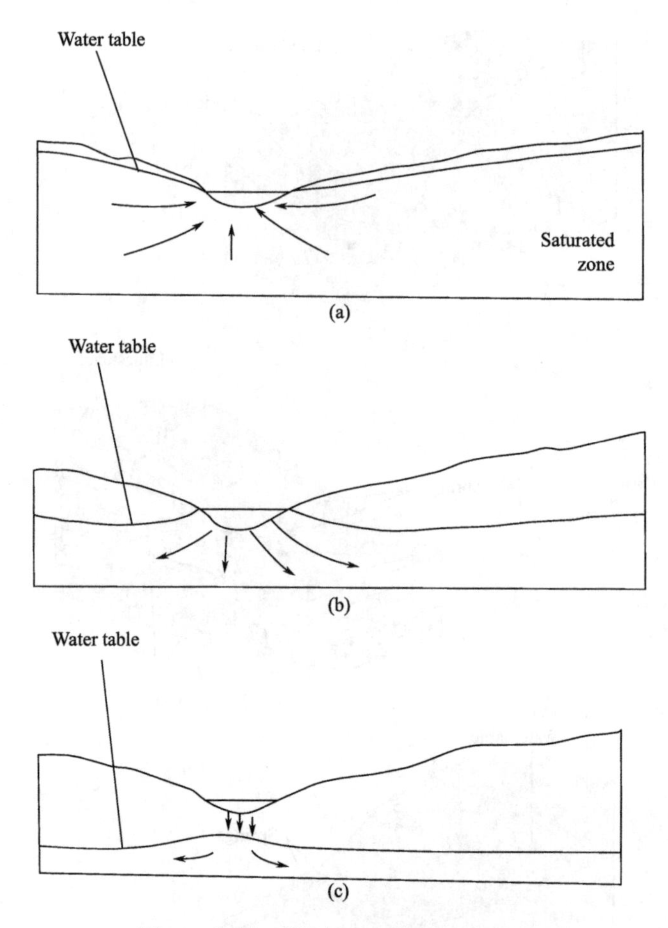

Figure 6-15 Gaining and losing streams

(a) Gaining stream; (b) Losing stream; (c) Losing stream (disconnected)

6.4 Movement of ground water

Compared with the rapid water flow in surface streams, most groundwater moves relatively very slowly through subsurface rock. Groundwater moves in response to variations in water pressure and elevation, causing water within the upper part of the saturated zone to move downward following the slope of the water table (Figure 6-16).

The groundwater circulation in the saturated zone is not confined to a shallow layer beneath the water table. Groundwater may move hundreds of meters vertically downward before rising again to discharge as a spring or seep into the beds of rivers and lakes at the surface due to the combined effects of gravity and the slope of the water table.

How fast groundwater moves also depends on the permeability of the rock or other materials it transfers. If rock pores are tiny and poorly connected, water moves slowly, and

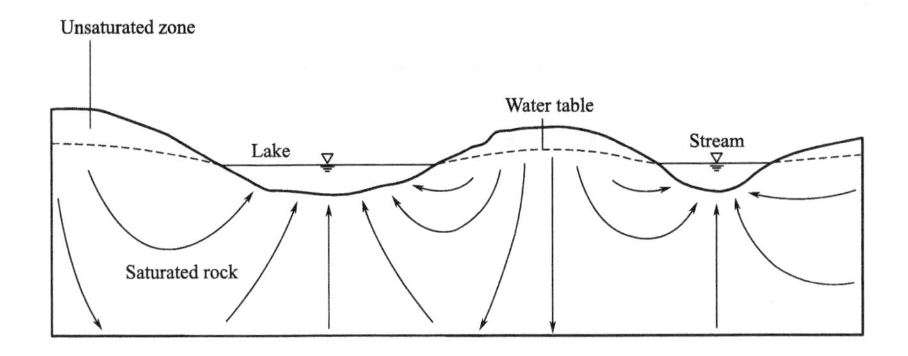

Figure 6-16 Movement of groundwater beneath a sloping water table in uniformly permeable rock. Near the surface, groundwater tends to flow parallel to the slanted water table

when openings are large and well connected, water flow is more rapid. One means of measuring groundwater velocity is to include a tracer, such as a colorant, into the water and then follow the color to emerge in a well or spring some extent away. Such investigations have revealed that the velocity of groundwater changes extensively, averaging a few centimeters to many meters a day. Nearly impermeable rocks may permit water to move only a few centimeters per year. However, highly permeable substances, such as unconsolidated gravel or cavernous limestone, may allow flow rates of hundreds or even thousands of meters per day.

6.4.1 Darcy's Law and Fluid Potential

In 1856, Henry Darcy, a French hydraulic engineer, found that the velocity at which water moves depends on the hydraulic head and the permeability of the substance that the water is moving through. In that year, he published his report on the water supply of the city of Dijon, France. In the report, Darcy described a laboratory experiment that he had carried out to analyze water flow through sands. The hydraulic head of a drop of water is equal to the water pressure plus the elevation of the drop: Hydraulic head = elevation head + pressure head+velocity head (in groundwater movement, the velocity head is tiny and generally negligible).

Consider an experimental device like that shown in Figure 6-17. A circular cylinder of cross-section A is packed with sand, stoppered at each end, and outfitted with inflow and outflow pipes and a pair of manometers. Water is carried into the cylinder and allowed to flow through it until all the pores are saturated with water, and the inflow rate Q is equivalent to the outflow rate. If we set an optional datum at elevation $z = 0$, the elevations of the manometer intakes are z_1 and z_2, and the elevations of the fluid levels are h_1 and h_2. The distance between the manometer intakes is Δl. We will define v, the specific discharge through the cylinder, as

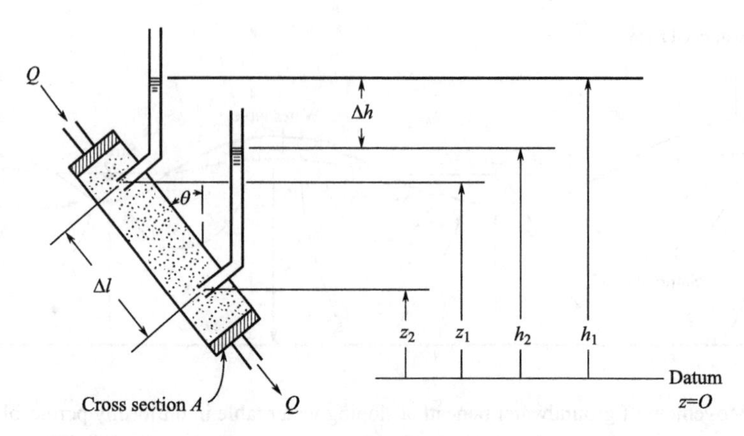

Figure 6-17 Experimental apparatus for the illustration of Darcy's law

$$v=\frac{Q}{A} \tag{6-1}$$

If the dimensions of Q are [L^3/T] and those of A are [L^2], v has the dimensions of a velocity [L/T].

The experiments carried out by Darcy showed that v is directly proportional to h_1-h_2 when Δl is held constant, and inversely proportional to Δl when h_1-h_2 is held constant. If we define $\Delta h = h_2-h_1$ (an arbitrary sign convention that will stand us in good stead in later developments), we have $v \propto -\Delta h$ and $v \propto 1/\Delta l$. Darcy's law can now be written as

$$v=-K\frac{\Delta h}{\Delta l} \tag{6-2}$$

or, in differential form,

$$v=-K\frac{dh}{dl} \tag{6-3}$$

In Eq. (6-3), h is called the hydraulic head and dh/dl is the hydraulic gradient; K is a constant of proportionality. It must be a property of the soil in the cylinder, for were we to hold the hydraulic gradient constant, the specific discharge would surely be larger for some soils than for others. In other words, if dh/dl is held constant, $v \propto K$. The parameter K is known as the hydraulic conductivity. It has high values for sand and gravel and low values for clay and most rocks. Since Δh and Δl both have units of length [L], a quick dimensional analysis of Eq. (6-2) shows that K has the dimensions of a velocity [L/T].

An alternative form of Darcy's law can be obtained by substituting Eq. (6-1) in Eq. (6-3) to yield

$$Q=-K\frac{dh}{dl}A \tag{6-4}$$

This is sometimes compacted even further into the form

$$Q=-KiA \tag{6-5}$$

where i is the hydraulic gradient.

We have transcribed that the specific discharge v has the dimensions of velocity or flux. For this logic, it is sometimes recognized as the Darcy velocity. The specific discharge is a macroscopic idea, and it is easily measured. It must be differentiated from the microscopic speeds associated with the actual tracks of singular particles of water as they zigzag their way through the particles of sand (Figure 6-18). The microscopic velocities are real, but they are probably impossible to measure. We will not refer to v as a velocity; instead, we will utilize the correct term, specific discharge.

Figure 6-18 Macroscopic and microscopic concepts of groundwater flow

In Figure 6-19 (a), points A and B are on the water table, so the pressure is zero (there is no water over points A and B to generate pressure). Point A is higher than B, so A has a higher hydraulic head than B. The elevation difference is equal to the difference in the head, which is A labeled h. Water will flow from point A to point B (as shown by the dark blue arrow) because water moves from a region of a high hydraulic head to an area of the low head. The distance the water moves from A to B is labeled L.

In Figure 6-19 (b), the two points have equal elevation, but the pressure on point C is higher than on point D (more water creates pressure over point C than point D). The head is higher at point C than at point D, so the water flows from C to D. In Figure 6-19 (c), point F has a lower elevation than point G, but F can also has a higher pressure than G. The pressure difference is greater than the elevation difference. Hence, F has a higher head than G, and water moves from F to G. Note that underground water may move downward, horizontally, or upward in response to differences in the head but always moves in the direction of the downward slope of the water table above it. One of the first aims of groundwater geologists, particularly in groundwater pollution investigations, is to find the slope of the local water table to determine the direction (and velocity) of groundwater movement.

The velocity of groundwater flow is controlled by the permeability of the sediment or rock and the hydraulic gradient. Darcy's Law states that the speed equals the permeability multiplied by the hydraulic gradient. It gives the Darcian velocity (or the velocity of water flowing

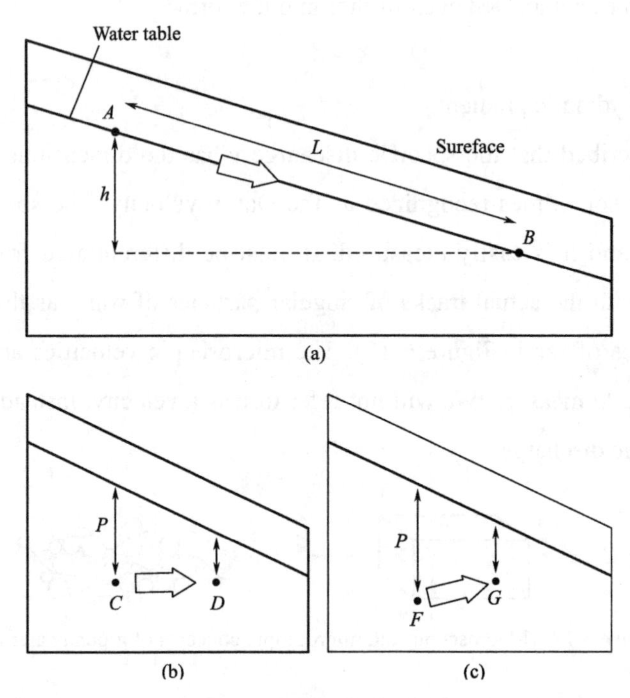

Figure 6-19 Groundwater moves in response to hydraulic head (elevation plus pressure). Dark blue arrows show water movement

(a) Points *A* and *B* have the equal pressure, but *A* has a higher elevation; therefore, water flows from *A* to *B*; (b) Point *C* has a higher pressure (arrow marked *P*) than *D*; thus, water moves from *C* to *D* at the same elevation; (c) Pressure also moves water upward from *F* to *G*

through an open pipe). To determine the actual velocity of ground water, since ground water only flows through the openings in sediment or rock, the Darcian speed must be divided by the porosity.

Groundwater velocity = permeability/porosity × hydraulic gradient

$$V=\frac{K}{n}\frac{\Delta h}{L} \tag{6-6}$$

(*K* is the hydraulic conductivity; it is a measure of permeability and is specific to a particular aquifer. The porosity is represented by *n* in the equation.)

As Hubbert (1956) has denoted, the constant proportionality in Darcy's law, which has been named the hydraulic conductivity, is a porous medium and fluid function. Consider once again the laboratory apparatus of Figure 6-19. If Δh and Δl are kept constant for two runs using the identical sand, but water is the fluid in the first race and molasses in the second, it would be no wonder to discover the specific discharge v much lower the second run than in the first. In light of such research, it would be enlightening to explore a parameter that can define the conductive properties of a porous medium alone from the fluid flowing through it.

To this end, investigations have been carried out with ideal porous media consisting of uniform glass beads of diameter d. When various fluids of density ρ and dynamic viscosity μ are run through the apparatus under a constant hydraulic gradient dh/dl, the following proportionality relationships are observed:

$$v \propto d^2$$
$$v \propto \rho g \tag{6-7}$$
$$v \propto \frac{1}{\mu}$$

Together with Darcy's original observation that $v \propto -dh/dl$, these three relationships lead to a new version of Darcy's law:

$$v = -\frac{Cd^2\rho g}{\mu}\frac{dh}{dl} \tag{6-8}$$

The parameter C is still another constant of proportionality. Natural soils must include the influence of other media properties that affect the flow, apart from the mean grain diameter: for example, the arrangement of grain sizes, the sphericity and roundness of the grains, and the nature of their disposition.

Comparison of Eq. (6-8) with the original Darcy equation [Eq. (6-3)] shows that

$$K = \frac{Cd^2\rho g}{\mu} \tag{6-9}$$

In this equation, ρ and μ are functions of the fluid alone and Cd^2 is a function of the medium alone. If we define

$$k = Cd^2 \tag{6-10}$$

Then

$$K = \frac{k\rho g}{\mu} \tag{6-11}$$

The parameter k is known as the specific or intrinsic permeability. If K is always called hydraulic conductivity, it is safe to drop the adjectives and refer to k as simply the permeability. That is the convention that will be followed in this text. Still, it can lead to some confusion, especially when dealing with older texts and reports where the hydraulic conductivity K is sometimes called the permeability coefficient.

Groundwater movement is shown in diagrams concerning equipotential lines (lines of constant hydraulic head). Groundwater moves from regions of the high head to areas of the low head. Figure 6-20 illustrates how flow lines, which show groundwater flow, cross equipotential lines at right angles as water flows from high to low head.

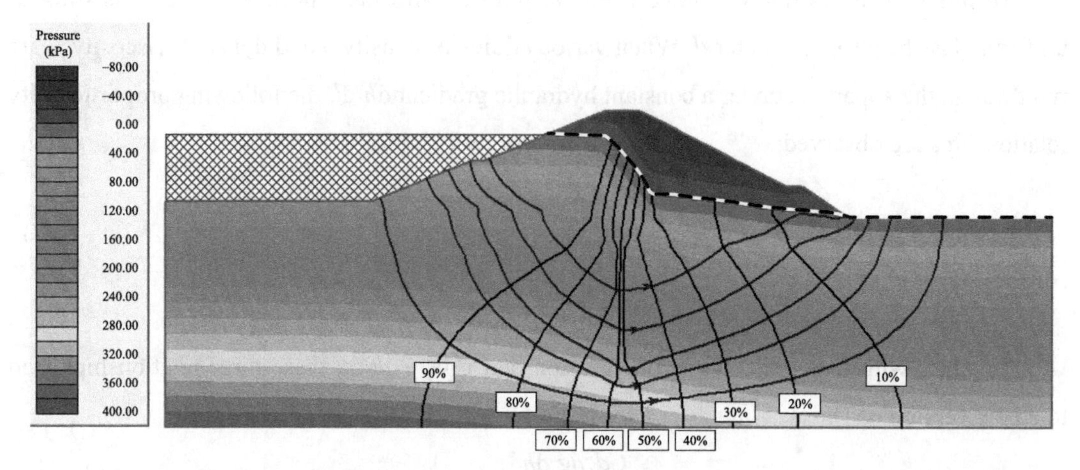

Figure 6-20 Dark arrows are flow lines, which show the direction of groundwater flow. Flow is perpendicular to equipotential lines (black lines with percentage numbers), showing equal hydraulic head regions

6.4.2 Equations of groundwater Flow

In almost every range of science and engineering, the analysis techniques are based on an understanding of physical processes. In most cases, it is possible to describe these processes mathematically. Groundwater flow is no exception, and the fundamental law of flow is Darcy's law. A partial differential equation of flow is the result when it is put together with an equation of continuity that describes the conservation of fluid mass during flow through a porous medium (Freeze & Cherry,1979).

In that so numerous of the standard routines of analysis in groundwater hydrology are based on boundary-value problems that include partial differential equations, it is helpful to have an essential judgment of these equations as one proceeds to learn the various techniques. Fortunately, it is not an absolute requirement. In most cases, the methods can be explained and understood without returning at every step to fundamental mathematics. The hydrogeologist must work with the flow equations daily; the practicing hydrogeologist can usually avoid advanced mathematics.

Suppose a unit volume of porous media such as that is shown in Figure 6-21. Such an element is usually called an elemental control volume.

The law of conservation of mass for steady-state flow through a saturated porous medium requires that fluid mass flows into any elemental control volume equal to the fluid mass flow rate out of any elemental control volume. The equation of continuity that translates this law into

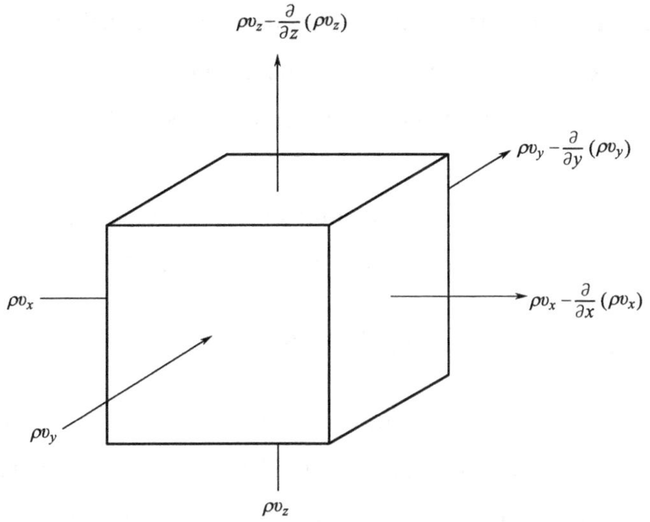

Figure 6-21 Elemental control volume for flow through porous media

mathematical form can be written as

$$-\frac{\partial(\rho v_x)}{\partial x}-\frac{\partial(\rho v_y)}{\partial y}-\frac{\partial(\rho v_z)}{\partial z}=0 \tag{6-12}$$

A quick dimensional analysis on the ρv terms will show them to have the dimensions of a mass rate of flow across a unit cross-sectional area of the elemental control volume. If the fluid is incompressible, $\rho(x, y, z)$ = constant and the ρ can be removed from Eq. (6-12). Even if the fluid is compressible and $\rho(x, y, z) \neq$ constant, it can be shown that terms of the form $\rho\ \partial v_x/\partial x$ are much greater than terms of the form $v_x\ \partial \rho/\partial x$, both of which arise when the chain rule is used to expand Eq. (6-12). In either case, Eq. (6-12) simplifies to

$$-\frac{\partial v_x}{\partial x}-\frac{\partial v_y}{\partial y}-\frac{\partial v_z}{\partial z}=0 \tag{6-13}$$

Substitution of Darcy's law for v_x, v_y, and v_z in Eq. (6-13) yields the equation of flow for steady-state flow through an anisotropic saturated porous medium:

$$\frac{\partial}{\partial x}\left(K_x\frac{\partial h}{\partial x}\right)+\frac{\partial}{\partial y}\left(K_y\frac{\partial h}{\partial y}\right)+\frac{\partial}{\partial z}\left(K_z\frac{\partial h}{\partial z}\right)=0 \tag{6-14}$$

For an isotropic medium, $K_x = K_y = K_z$, and if the medium is also homogeneous, then $K_{(x, y, z)}$ = constant. Eq. (6-14) then reduces to the equation of flow for steady-state flow through a homogeneous, isotropic medium:

$$\frac{\partial^2 h}{\partial x^2}+\frac{\partial^2 h}{\partial y^2}+\frac{\partial^2 h}{\partial z^2}=0 \tag{6-15}$$

Eq. (6-15) is one of the most basic partial differential equations known to mathematicians.

It is called Laplace's equation. The solution of the equation is a function $h_{(x, y, z)}$ that describes the value of the hydraulic head h at any point in a three dimensional flow field. A solution to Eq. (6-15) allows us to produce a contoured equipotential map of h, and with the addition of flowlines, a flow net.

For steady-state, saturated flow in a two-dimensional flow field, say in the xz plane, the central term of Eq. (6-15) would drop out and the solution would be a function $h_{(x, z)}$.

6.4.3 Seepage calculation using FLAC3D

FLAC3D (Fast Lagrangian Analysis of Continua in 3 Dimensions) models the flow of fluid through a permeable solid, such as soil. The flow modeling may be done by itself, independent of the usual mechanical calculation of FLAC3D, or it may be done in parallel with the mechanical modeling to capture the effects of fluid and solid interaction (Itasca, 2014). One type of fluid and solid interaction is consolidation, in which the slow dissipation of pore pressure causes displacements to occur in the soil. This type of behavior involves two mechanical effects. First, changes in pore pressure cause changes in effective stress, which affect the response of the solid. (For example, a reduction in effective stress may induce plastic yield.) Second, the fluid in a zone reacts to mechanical volume changes by a change in pore pressure.

The basic flow scheme manages both fully saturated flow and flow in which a phreatic surface exhibits. In this case, pore pressures are zero above the phreatic surface, and the air phase is passive. This thought is applicable to coarse materials when capillary effects can be neglected. To describe the growth of an internal transition between saturated and unsaturated zones, the flow in the unsaturated region must be simulated so that fluid may migrate from one area to the other. A simplistic law that relates the apparent permeability to saturation is used. The transient response in the unsaturated region is only approximate (due to the simple rule used), but the steady-state phreatic surface should be accurate.

This example is the traditional problem of steady-state seepage flow over a homogeneous embankment with vertical slopes exposed to different water levels and resting on an impervious base. The total discharge, Q, and the seepage face, s length are compared to the exact solutions. Figure 6-22 shows the geometry and boundary conditions of the problem.

The fluid is homogeneous, Darcy's law governs flow, and it is assumed that the soil's pores beneath the phreatic surface are filled with water, and the pores above it are filled with air. The width of the dam is L, the head and tailwater elevations above the impervious base are h_1 and h_2,

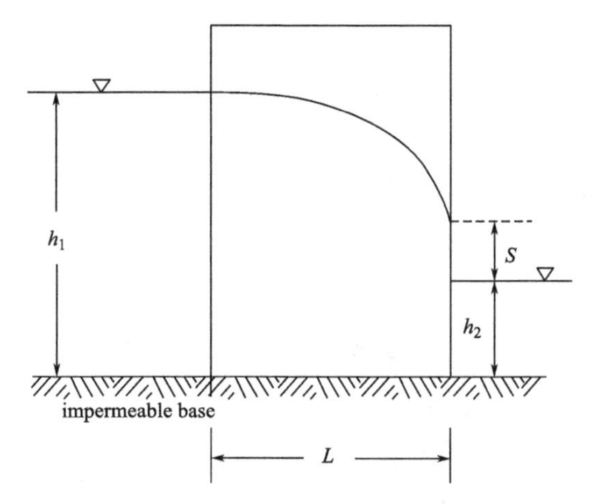

Figure 6-22 Problem geometry and boundary conditions

respectively.

The exact solution for the total discharge through a dam section of unit thickness was shown by Charny (Harr, 1991) to be given by Dupuit's formula:

$$Q=k\rho_w g \cdot \frac{h_1^2-h_2^2}{2L} \tag{6-16}$$

where k is mobility coefficient, ρ_w is water density and g is gravity. The length, s, of the seepage face (elevation of the free surface on the downstream face of the dam above h_2) was obtained by Polubarinova-Kochina (1962), and is given in Figure 6-23 as a function of the characteristic dimensions of the problem.

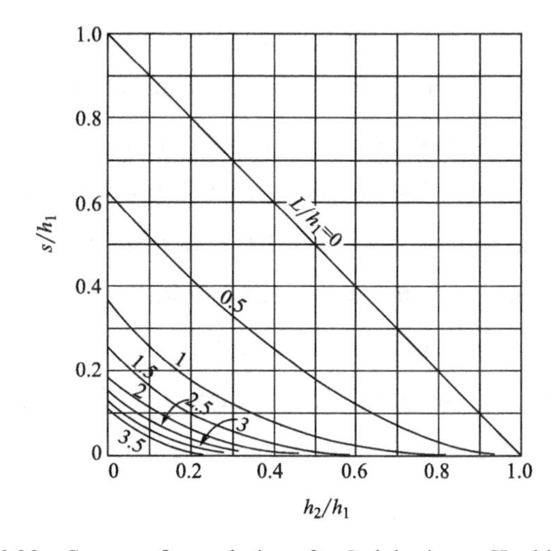

Figure 6-23 Seepage face solution after Polubarinova-Kochina (1962)

The FLAC3D simulation is conducted for a particular set of parameters:

L =9m

h_1 =6m

h_2 =1.2m

Several material properties are used:

permeability (k)	10^{-10}m^2
porosity (n)	0.3
water density (ρ_w)	1000kg/m^3
water bulk modulus (K_f)	1000Pa
soil dry density (ρ)	2000kg/m^3
gravity (g)	10m/s

The grid and boundary situations are the same for both cases. The grid comprises 30 zones in width, 20 zones in height, and 1 zone in thickness. The boundary conditions correspond to a static pore-pressure pattern up to level h_1 on the upstream face and up to h_2 on the downstream face, zero pore pressure from level h_2 to h_1 on the downstream face no-flow conditions across the remaining boundaries. The variations between the two cases are the initial pore pressure and saturation distributions. The saturation and pore pressure are zero above h_2. Below that level, the saturation is 1, and the pressure is hydrostatic. The numerical simulation is carried out until steady-state conditions are discovered.

To speed the calculation to steady state, the water bulk modulus is given a small value (K_f=10^3Pa) compatible with free surface stability (The criterion used is K_f=$0.3\rho_w g L_z$, where L_z is the maximum vertical zone dimension in the vicinity of the phreatic surface).

The final flow pattern is similar for initial conditions. The numerical value of seepage length is defined as the distance on the dam's downstream face between the tailwater elevation and the point where the magnitude of the flow vector vanishes. The analytical value of seepage length is determined from Figure 6-24. For this particular problem, h_2/h_1=0.2, L/h_1=1.5, and the value of s/h_1 is thus 0.1. The figures show that the numerical value of seepage length compares well with the analytical solution sketched there as a bold line. A FISH function, qflac, is used to determine the discharge, Q, per unit thickness of the dam: the steady-state numerical value is 1.914×10^{-6}m^2/s. The value is close to the analytic value of 1.920×10^{-6}m^2/s, as determined from Eq. (6-16) for this particular problem.

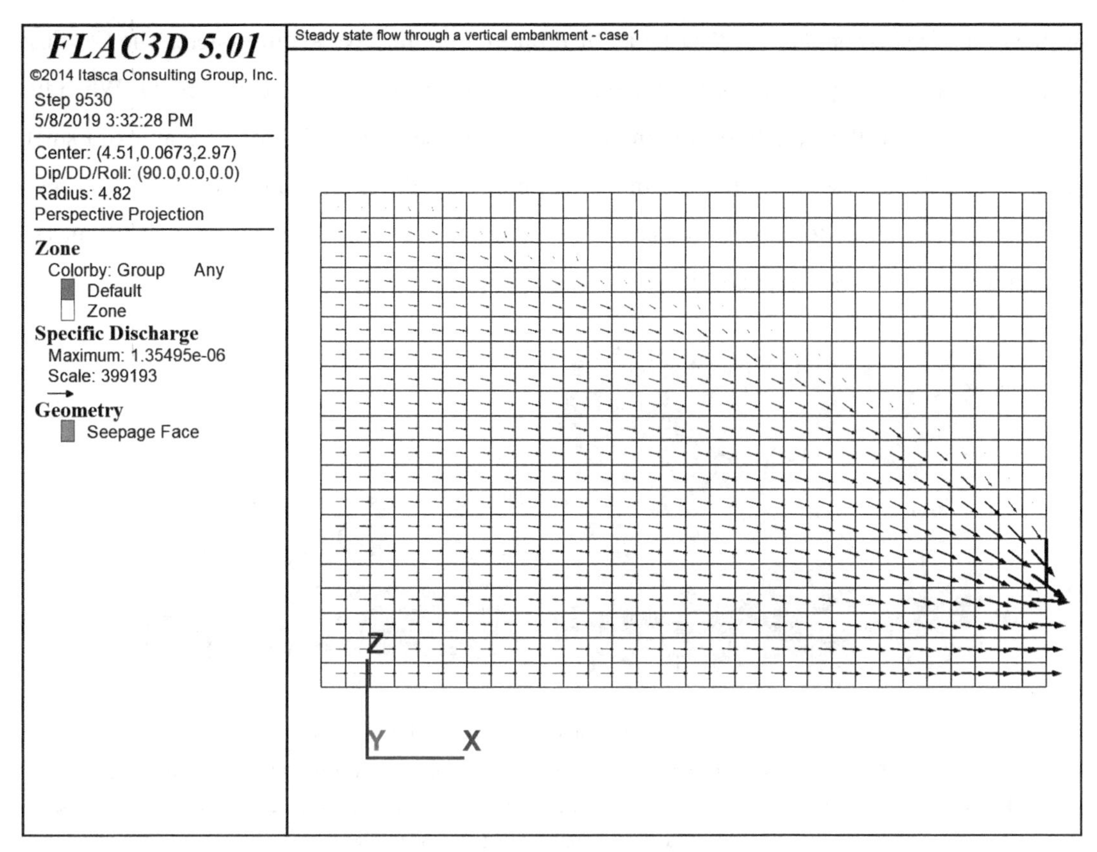

Figure 6-24 Steady-state flow vectors and seepage face solution

6.5 Effects of groundwater action

6.5.1 Caves, Sinkholes, and Karst Topography

Caves (or caverns) are naturally developed underground chambers(Figure 6-25a). Most caves form when slightly acidic groundwater dissolves limestone along joints and bedding planes, opening up cavern systems as calcite is carried away in solution. Natural groundwater is generally lightly acidic because of dissolved carbon dioxide (CO_2) from the atmosphere or soil gases.

Geologists quarrel whether limestone caves form above, below, or at the water table. Most caves are probably formed by groundwater flowing below the water table, as shown in Figure 6-25(a). If the water table drops or the ground is elevated above the water table, the cave may fill with calcite precipitation again.

Groundwater with a high intensity of calcium (Ca^{++}) and bicarbonate (HCO_3^-) ions may drip slowly from the ceiling of an air-filled cave. As a water drop hangs on the top of the cave, some of the dissolved carbon dioxide (CO_2) may be dropped into the cave's atmosphere. The CO_2 loss causes a small quantity of calcite to precipitate out of the water onto the cave ceiling. When the water drop falls to the cave floor, the influence may induce more CO_2 loss, and an extra small amount of calcite may precipitate on the cave floor. Therefore, a falling water drop can pour small amounts of calcite on both the cave ceiling and the cave floor, and each subsequent fall adds more calcite to the first deposits.

Figure 6-25 Solution of limestone to form caves

(a) The Monk Cave formed in a soluble limestone rock. The width is 21 meters at the hole of the cave, and height is 40m; (b) Falling water table allows cave system, now greatly enlarged, to fill with air. Calcite precipitation forms stalactites, stalagmites, and columns above the water table

Deposits of calcite (other minerals) built up in caves by dripping water are called peleothems or dripstones. Stalactites are icicle-like pendants of dripstone hanging from cave ceilings. They are generally slender and are commonly aligned along cracks in the top, which act as conduits for groundwater. Stalagmites are cone-shaped masses of dripstone formed on cave floors, generally directly below stalactites. Splashing water precipitates calcite over a large area on the cave floor, so stalagmites are usually thicker than the stalactites above them. As a stalactite grows downward and a stalagmite grows upward, they may eventually join to form a column (Figure 6-25b). Figure 6-26 shows some of the intriguing features included in caves.

In some caves, water flows in a thin film over the cave surfaces rather than dripping from the ceiling. Sheetlike or ribbonlike flowstone deposits develop from calcite precipitated by pouring water on cave walls and floors.

The grounds of most caves are covered with sediment, some of which is remaining clay, the fine-grained particles left behind as insoluble debris when a limestone, including clay, dissolves (Some limestone contains only about 50% calcite). Other sediments, including most coarse-grained material found on cave floors, may be carried by streams, mainly when surface water drains into a cave from openings on the ground surface.

Figure 6-26 The entrance and stalactites in Yuquan Cave in Zigui, Hubei

Solution of limestone underground may offer features that are visible on the surface. Extensive cavern systems can weaken a region so that houses collapse and form depressions above the land surface. Sinkholes are closed depressions found on land surfaces underlain by limestone (Figure 6-27). They include either the collapse of a cave roof or by solution as descending water enlarges a crack in limestone. Limestone regions in Florida, Missouri, Indiana, and Kentucky are heavily dotted with sinkholes. Sinkholes can also form in the areas underlain by gypsum or rock salt, which are also soluble in water.

(a)

(b)

Figure 6-27 Sinkholes in different countries (Plummer et al., 2016)

(a) Sinkholes formed in limestone near Timaru, New Zealand; (b) A collapse sinkhole formed suddenly in Winter Park, Florida, in 1981

An area with many sinkholes and caves under the land surface has karst topography (Figure 6-28). Karst areas are characterized by a lack of surface streams, although one major river may flow at a level lower than the karst area.

Streams sometimes go down sinkholes to flow through caves below the surface. In this specialized instance, a proper underground stream exists. However, such streams are rare as most groundwater flows very slowly through pores and cracks in deposits or rock. Wells tap groundwater in the rock pores and cracks, not underground streams. If a well did tap an actual underground river in a karst region, the water would seemingly be too polluted to drink, especially if it had washed down the surface into a cavern without being filtered through soil and rock.

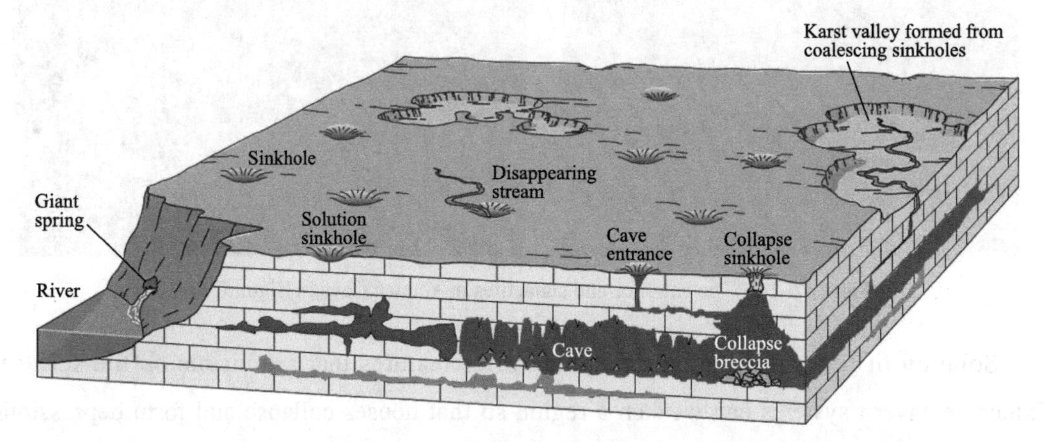

Figure 6-28 Karst topography is marked by underground caves and numerous surface sinkholes. A significant river may cross the region, but small surface streams generally disappear down sinkholes

6.5.2 Balancing withdrawal and recharge

A local supply of groundwater will last if it is withdrawn for use at a rate equal to or less than the rate of recharge to the aquifer. However, if groundwater is removed faster than it is being recharged, the supply is reduced and will be gone one day. Heavy use of groundwater causes a regional water table to drop and even land subsidence.

Land subsidence generally occurs slowly and is difficult to be detected. Once a settlement has occurred, even if the cause of the settlement is eliminated, the settled ground cannot be completely restored. In the three urban areas of Suzhou, Wuxi, and Changzhou in Jiangsu Province of China, the groundwater level in these areas has been falling continuously due to the over-exploitation of the second layer of groundwater for a long time (Figure 6-29). It has exceeded the warning level of groundwater, forming a regional groundwater falling funnel centered on the three urban areas of Suzhou, Wuxi, and Changzhou. The decrease rate is 1~2m/a, and the lowest water level has been reduced to −69.62m, affecting a range of nearly 5000km^2, resulting in severe area land subsidence in Suzhou, Wuxi, and Changzhou, with accumulative subsidence of 1100mm, 1048mm and 870mm, respectively. It has grown at 67.3mm, 31.4mm, and 52.2mm per year, respectively. Only in Suzhou city in 1980, it was found that over 50 deep well casings were titled or disconnected; over 30 wells rose 100~300mm. The river's water surface increased, the water quality deteriorated sharply, the water level changed, and the flooded area expanded yearly. Since the 1980s, land subsidence has been growing year by year. During the seven years from 1983 to 1989, the direct economic losses caused by the land subsidence reached 260 million yuan in anti-flood dike construction, foundation height of new buildings, river dredging, and repair of municipal facilities. Besides, Suzhou is an ancient cultural city with gardens and cultural relics all over the city. The direct economic losses caused by land subsidence are inestimable.

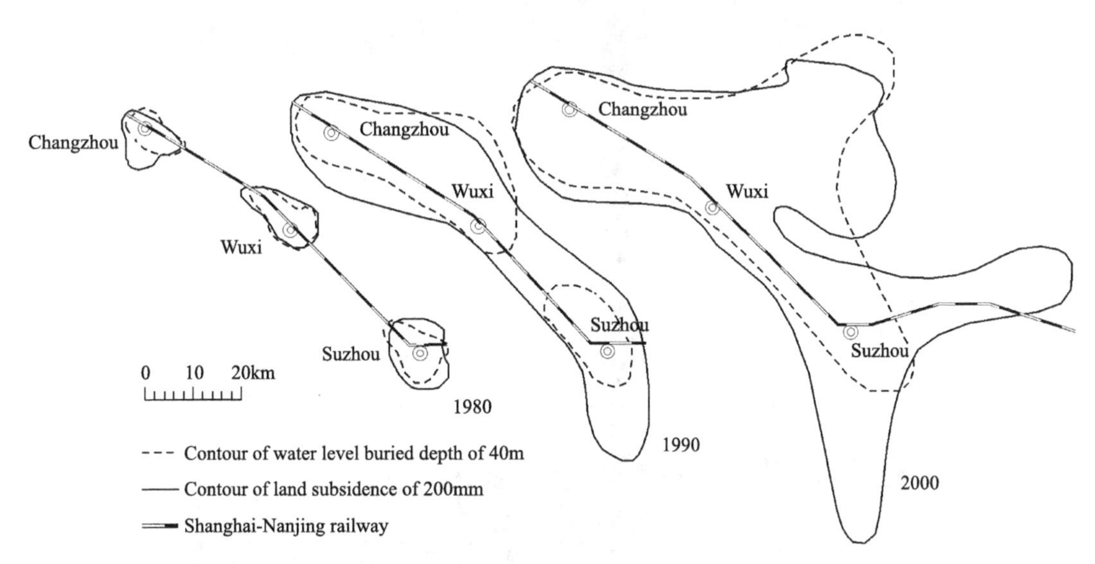

Figure 6-29 The contour of water level buried depth of 40m and the contour of ground subsidence of 200mm in Su-Xi-Chang

In parts of western Texas and eastern New Mexico, pumping groundwater from the Ogallala aquifer has brought the water table down 30m in recent decades. Lowering the water table means that wells need to be deepened and more power needs to be used to pump water to the ground. Furthermore, as the water is extracted, the soil surface can be deposited because the water no longer supports the rock and sediment. Mexico City has decreased by more than 7m and parts of the central valley of California by 9m due to groundwater extraction (Figure 6-30). Such subsidence can break up foundations, roads, and pipelines. Groundwater crowding also results in compaction and loss of porosity in rock and soil and can permanently ruin suitable aquifers.

To avoid the problems of groundwater loss, subsidence and compaction, many cities use artificial recharge to increase the natural recharge rate. Natural floodwater or treated industrial or domestic wastewater is stored in surface infiltration ponds to increase the rate of percolation of water into the ground. The clean, recycled water of wastewater treatment plants is often used for this purpose. In some cases, particularly in areas where groundwater is confined, water is actively pumped into the soil to replenish the groundwater supply. It costs more than filling surface ponds, but reduces the amount of water lost by evaporation.

Figure 6-30 Subsidence of the land surface caused by groundwater extraction near Mendota, San Joaquin Valley, California. Signs on the pole indicate the positions of the land surface in 1925, 1955, and 1977. The ground sank 9 meters in 52 years (USGS & Plummer et al., 2016)

6.5.3 Other effects

Groundwater is essential in preserving fossils such as petrified wood, which develops when porous buried wood is either filled in or displaced by inorganic silica carried in by groundwater (Figure 6-31). The result is a hard, permanent rock, commonly preserving the growth rings and other details of the wood. Calcite or silica carried by groundwater can also replace the original material in marine shells and animal bones. Sedimentary rock cement, usually silica or calcite, is brought into place by groundwater. When a considerable amount of cementing material precipitates locally in a rock, a hard, rounded mass called a concretion develops, typically around an organic nucleus such as a leaf, tooth, or other fossils (Figure 6-32).

Geodes are partially hollow, globe-shaped bodies found in some limestones and locally in other rocks. The outer shell is amorphous silica, and well-formed crystals of quartz, calcite, or other minerals project entering toward a central cavity (Figure 6-33). The origin of geodes is complex but related to groundwater. Crystals in geodes may have filled original cavities or have replaced fossils or other crystals.

Figure 6-31 Petrified wood in the Painted Desert. The log was displaced by silica carried in solution by groundwater, and small amounts of iron and other elements color the silica in the log

Alkali soil may develop in arid and semiarid climates because of the precipitation of significant sodium salts by evaporating groundwater. Such soil is generally unfit for plant growth. Alkali soil typically forms on the ground surface in low-lying areas.

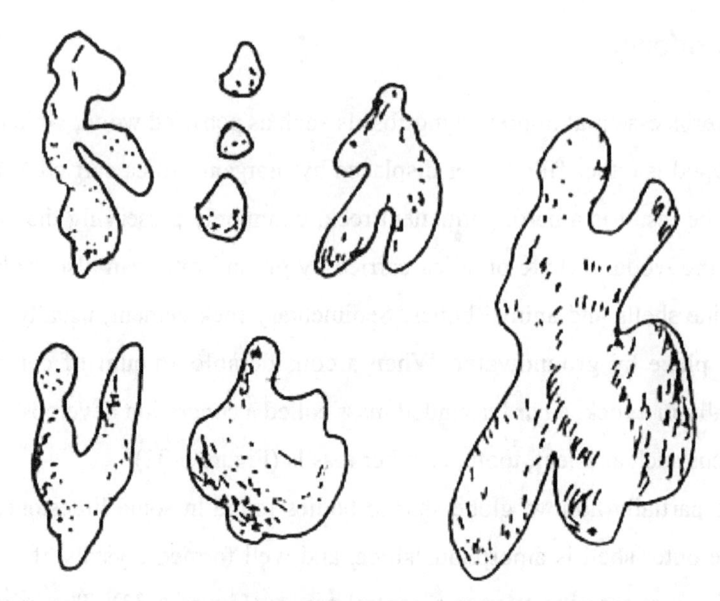

Figure 6-32 Calcareous concretions that have weathered out of shale. Concretions contain more cement than the surrounding rock and therefore are very resistant to weathering

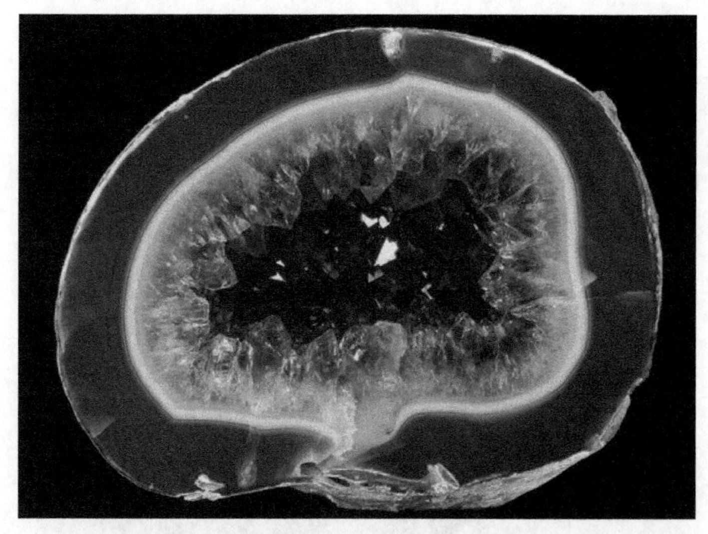

Figure 6-33 Concentric layers of amorphous silica are lined with well-formed amethyst (quartz) crystals growing inward toward a central cavity in a geode

6.6 Reservoir leakage

Reservoir leakage refers to the phenomenon that reservoir water leaks outside of reservoir basins along with rock pores, fissures, faults, and karst caves, or downstream through dam foundation (shoulder). The function of the reservoir is to store water for profit. Under certain geological conditions, leakage will occur during and after the reservoir impoundment. For any reservoir, in the absence of effective engineering treatment measures, severe leakage will directly affect the benefits of the reservoir. In many cases, leakage in the dam area often results in seepage deformation of the dam foundation, which endangers the dam's safety. Therefore, the leakage of reservoir and dam areas is an important engineering geological problem, which is also the most frequently encountered problem.

6.6.1 Geological conditions of reservoir leakage

Reservoir leakage can be divided into temporary leakage and permanent leakage. Transient leakage refers to the loss of reservoir water due to the pores, fissures, and voids of the rock and soil layer below the reservoir water level and the gradual elevation of reservoir water level in the initial stage of reservoir impoundment. In general, this part of the water leakage does not leak out of the reservoir but stops after a period of time.

Permanent leakage refers to the continuous leakage phenomenon of reservoir water flowing to adjacent valleys, depression, or distant low-lying drainage areas through the pores, fissures, faults, karsts, caves, and other leakage channels in the rock and soil body of the reservoir bank or the bottom of the reservoir basin. This kind of leakage will directly affect the reservoir's storage and may cause such adverse phenomena as immersion, swamping, and salinization. Permanent leakage in the reservoir area must have suitable topography, structure, lithology, and hydrogeological conditions. To judge whether leakage will occur in the reservoir area, a comprehensive analysis should be conducted from the following aspects.

1. Geomorphic conditions

Reservoir leakage is closely related to different geomorphic units. If overlapping mountains and continuous peaks characterize the terrain around the reservoir area, the possibility of reservoir leakage is tiny. On the contrary, the reservoir bank mountain is thin, and there are adjacent valleys that are deeply cut down, the reservoir leakage is more likely to occur. Suppose the reservoir is built at a sharp bend in a valley located in a bedrock mountainous area. In that case, the ridge between the curves may be narrow in places, and such topographical conditions may result in reservoir leakage. The deeper the adjacent valley is cut, the more significant the difference between the elevation of the adjoining valley bottom and the reservoir's water level, and the more severe the leakage (Figure 6-34a). On the contrary, if the adjacent valley is not cut deep and the valley's bottom elevation is higher than the reservoir's water level, there will be no leakage to the adjoining valley. (Figure 6-34b).

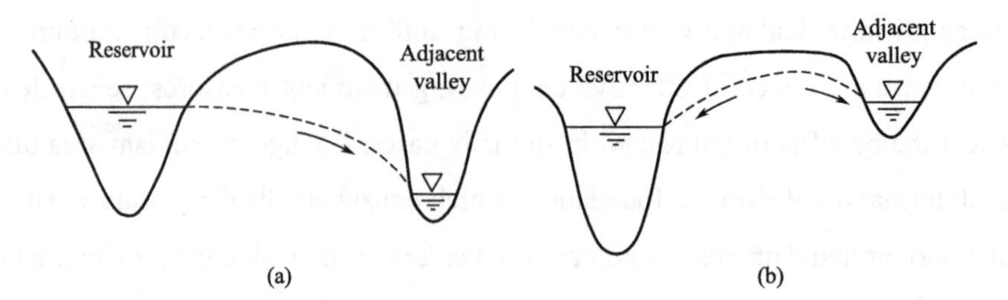

Figure 6-34 Relationship between elevation of adjacent valley and reservoir leakage

Valleys are generally cut shallower in plain areas, and the reservoir area is often far away from the adjacent valley. Hence, it is not easy for the reservoir water to leak through the interior massif to the adjoining valley. However, in the river bend area developed, the interfluvial block is relatively thin, where leakage is likely to occur.

2. Lithology conditions

Different lithology has a decisive influence on reservoir leakage. According to the lithology of rock, there are two kinds of rock layers: permeable and impermeable layers. In the analysis of reservoir leakage, the permeability of the rock layer should be taken as the focus because the strong permeable layer can lead to reservoir leakage, and the impermeable layer can prevent leakage.

The rock layer, which is capable of preventing leakage, is a weakly permeable or almost impermeable rock layer, such as claystone, shale, and clay deposits layer in clay-like rocks and all kinds of a solid rock layer that is intact and dense. Suppose an impermeable layer exists around the reservoir basin or reservoir. In that case, it can prevent leakage so that the reservoir water will not leak out of the reservoir. For all rock formations that can act as a water barrier, the thickness, distribution range, attitude, and degree of crack development must be ascertained to determine whether it can play a role in preventing seepage and the degree of seepage prevention.

Water leaking from a reservoir will come into chemical equilibrium with the rocks through which it flows and may dissolve them. Sediments containing the minerals halite, gypsum, anhydrite, and calcite are particularly susceptible to dissolution by flowing groundwater. The failure of the Macmillan reservoir in the U.S.A. resulted from water escaping through a gypsum layer that had been dissolved by leakage. Subramanian and Carter (1969) describe how the dissolution of calcareous sandstones undermined the Clubbiedean Dam in Scotland. Various formulas have been proposed for assessing the solubility of rock in-situ (Goudie, 1970; Wigley, 1977; James and Kirkpatrick, 1980), although none is precise. Leakage can also erode from joints, other fractures, and the cavities of karstic limestones, clayey material that often infills these voids.

3. Geologic structures

The fault zones or fault intersection zones, anticline and syncline structures, and rock attitudes are closely related to the reservoir leakage.

Faults, especially fault fracture zones with uncemented or incomplete cementation, are the main channels for reservoir leakage. Some faults penetrate the upstream and downstream of the dam (F_1 in Figure 6-35), while others extend from the reservoir area to the low valley outside the reservoir (F_2 in Figure 6-35), causing reservoir leakage.

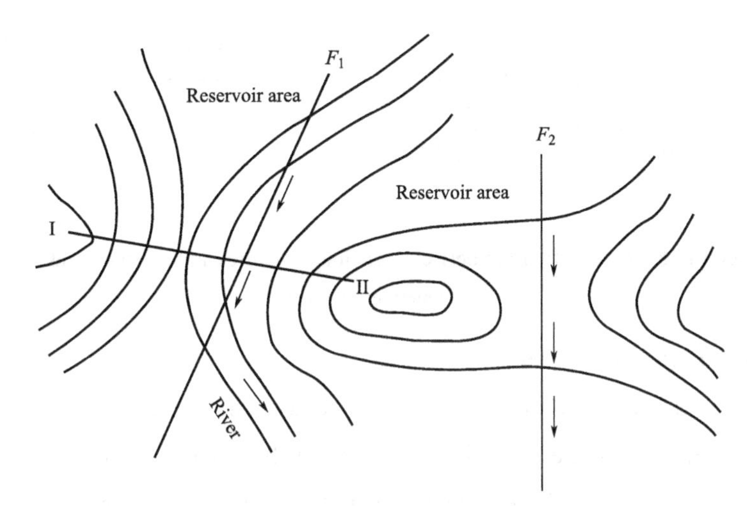

Figure 6-35 Schematic diagram of reservoir leakage along faults

The relationship between anticline structure and syncline structure and reservoir leakage should be analyzed from two aspects. First, the dense joint zone or interlaminar shear zone associated with the core of anticline and synclinal may become the leakage channel. Second, the cooperation and attitude of permeable layer and waterproof layer may affect reservoir leakage. For example, Figure 6-36(a) is an anticline structure. The permeable limestone layer has a slight dip angle and is exposed by adjacent valleys. In this case, the reservoir water can leak along the dissolution channel to the adjoining valley. Figure 6-36(b) shows a different situation. The permeable limestone layer has a steep dip angle and is not exposed by the adjacent valley, while the shale beneath it is impervious to water. Although the adjacent valley exposes the upper sandstone, the reservoir water does not leak to the adjoining valley due to its weak permeability. Figure 6-37 shows the relationship between synclinal structure and reservoir leakage. Figure 6-37(a) is a synclinal structure with an aquiclude layer. The shale is impervious to water, which generally does not produce leakage. Figure 6-37(b) is also a synclinal structure, but the permeable limestone has no impermeable layer and is connected with the adjacent valley. In this case, the reservoir water generally leaks along the karst channel.

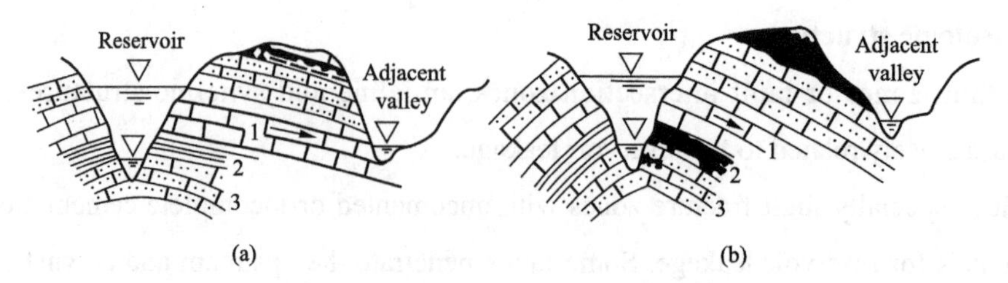

Figure 6-36 Relationship between anticline structure and reservoir leakage

1—limestone; 2—shale; 3—silty clast limestone

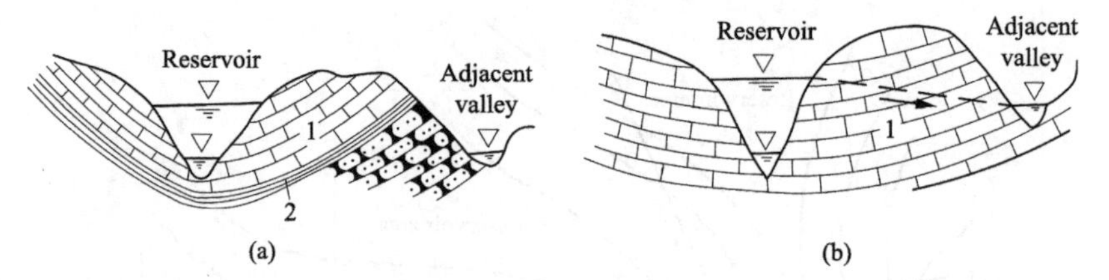

Figure 6-37 Relationship between syncline structures and reservoir leakage

1—limestone; 2—shale

4. Hydrogeological conditions

The hydrogeological condition of the reservoir area is one of the essential conditions for the reservoir leakage, especially the groundwater divide of the reservoir bank, and the elevation of the groundwater divide is of decisive significance to the reservoir leakage.

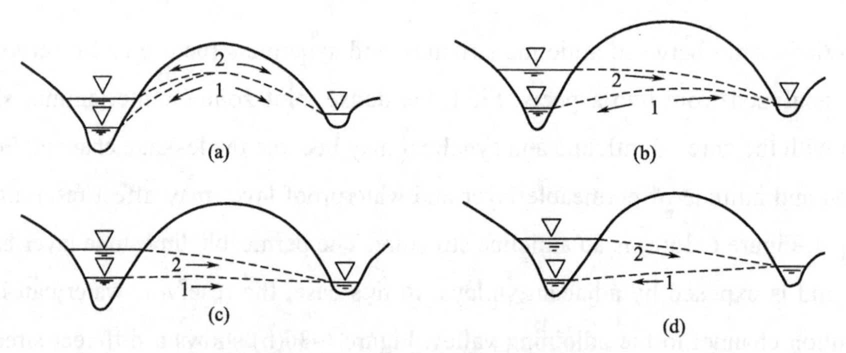

Figure 6-38 Several cases of groundwater divide

1—Before impoundment; 2—After impoundment

When it is confirmed that there are springs replenished by phreatic water on both slopes of the watershed, it indicates groundwater divides in the mountain of the watershed. According to the relationship between the elevation of the ridgeline of the groundwater divide and the normal water level of the reservoir, it can be judged whether the reservoir water may leak to the adjacent valley. There are four cases:

(1) The groundwater divide before the construction of the reservoir is higher than the

normal water level of the reservoir. The reservoir water will not leak to the adjacent valley after constructing the reservoir (Figure 6-38a).

(2) The groundwater divide before the construction of the reservoir is lower than the normal water level of the reservoir. The reservoir water will leak to the adjacent valley after constructing the reservoir (Figure 6-38b).

(3) Before the construction of the reservoir, the groundwater leaks from the reservoir to the adjacent valley, and the water head will become larger, and the leakage will become more severe after the water storage (Figure 6-38c).

(4) Before constructing the reservoir, the water from the adjacent valley flows to the reservoir valley through the underground. After the construction of the reservoir, the water level of the adjoining valley is lower than the reservoir's water level so that the reservoir water will leak into the adjoining valley (Figure 6-38d).

Sometimes, although the groundwater divide is slightly lower than the normal water level of the reservoir (Figure 6-38b), the groundwater divide may eventually be slightly higher than the reservoir level due to the jacking action of the reservoir water after water storage, and the reservoir water will not leak out. When the groundwater divide is very thick, and the permeability of rock and soil is small, the reservoir water will not leak out.

When confined water exists in the groundwater divide area, if the exposed elevation of the permeable layer of confined water in the adjacent valley is lower than the normal water level of the reservoir, the reservoir water may leak to the adjoining valley along with the permeable layer.

There are many recorded examples of reservoirs with almost uncontrollable water leakage from their base and sides. To avoid such losses, the valleys selected as sites for reservoirs should have at least one of the following geological characteristics; either a floor and margins that contain formations of low permeability, or a natural water level in the valley sides that is higher than the level proposed for the reservoir. The effect that impermeable rocks may have upon restraining leakage is illustrated in Figure 6-39. To be effective, they must create natural barriers that prevent the rapid loss of water away from the sides and base of the reservoir.

No such barriers are required to retain water in a reservoir if the groundwater level (i.e., its total head) is greater than that of the proposed reservoir. This situation is illustrated in Figure 6-40(a); no leakage can occur because the natural direction of groundwater flow is from the ground to the reservoir. Only when the reservoir head exceeds that of the groundwater can leakage commence (Figure 6-40b).

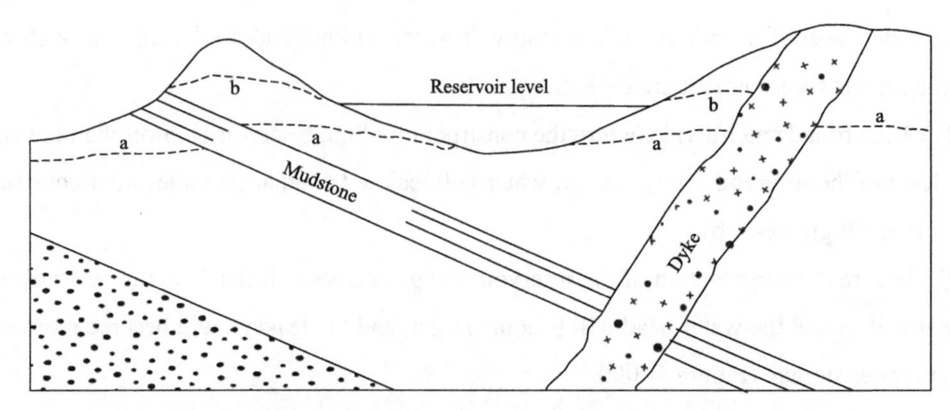

Figure 6-39 Water-tight reservoir assured by sedimentary and igneous aquicludes

a—original water level; b—water level after impounding

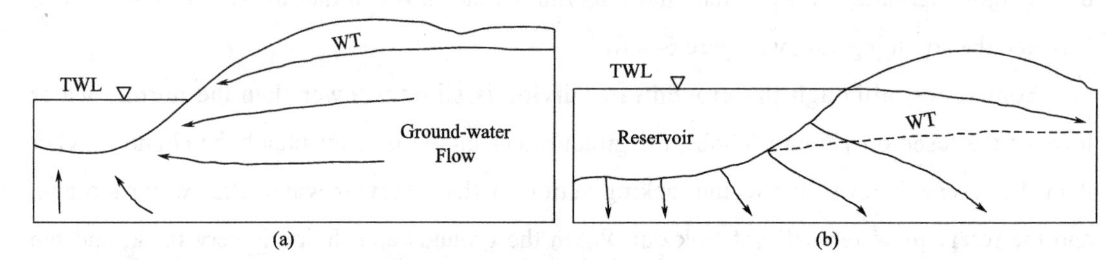

Figure 6-40 Water-tight and leaky reservoirs

(a) The total head of water in the ground exceeds that in the reservoir, and there is no leakage of reservoir water; (b) The reverse situation results in leakage. WT—water table; TWL—top water level

6.6.2 Measures to prevent leakage of reservoirs

Kennard and Knill (1969) demonstrated how the accurate investigation of the head in a valley side permitted a large reservoir to be successfully filled even though the valley contained cavernous limestone and unsealed abandoned mine workings.

Hydrogeological investigations of surrounding water levels are essential to an assessment of likely reservoir leakage. Care must be taken to accurately interpret the significance of water levels encountered in boreholes and distinguish perched water tables from the main water table and other piezometric levels (Figure 6-7). Water levels should therefore be monitored diligently during the drilling of boreholes. Water levels vary; they may be sufficiently high to either prevent or greatly retard reservoir leakage but may later fall to lower levels during dry periods and permit excessive reservoir leakage at a time when the reservoir of water is most needed. For this reason, the fluctuation of water levels should be recorded.

Accurate mapping of geological structure and rock types is necessary to reveal the presence of zones and horizons that may either prohibit or permit excessive leakage from a reservoir. The vertical section of Figure 6-41 illustrates the Dol-y-Gaer dam, built on Carboniferous Limestone, whose extent could easily be mapped and through which serious leakage later occurred. Much of the lost water re-appeared downstream, where it was brought to the surface by the presence of relatively impermeable Devonian strata. Remedial measures failed to control the leakage, and the Pontsticill dam was constructed downstream to impound much of the water leaking from the upper reservoir.

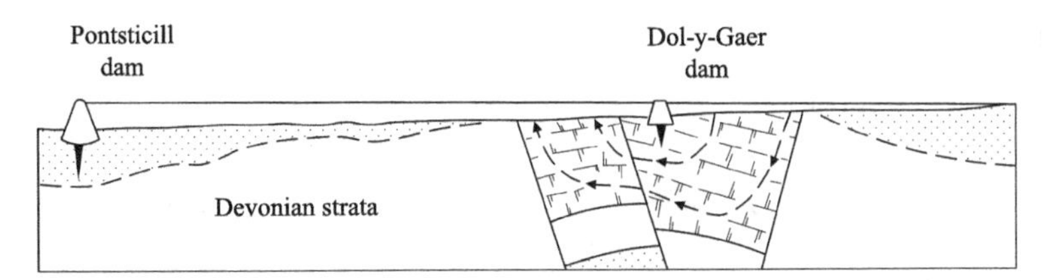

Figure 6-41 Geological section to illustrate leakage at the Taf Fechan reservoirs, S. Wales (modified from Blyth, 2005)

The presence of a relatively impermeable layer in the floor of a valley needs no guarantee against leakage when the head of water in the strata beneath the layer is lower than that of the reservoir. Seepage will eventually reach these lower levels and may lead to failure of the sealing layer. In arid regions, where water levels can be low, an artificial impermeable barrier may be created by injecting cement and clay mixtures into the ground to reduce unacceptable reservoir leakage (Figure 6-42). The cost of such work prohibits its extensive use.

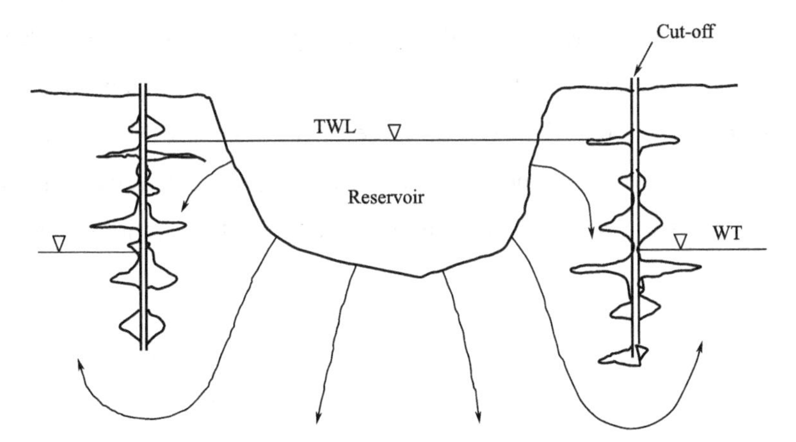

Figure 6-42 Partially penetrating cut-off to reduce (but not stop) reservoir leakage, shown by flow lines (Blyth & de Freitas, 2005)

Special care should be taken to identify buried valleys as their course need not coincide with existing rivers. The Ming Tombs reservoir in the suburbs of Beijing has an ancient river channel (Figure 6-43) near Dagongmen on the right bank of the reservoir. The ancient river channel

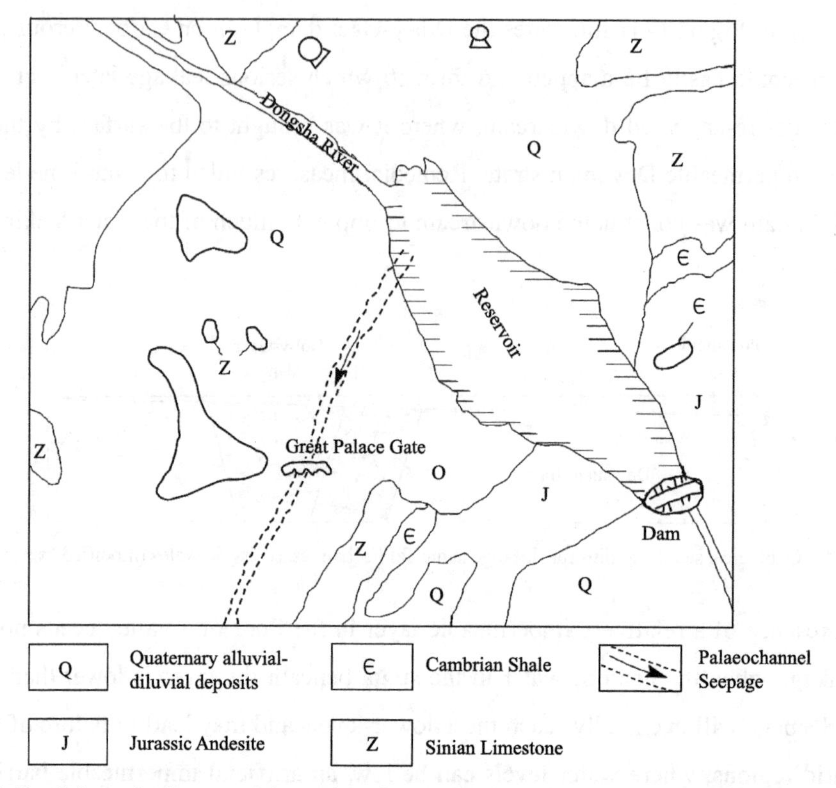

Figure 6-43 Schematic diagram of the ancient river reservoir on the right bank of Ming Tombs Reservoir

is composed of a solid porous gravel layer. When the reservoir reaches a specific elevation, concentrated leakage and a large amount of water leakage will form. It is a fundamental reason for the reservoir not operating at full capacity for a long time after completing the project. A 1300-meter-long cut-off is built at the end of the reservoir to cut off the channel connecting the reservoir with the ancient river channel near Dagongmen to prevent leakage. After the treatment, the normal storage of the reservoir can be ensured to meet the requirements of reservoir operation.

Reference

[1] BLYTH F G H, DE FREITAS M. A geology for engineers[M]. 7th ed. Amsterdam: Elsevier Press, 2005.

[2] CARLSON D H, PLUMMER C C, HAMMERSLEY L. Physical geology: earth revealed[M]. 9th ed. New York: McGraw-Hill Education. 2011.

[3] DE VALLEJO L G, FERRER M. Geological engineering[M]. Florida: CRC Press. 2011.

[4] FREEZE R A, CHERRY J A. Groundwater[M]. Englewood Cliffs: Prentice-Hall. Inc.,

1979.

[5] GOUDIE A S. Input and output considerations in estimating rates of chemical denudation[J]. Earth Science Journal, 1970, 4, 60-66.

[6] HARR M E. Groundwater and Seepage[M]. New York: Dover, 1991.

[7] HEATH R C. Basic ground-water hydrology[M]. US Geological Survey, 1987.

[8] HUBBERT M K. Darcy's law and the field equations of the flow of underground fluids[J]. International Association of Scientific Hydrology. Bulletin, 1956, 2(1): 23-59. DOI: 10.1080/02626665709493062.

[9] Itasca Consulting Group Inc. FLAC3D-Fast Lagrangian Analysis of Continua in 3 Dimensions. Ver. 5.01 User's Manual[Z]. Minneapolis: ICG, 2014.

[10] JAMES A N, Kirkpatrick L M. Design of foundations of dams containing soluble rocks and soils[J]. Quarterly Journal of Engineering Geology, 1980, 13(3): 189-198. DOI: 10.1016/0148-9062(81)90851-2.

[11] KENNARD M F, KNILL J L. Reservoirs on limestone, with particular reference to the Cow Green Scheme[J]. Journal of the Institution of Water Engineers, 1969, 23: 87-113.

[12] PLUMMER C C, CARLSON D H, HAMMERSLEY L. Physical geology[M]. 15th ed. New York: McGraw-Hill Education, 2016.

[13] POLUBARINOVA-KOCHINA P Y. Theory of Groundwater Movement[M]. Princeton: Princeton University Press, 1962.

[14] SUBRAMANIAN S A, Carter A. Investigation and treatment of leakage through Carboniferous rocks at Clubbieden Dam[J]. Midlothian: Scottish Journal of Geology, 1969, 5(3): 207-223. DOI: 10.1144/sjg05030207.

[15] WIGLEY T M L. WATSPEC: A computer program for determining the equilibrium specification of aqueous solutions[CP]. British Geomorphological Research Group, Technical Bulletin, No. 20, 1977.

Exercises

Choice Questions

(1) Porosity is:

(a) the percentage of a rock's volume that is openings

(b) the capacity of a rock to transmit a fluid

(c) the ability of a sediment to retard water

(d) none of the preceding

(2) Permeability is:

(a) the percentage of a rock's volume that is openings

(b) the capacity of a rock to transmit a fluid

(c) the ability of a sediment to retard water

(d) none of the preceding

(3) The subsurface zone in which all rock openings are filled with water is called the:

(a) saturated zone (b) water table

(c) unsaturated zone (d) aquiclude

Review Questions

(1) Discuss the difference between porosity and permeability.

(2) What is the water table? What is the perched water table?

(3) Describe the classification of groundwater based on buried conditions.

(4) Sketch main four different origins for springs.

(5) What conditions are necessary for an artesian well?

(6) What distinguishes a geyser from a hot spring? Why does a geyser erupt?

(7) What is karst topography? How does it form?

(8) How to judge whether leakage will occur in a reservoir area?

7
Landforms Made by Running Water

7.1 Introduction

Water is the most widely distributed and most important substance on the earth's surface. Oceans, rivers, lakes, swamps, groundwater, glaciers, and moisture in the atmosphere together constitute the hydrosphere on the earth. Surface water flow and groundwater flow are the most extensive and most vigorous external geological factors. They are constantly eroding, transporting, and depositing soil and rock materials in the process of flowing to lakes, seas, and other low-lying places. Due to the combined influence of this process and internal forces and geological processes, various landforms have been shaped, forming different types of loose Quaternary sediments, which also promotes some undesirable geological processes.

The study of the resulting landforms by flowing water is termed fluvial geomorphology. Fluvial geomorphology includes the action of both channelized and unchannelized flow moving downslope due to gravity. Flowing water is more influential in shaping the surface of our planet than any other exogenic geomorphic process, primarily because of the sheer number of streams on Earth. Water flowing downslope over the land surface, mainly when concentrated in channels, modifies existing landforms and creates others through erosion and deposition. Nearly every region of Earth's land surface exhibits at least some topography that the effect of flowing water has shaped, and many areas indicate extensive evidence of stream action. Flowing water is the primary geomorphic agent in arid as well as humid environments. Rivers vary greatly in appearance with changes both from source to mouth and between individual rivers. It is this morphological diversity that is the fascination for many people (Holden,2017).

Because of the widespread occurrence of stream systems and their crucial role in providing fresh water for people and agricultural, industrial, and commercial activities, a substantial portion of the world's population lives close to streams. Understanding stream processes, landforms, and hazards are fundamental for maintaining human safety and quality of life. As the primary place for surface water activities, rivers provide essential sites for water conservancy and hydropower projects. In the hydrosphere, the most direct and indirect influence of water conservancy and hydropower projects is on the river section of the reservoir and its related groundwater. After the dam is completed, the water level of the reservoir increases, which affects the river section and the groundwater connected to it. The geological effects produced by the flow of rivers have shaped various types of river valley landforms and accumulated in the rivers to form complex quaternary strata. These are the primary conditions for constructing reservoirs and dams and have an essential impact on water conservancy and hydropower projects.

Stream is the general term for natural, channelized flow. In the Earth sciences, the term stream pertains to water flowing in a channel of any size, although in general usage, we describe large streams as rivers and use local terms, such as creek, brook, run, draw, and bayou, for smaller streams (Christopherson & Brikeland, 2017). Whether dominated by erosion or deposition, the long-term effects of streamflow are sometimes also quite dramatic (Figure 7-1).

Figure 7-1 Ox liver and horse lung gorge in Xiling Gorge

Xiling Gorge starts from Zigui Xiangxi Estuary in the west and ends at Nanjin Gate in the east of Yichang. It is about 75km long and is the longest canyon in the Three Gorges of the Yangtze River. The gorge belongs to the world-famous Huangling anticline in terms of structure. The two wings of the anticline crossed by the Yangtze River are built with steep cliffs and flooded beaches.

7.2 Erosion, Transportation, and Deposition

7.2.1 Composition of river valleys

The trough terrain through which a river flows is called a river valley. A river valley is composed of valley bottom and valley slope (Figure 7-2). The valley bottom includes the river bed and flood plain. The river bed refers to the river occupied in the mean-flow period or called the channel. The flood plain is the part of the valley floor of a river beyond its flooded bed during flood time but is exposed to the water when it is dry. Valley slope is the bank slope on both sides of a valley. A terrace is a platform along a river with a steep slope under a valley slope that perennial floods cannot submerge.

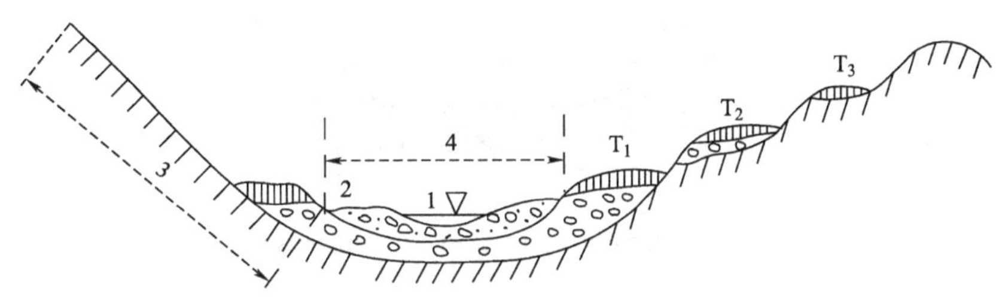

Figure 7-2 Composition of river valleys
1—River bed; 2—Flood plain; 3—Valley slope; 4—Valley bottom;
T_1—First terrace; T_2—Second terrace; T_3—Third terrace

Most of the world's land surface has been sculpted by running water, which acts to shape landforms through three closely related processes—erosion, transportation, and deposition. From bedrock or regolith, mineral materials are removed from slopes and stream channels by erosion, carving out gorges and valleys. These particles are then transported by water, either in solution as ions or sediment of various sizes. The residue is finally deposited downstream, where it builds up into plains, levees, fans, and deltas. The landforms shaped by the progressive removal of bedrock are called erosional landforms. Fragments of soil, regolith, and bedrock removed from the parent rock mass are transported and deposited elsewhere, where they take shape as an entirely different set of surface features—the depositional landforms.

7.2.2 Stream erosion

Fluvial erosion is the removal of rock material by flowing water (Figure 7-3). Fluvial erosion may take the chemical removal of ions from rocks or the physical removal of rock fragments (clasts). Physical removal of rock fragments includes breaking off new pieces of bedrock from the channel bed or sides, moving them, and picking up and removing preexisting clasts that were temporarily resting on the channel bottom. The breaking off of new pieces of bedrock proceeds very slowly where highly resistant rock types are found. Net erosion results in the lowering of the affected part of the landscape and is termed degradation. Net deposition of sediments results in a building up, or aggradation, of the landscape (Gabler et al., 2009).

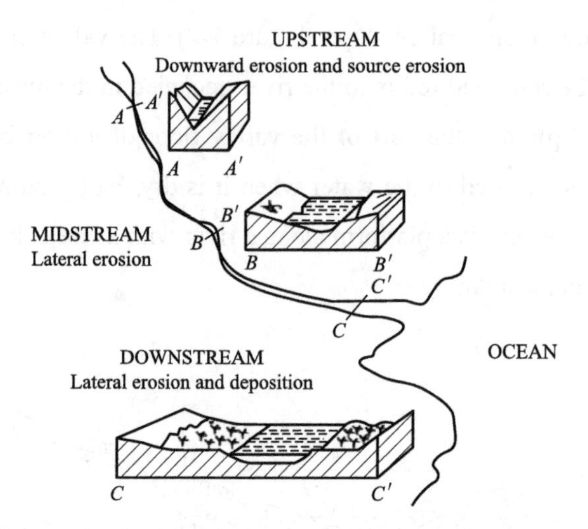

Figure 7-3 Geological processes in different parts of the river

Hydraulic action refers to the physical, instead of chemical, stream water removing pieces of rock. As stream water flows downslope under gravity, it exerts stress on the streambed. Whether this stress results in entrainment and removal of a preexisting clast currently resting on the channel bottom or even the breaking off of a new piece of bedrock from the channel depends on several factors. The factors include water volume, flow velocity, flow depth, stream gradient, friction with the streambed, the strength and size of the rocks over which the stream flows, and the degree of stream turbulence. Turbulence is a chaotic flow that mixes and churns water, often with a significant upward component that dramatically increases erosion rate and the load-carrying capacity of streams. Turbulence is controlled by channel roughness and the gradient over which the stream is flowing. Plunge pools at the base of waterfalls and rapids reveal how decisive the turbulence-enhanced hydraulic action is when directed toward a localized point.

As soon as a stream begins carrying rock fragments as a load, it starts to erode the ground by abrasion, a process even more potent than hydraulic action. As rock particles bounce, scrape, and drag along the bottom and sides of a stream channel, they break off additional rock fragments. Because solid rock particles are heavier than water, the impact of having a clastic load thrown against the channel bottom and sides by the current is much more effective than the impact of water alone. Under certain conditions, stream abrasion makes distinctive round depressions called potholes in the rock of a bedrock streambed. Potholes originate in particular circumstances, such as below waterfalls or swirling rapids, or at the points of structural weakness, including joint intersections in the streambed (Figure 7-4). Potholes range in diameter and depth from a few centimeters to many meters. If you peer into a pothole, you can often see one or more round stones at the bottom. These are the abraders or grinders. They are swirling whirlpool movements of the stream water cause such rocks to grind the bedrock and enlarge the pothole by abrasion while finer sediments are carried away in the current.

Figure 7-4 Potholes in a streambed

As rock fragments are transported downstream during abrasion, they are gradually reduced in size, and their shape changes from angular to rounded. This wear and tear of sediments as they tumble and bounce against one another, and the stream channel is called attrition. Attrition explains why gravels found in streambeds are rounded and why the load carried in the lower reaches of most large rivers is composed primarily of fine-grained sediments and dissolved minerals.

Stream erosion widens and lengthens stream channels and the valleys they occupy. Lengthening occurs primarily at the source through headward erosion, accomplished partly by surface runoff flowing into a stream and partly by springs undermining the slope. Extending a river's course in an upstream direction is particularly important where erosional gullies are rapidly dissecting agricultural land. Such gullying may be counteracted by soil conservation practices to reduce erosional soil loss. Channel lengthening, which results in a decrease in stream gradient, also occurs if the path of the stream channel becomes more winding and sinuous.

According to the directions of stream erosion, there are two types: downward erosion and lateral erosion. The river water and its entrained gravel, in the process of constantly flowing from a high place to a low place, continuously impact, scour, grind and dissolve the bedrock, and deepen the river valley, which is called the downward erosion of the river. The result of this effect is that the valley becomes deeper and steeper. Lateral erosion refers to the process of river water scouring and destroying both banks of the river, causing the river bed to swing from side to side, the valley slopes receding, and the continuous widening of the river valley. The result of lateral erosion is to widen the river bed and valley bottom, complicating the river valley's morphology, forming meanders, convex banks, ancient river beds, and oxbow lakes. Lateral erosion mainly occurs in the middle and lower reaches of the river.

7.2.3 Stream transportation

A stream directly erodes part of the sediment it carries, and most of the chemical residue flows into the base flow of the channel. Still, a far more significant proportion of its load is delivered by surface runoff and mass movement. Regardless of the sediment source, streams transport their load in several ways. Some minerals are dissolved in water and are therefore transported through solution transport processes. The finest solid particles are held in suspension, buoyed by vertical turbulence. Such small grains can remain suspended in the water column for long periods, as long as the force of upward turbulence is stronger than the downward settling tendency of the particles. Some grains are too large and heavy to be carried in suspension to rebound along the bottom of the channel in a process known as saltation. Particles too large and heavy to move by saltation can slide and roll along the bottom of the channel through the traction transport process.

So there are three main types of stream load. Ions of rock material held in solution constitute the dissolved load. The suspended load consists of the small clastic particles being moved in suspension. Larger particles that saltate or move in traction along the streambed comprise the bed load (Figure 7-5). The total amount of load that a stream carries is expressed in terms of the weight of the transported material per unit time. Generally, most of the stream load is carried in suspension. A large river such as the Mississippi has as much as 90 percent of its load in suspension. The capacity to move the larger particles of the bed load (Stream capacity) increases with velocity because faster-moving water drags more powerfully against the bed. The ability to move bed load will increase with the third to the fourth power of the velocity. In other words, if a stream's speed is doubled in times of flood, its capability to transport bedload will increase from 8 to 16 times.

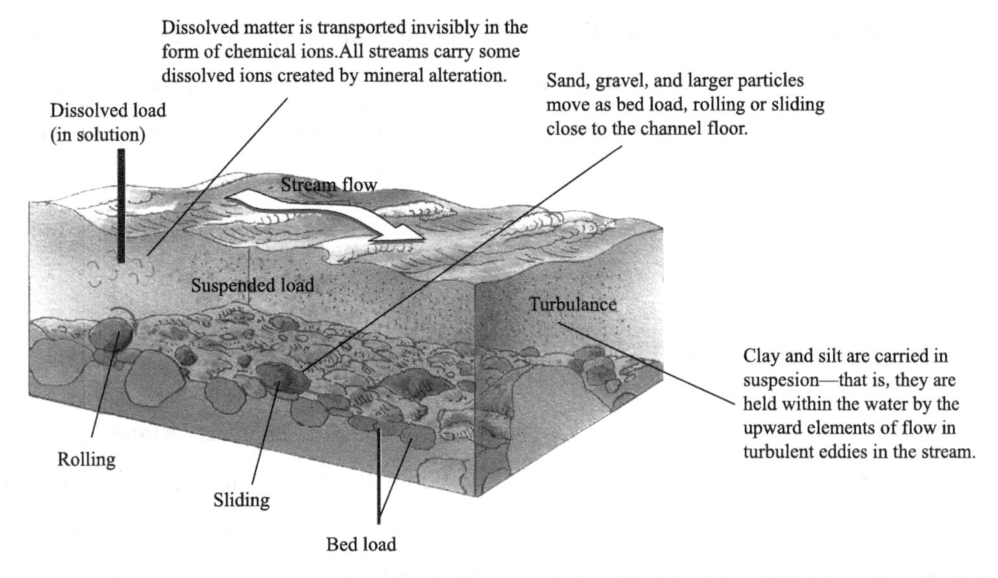

Figure 7-5 Streams carry their load as dissolved, suspended, and bed load (modified from Strahler, 2013)

The relative proportion of each load type present in a given stream varies with drainage basin characteristics such as climate, vegetation cover, slope, rock type, and the infiltration capacities and permeabilities of the rock and soil types. Dissolved loads will be larger than average in basins with high amounts of infiltration and baseflow, and therefore limiting surface runoff because slow-moving groundwater that feeds the base flow acquires ions from the rocks through which it moves. Humid regions experience considerable weathering, which produces plenty of fine-grained sediment, and thus moist region streams tend to have large amounts of the suspended load. Rivers that are carrying a high suspended load look characteristically muddy.

The Huang He in northern China (Figure 7-6), also known as the "Yellow River" because

Figure 7-6 The Yellow River landscape

of the color of its silty suspended load, carries a vast amount of sediment in suspension, with more than 1 million tons of suspended load per year. Compared to the "muddy" Mississippi River, the Huang He transports five times the suspended sediment load with only one-fifth the discharge. Streams dominated by bedload tend to occur in arid regions because of the limited weathering rate in dry climates. Little weathering leaves considerable coarse-grained sediment in the landscape available for transportation by the stream system.

7.2.4 Stream deposition

A stream carries its load downslope toward a valley, a lake, or an ocean. Along that journey, whenever a stream's load exceeds its capacity, it deposits some of its load. This deposited sediment is called alluvium. Deposition typically occurs where the velocity of streamflow decreases. For example, deposition occurs along stream banks when the streamflow slows down inside a bend in the channel. During flooding, fast-moving floodwaters slow down and spread over the valley floor, depositing alluvium in layers. Fine sediment, rich in organic matter, can improve soil fertility—although, sometimes, flooding can also leave behind sterile layers of sand or gravel.

The alluvium is the general name given to fluvial deposits, regardless of the type or size of material. The alluvium is recognized by the characteristic sorting and rounding of sediments that streams perform. A stream sorts particles by size, transporting the dimensions that it can and depositing larger ones. As velocity fluctuates due to changes in discharge, channel gradient, and roughness, particle sizes that can be picked up, transported, and deposited vary accordingly (Figure 7-7). The alluvium deposited by a stream with fluctuating velocity will exhibit alternating layers of coarser and finer sediment.

Dams represent a particular case for sediment deposition and may exert negative influences on fluvial systems. Where earthen or concrete barriers block the streamflow, transported sediment quickly settles at the dam's base. Sediment that would otherwise continue downstream and settle out during floods continuously fills in the reservoir. Eventually, the deposit may displace enough water to render the original dam inoperable for water storage or electrical generation. Costly dredging is the only remedy for removing depositional sediment.

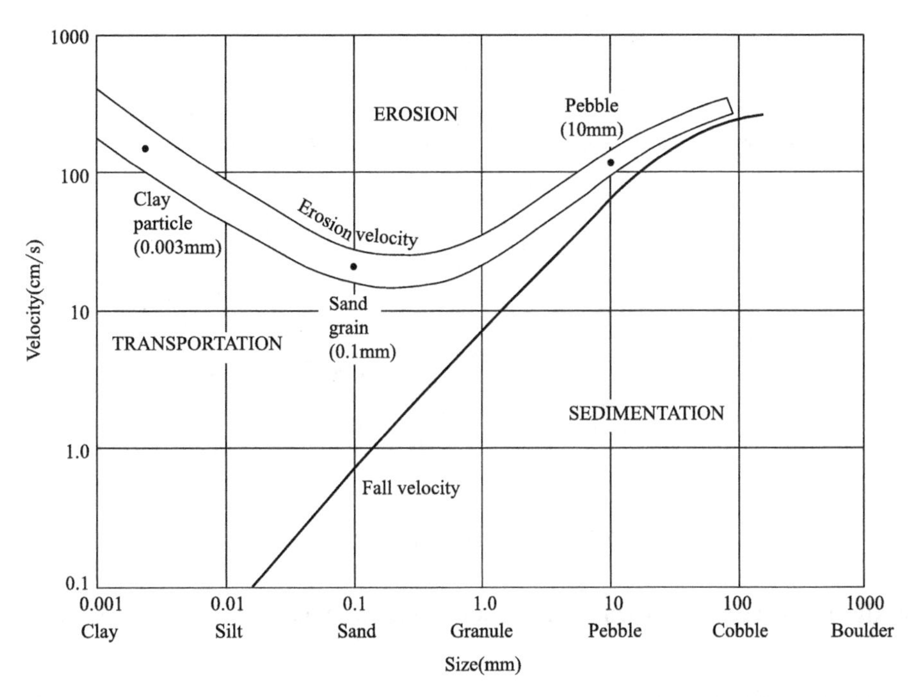

Figure 7-7 Velocity characteristics of erosion, transportation, and deposition for different sized particles

7.3 Fluvial Landforms

Now that you've seen how landscapes are shaped by water over time. Let's take a closer look at several of the landforms associated with them, called fluvial landforms.

7.3.1 Meandering streams

Meandering, as described earlier, is a characteristic of graded rivers carrying substantial loads of sediment of varying sizes. On the outside of a meander bend, water moves with incredible velocity, and the channel is deeper. The greater velocity erodes the floodplain sediment, creating a cut bank and causing the bend to grow outward and move in a downstream direction (Figure 7-8). On the inside of the curve, the flow is slower, and sediment accumulates on a point bar. Sometimes the streamflow cuts off a meander loop by eroding through the narrow portion of the meander neck. After the cutoff, silt and sand are deposited across the ends of the former channel, producing an ox-bow lake. Gradually, the lake fills in with sediment, dries into a swamp, and finally becomes a meander scar. Cutoffs are sometimes human-induced to straighten the stream channel and make it more convenient for transport.

A meander grows outward and migrates downstream as the riverbank is eroded on the outside of the meander, forming a cut bank, while deposition occurs at the inside bank, creating a point bar. The

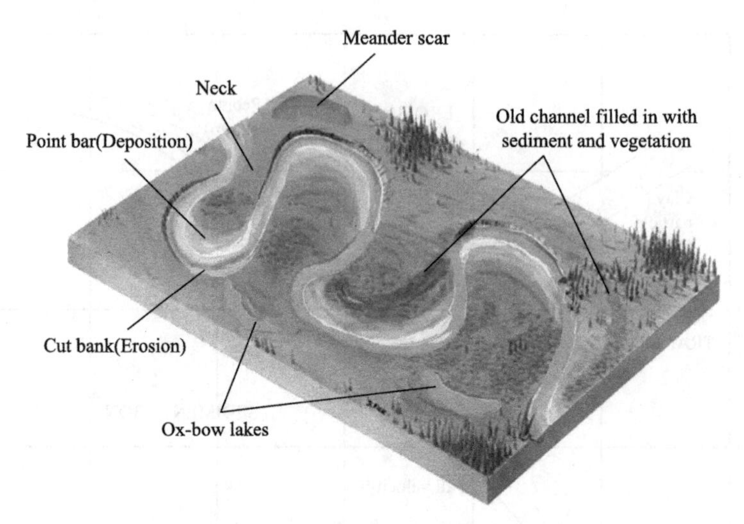

Figure 7-8 Growth and movement of meanders

river quickly takes the shortcut when meanders touch, leaving a meander scar or an ox-bow lake.

7.3.2 Floodplain

The higher part of the valley bottom above the river surface during the normal period but submerged during large floods is called the river floodplain (Figure 7-9). The terrain, flat or slightly undulating, is long and narrow on the verge of the riverbed. The range of the floodplain varies in width and is often several to several tens of times larger than the width of the river bed. The floodplains of mountain rivers are generally not well developed, or only small floodplains with a narrow width are developed. In plain areas, rivers and floodplains are not only widely grown in the area but may also develop into extensive floodplains or alluvial plains. The genetic floodplain is a geomorphological term for a relatively flat alluvial landform, constructed largely by the flow regime of the present river and subject to flooding. The hydraulic floodplain is an engineering

Figure 7-9 The floodplain developed in the middle and lower reaches of the river (Ziyang, Sichuan)

concept for any surface subject to river flooding within a given return period (Thomas, 2016).

The floodplain is the inevitable product of the lateral erosion of the river and the lateral migration or swing of the river bed. Under the action of lateral erosion of the river, the valley's slope gradually recedes, the bottom of the valley widens, and the rudimentary riverbed shoal is formed on the convex bank of the river bend. It further develops and stabilizes, and the rudimentary floodplain is formed. As the lateral erosion continues, the rudimentary river floodplain expands and begins to accept the deposition of fine-grained materials during flooding, forming a thin cover layer. This kind of accumulated body with a binary structure on the bottom is the floodplain (Figure 7-10). In addition, after the straightening of the river or the disappearance of non-mainstream channels such as the branching river, floodplains can also be formed from the original abandoned river channels after a large amount of material accumulation and siltation.

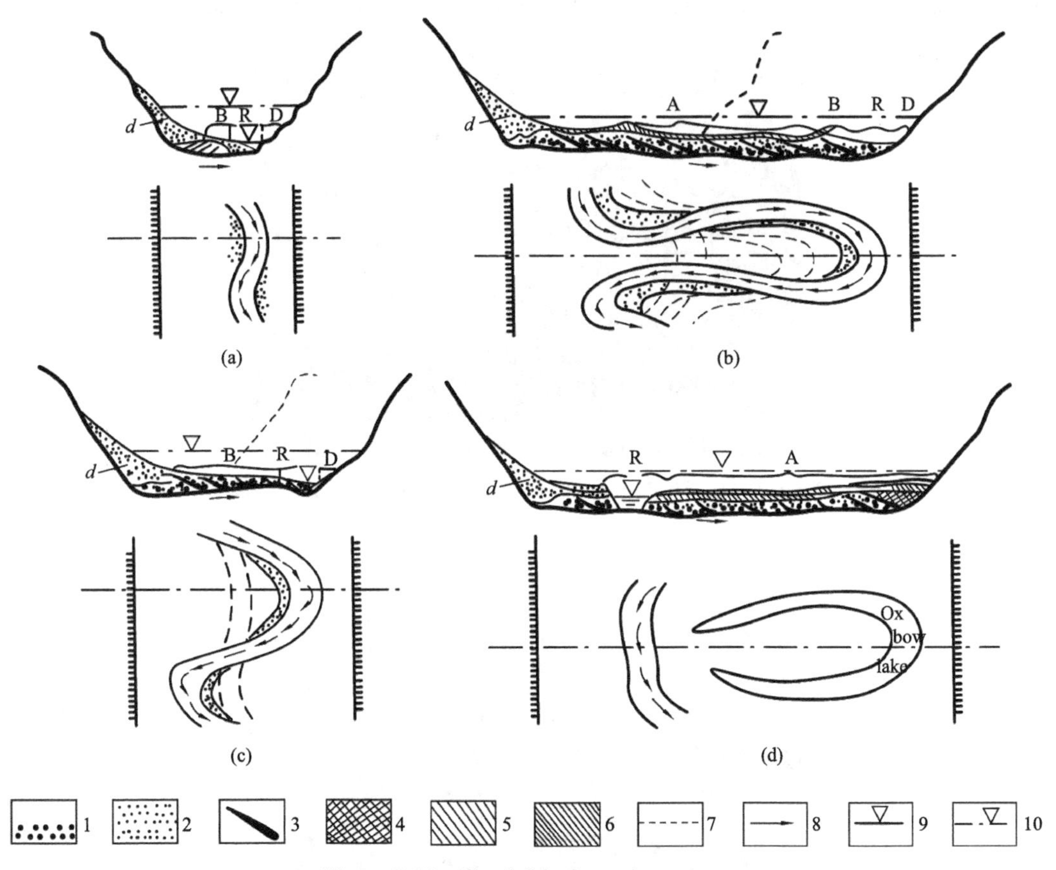

Figure 7-10 Floodplain formation process

(a) Forming a rudimentary river floodplain; (b) Forming a primitive river floodplain; (c) Forming a floodplain; (d) Forming an oxbow lake. 1 to 3—river facies alluvium (1—gravel and pebbles; 2—sand; 3—silt); 4—oxbow lake facies alluvium; 5、6—river floodplain, phase alluvial deposits (accumulation in sequence); 7—the location of the early scouring bank; 8—the movement direction of the river bed; 9— the level water level; 10—the flood level; A— river floodplain; B— riverbed shoal; D—bedrock shoal; *d*— slope deposit; R—riverbed

When streams leave the confines of their channels during floods, the channel cross-sectional width is suddenly enlarged so much that the flow velocity must slow down to counterbalance it. The resulting decrease in stream competence and capacity causes sediment deposition on the flooded land adjacent to the channel. This sedimentation is significant, rightly next to the channel where aggradation constructs channel-bounding ridges known as natural levees. Still, some alluvium will be left behind wherever load settles out of the receding floodwater.

Floodplains often constitute the extensive, low-gradient land areas composed of alluvium that lie adjacent to many stream channels (Figure 7-11). Floodplains are aptly named because they are inundated during floods and at least partially composed of vertical accretion deposits, the sediment that settles out of slowing and standing flood-water. Most floodplains also contain lateral accretion deposits (Figure 7-12). These are generally channel bar deposits that get left behind as a channel gradually shifts its position in a sideways fashion (laterally) across the floodplain.

Figure 7-11 The Yellow River floodplain in Zhengzhou, Henan

Figure 7-12 When the Yellow River flows through the Linhe District of Bayannaoer City, Inner Mongolia, it presents the characteristics of a typical braided river. Braided rivers are a type of river with multiple channels, and the channels are extensive and shallow as a whole, and most of the river banks are very unstable

7.3.3 River terraces

River terraces are terraced terrains distributed along the river banks and above the general flood level. They are formed by the erosion and sedimentation of the river alternately. The terrace comprises ground, steep slope, front edge, rear edge, and toe (Figure 7-13).

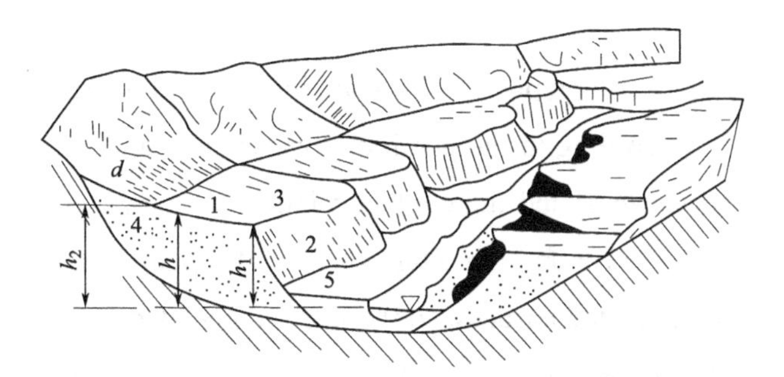

Figure 7-13 Schematic diagram of terrace morphological elements

1— a terrace ground; 2— a steep slope; 3— a front edge; 4— a rear edge; 5—a toe; h_1— front edge height; h—terrace height; h_2— a rear edge height; d— a sloped skirt

The terraces are the bottoms of river valleys in different periods in the past, and most of them are slightly inclined towards the river bed and the lower reaches of the river. The surface is undulating due to the accumulation of colluvial deposits, slope deposits, and alluvial deposits. Steep hills are also called terrace slopes, which refer to the slopes below the terrace ground, with steep slopes and inclined to the riverbed. Multi-level terraces are often developed in river valleys. The order of marked terraces is graded from new to old, from bottom to top, and the latest terraces above the floodplain are called class Ⅰ terraces (Figure 7-14). Therefore, on the cross-section of the same river valley, the higher the terraces, the larger the series, and the earlier the formation age.

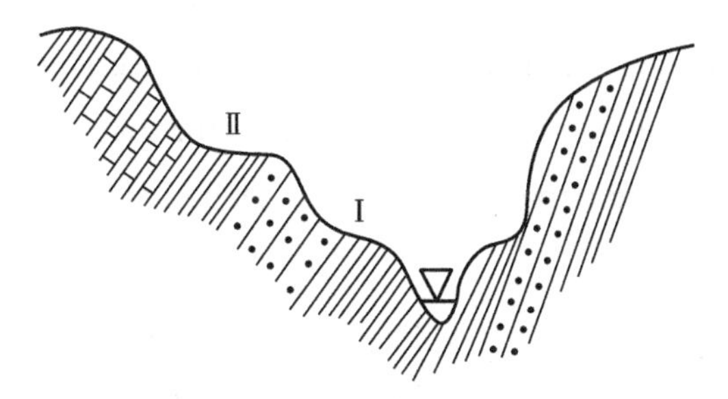

Figure 7-14 the structure of eroded terrace

Ⅰ— Class Ⅰ terrace; Ⅱ— Class Ⅱ terrace

Terraces may be rock floored (though a veneer of river deposits may remain) or alluvial (accumulation terraces). According to the structure of the constituent materials of the river terraces and the nature of the geological processes when they were formed, the terraces are mainly divided into the following three categories (Yang, 2004).

Eroded terraces are the river terraces composed of bedrock and rarely covered by alluvial deposits on the terraces (Figure 7-14). It is mainly formed by river erosion and developed in mountain valleys or upstream reaches of rivers with significant neotectonic movement and intense erosion.

Accumulative terrace. It is composed of fluvial alluvium and is expected in the middle and lower reaches of rivers. According to the depth of river erosion and the thickness of alluvium during the formation of terraces, and the contact relationship between each terrace, the river is divided into upper terrace river and inner terrace river. The late river did not cut through the early alluvium on the accumulated terraces, so the late accumulated terraces were superimposed on the earlier accumulated terraces as upper terraces. On the accumulated terraces that have been formed, later rivers cut through the early alluvium to the bedrock so that the subsequently formed new terraces are enclosed within the older terraces, which are called inner terraces (Figure 7-15).

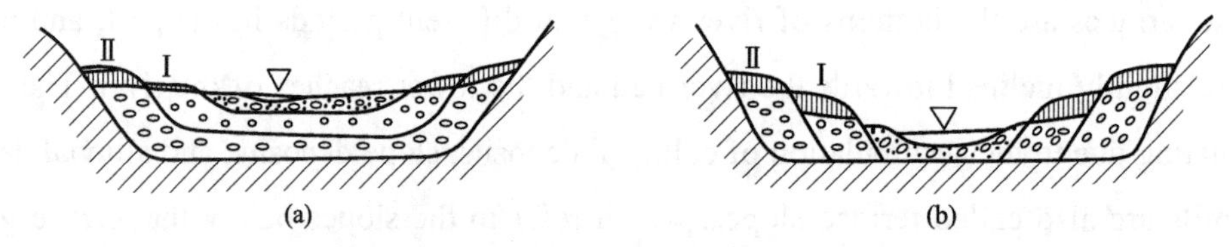

Figure 7-15 Structures of accumulative terraces

(a) Upper terraces; (b) Inner terraces

The base terrace is composed of two layers of materials from different periods. The upper part of the terrace is the river alluvium, and the lower part is the bedrock (Figure 7-16a). The base terrace's formation process is as follows: first, a broad valley is formed, small river alluvium is accumulated at the bottom of the valley; then the river erodes down and cuts through the alluvial layer into the bedrock; a base terrace is generated at last. This kind of terrace distribution is more common. In addition to the main types of terraces mentioned above, there are also embedded terraces (Figure 7-16b), buried terraces (Figure 7-16c), etc.

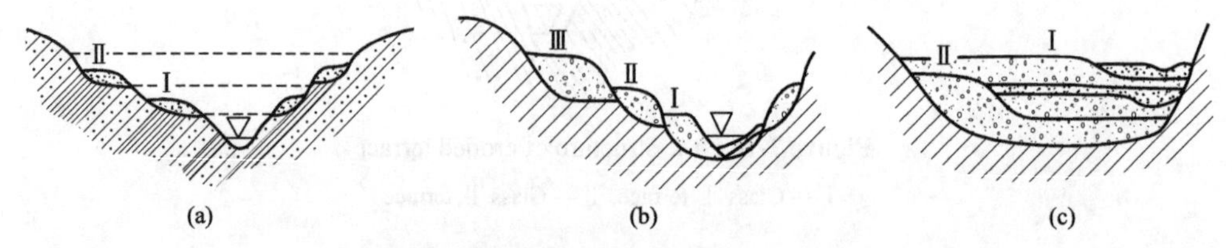

Figure 7-16 Structures of other terraces

(a) Base terraces; (b) Embedded terraces; (c) Buried terraces

7.3.4 Deltas

Where a stream enters a body of standing water, such as a lake or an ocean, it first drops its bedload, forming a bar across the river's mouth. Coarse suspended sediment, carried near the channel bottom, then begins to settle out. The flow commonly breaks into several distributaries as it crosses the bar and the newly deposited sediment. The flow of water and fine sediment is transported into the open water, maintaining its velocity for a short distance. Eventually, the flow mixes into the water body, and fine sediment settles out as well. If the freshwater flow mixes with saltwater, the salts cause even the finest clay particles to clot together and fall to the seafloor. The deposit built by this process is known as a delta (Figure 7-17).

Figure 7-17 Yellow River delta

Deltas can grow rapidly, ranging from 3m per year for the Nile River to 60m per year for the Mississippi. Some cities and towns that were at river harbors several hundred years ago are today several kilometers inland.

Human activities have recently been linked to the degradation of delta environments. In many locations, delta land is sinking due to the compaction of sediment and the removal of water, oil, and gas. Agricultural activities accelerate the loss of land with soil runoff. Upstream river water diversion projects remove from the river sediment that would otherwise replenish the delta landscape. Millions of people living in sinking deltas will need to develop adaptations as climate change heightens the threat of flooding from storms and a rising sea level. The Pearl River Delta is located in the south-central part of Guangdong Province, the lower reaches of the Pearl River, and is adjacent to the South China Sea (Figure 7-18). It is a composite of sediment brought by the Xijiang, Beijiang, Dongjiang, Tanjiang, Suijiang, and Zengjiang in the Pearl River Estuary. The Pearl River Delta gathers essential scientific and technological resources of Guangdong Province. It is the primary research and development base of the province's high-tech industries, the most prominent high-tech industrial belt in China, and a substantial domestic and international production base for high-tech industries.

Figure 7-18 A close-up view of the Pearl River Delta

Reference

[1] Christopherson R, Birkeland G H. Geosystems: An Introduction to Physical Geography[M]. 10th ed. Upper Saddle River, NJ: Pearson Education, 2017.

[2] GABLER R E, PETERSEN J F,TRAPASSO L M, SACK D. Physical geography[M]. 9th ed. Boston: Cengage Learning, 2008.

[3] STRAHLER A H, STRAHLER A. Introducing physical geography[M]. 6th ed. New York: Wiley, 2013.

[4] YANG L. Water Conservancy and Hydropower Engineering Geology[M]. Wuhan : Wuhan University Press, 2004.(In Chinese)

[5] Thomas D S. The dictionary of physical geography[M]. 4th ed. Hoboken: John Wiley & Sons, 2016.

[6] Holden J. An introduction to physical geography and the environment[M]. 4th ed.

Upper Saddle River, NJ: Pearson Education, 2017.

Exercises

Review Questions

(1) What is the meaning of the term stream in the Earth sciences?

(2) What are the components of a valley landform? Try to illustrate.

(3) What are the geological effects of rivers?What are the results?

(4) Describe the classification of fluvial erosion according to the directions of erosion.

(5) What are the differences among the fluvial transportation processes? Which moves the largest particles?

(6) Describe the composition of river valleys.

(7) Why is fluvial action so effective in arid climates, as precipitation is rare?

(8) What are stream terraces? How many types of river terraces are there? How are they formed? What are the basic features?

(9) How did sea-level changes during the Pleistocene cause stream terraces to form on land?

(10) What is a delta? How and why does it form?

8
Engineering Geological Mechanics of Rock Mass

8.1 Introduction

Before the 20th century, due to the limitation of the scale of production and the level of science, people only studied the rock's softness and hardness to distinguish the quality of the site and rarely doubted its overall stability when building buildings on the rock foundation. In the past hundred years, with the development of science and technology, the number of projects built on rock foundations has increased day by day, and the scale has become larger and larger. Strict requirements have been placed on rock foundations. Because of the catastrophic accidents in St. Francis Dam, the Malpasset Arch Dam, and the Vajont Reservoir, people realized that the rock foundation's quality depends not only on the strength of the rock itself but also on its own on the integrity of the rock and the groundwater. The role of rock mass is related to various factors, and thus the concept of rock mass is proposed.

A rock mass is a geological body with a specific composition and structure formed during geological history. The rock complex within the scope of the project's influence is usually called a rock mass. The scale of a rock mass can be determined by the content of the engineering geological problem being studied and the characteristics of the rock mass. The rock mass is a heterogeneous and anisotropic discontinuity. In some references, rock masses are often referred to as fractured rocks (Zhu & Zhao, 2003). In the rock masses, parts with low mechanical strength or relatively weak interlayers constitute a discontinuous surface of the rock mass, also called a discontinuity(structural plane). The discontinuity is a geological interface with a specific direction, scale, shape, and characteristics formed in the rock mass in the history of geological development. These geological interfaces can be rigid contact surfaces between rock blocks without any filling, such as cleavage planes, joint planes, bedding planes, schistose planes, etc. They can be a weak interlayer with fillings between the upper and lower sides. They can also be a fractured zone with a certain thickness (some called a width) or an ancient weathered crust.

According to their attitudes, a series of discontinuities combined cut the rock mass into rock blocks of different shapes, sizes, and compositions, collectively called rock mass structures. The rock mass structure can be composed of two basic units: discontinuity and structural body.

Various structural surfaces in the rock mass are the products of internal and external geology dynamics. By studying all discontinuities as a whole, it can be found that their combination forms are different. Summarizing the combination of structural planes, the concept of "rock mass structure" is born, scientifically reflecting rock mass's nature. In a large number of engineering practices and experimental studies, it is found that the rock mass structure is different, the physical and mechanical properties of the rock mass, the degree of difficulty, and the way of deformation and failure of the rock mass are also different. At the same time, the rock mass structure also controls the hydrogeological condition and weathering of the rock mass; more importantly, the rock mass structure is different, and the rock mass stability characteristics are entirely different (Figure 8-1).

Figure 8-1 Rock mass is including discontinuities such as joints, faults, layer, and etc

The type of anisotropy and inhomogeneity variation could be gradual within the intact rock or sudden as a discontinuity is crossed. There can be a variation on all scales: within grains or crystals, within the microstructure, within laboratory-sized samples of intact rock, within engineering structure-sized volumes of rock, and so on. However, a distinction can be made between the assumptions traditionally required for modeling and the natural properties of the rock. It can be remembered by two acronyms: CHILE and DIANE. A Continuous, Homogeneous, Isotropic, and Linearly-Elastic (CHILE) material is most commonly assumed for modeling purposes. Traditional stress analysis techniques are formulated in these four attributes, simply for necessity and convenience for obtaining closed-form solutions. In the past, limited computational methods precluded any more sophisticated analysis. Nowadays, however, especially in consulting and research organizations, computer codes are available that will routinely violate any of these traditional assumptions. It leads directly to the second acronym. A Discontinuous, Inhomogeneous, Anisotropic, Non-Elastic (DIANE) rock is the engineer's material to deal with. Therefore, we should consider the significance of the difference between the CHILE material being modeled and the DIANE rock being engineered and the likely error arising from the direct application of a model based on a CHILE material. Alternatively, the specific attributes of the DIANE rock can be modeled. Superb examples of the latter procedure are the development of block theory and distinct element methods in numerical analysis (Hudson & Harrison, 1997).

8.2 Structural characteristics of rock mass

8.2.1 Types of discontinuity

Geological investigations usually categorize discontinuities according to how they were formed. It is helpful for geotechnical engineering because discontinuities within each category have similar properties regarding both dimensions and shear strength properties which can be used in the initial review of stability conditions of a site. Discontinuities in rocks are cracks, fissures, fractures, joints, bedding planes, schistosity or foliation planes, and faults. Discontinuities are products of certain phenomena to which rocks were subjected in their geological past, and they are expected to be regularly distributed within the rock mass. The following are standard definitions of the most commonly encountered types of discontinuities.

1. Commonly encountered discontinuities

1) Fault

A discontinuity along which there has been an observable amount of displacement. Faults are rarely single planar units; normally, they occur as parallel or sub-parallel sets of discontinuities along which movement has taken place to a greater or less extent (Aydan, 2019).

2) Bedding plane

This is a surface parallel to the surface of deposition, which may or may not have a physical expression. Note that the original attitude of the bedding plane should not be assumed to be horizontal.

3) Foliation

Foliation is the parallel orientation of platy minerals or mineral banding in metamorphic rocks.

4) Joint

A joint is a discontinuity in which there has been no observable relative movement. In general, joints intersect primary surfaces such as bedding, cleavage, and schistosity. A series of parallel joints is called a joint set; two or more intersecting sets produce a joint system; two sets of joints approximately at right angles to one another are said to be orthogonal.

5) Cleavage

Parallel discontinuities formed in incompetent layers in a series of beds of varying degrees of competency are cleavages. In general, the term implies that mineral particles in parallel orientation do not control the cleavage planes.

6) Schistosity

This is the foliation in schist or other coarse-grained crystalline rock due to the parallel arrangement of mineral grains of the platy or prismatic type, such as mica.

7) Vein

Vein is an infilling of a discontinuity caused by circulation of mineralized fluid and deposition of minerals. Veins can cause healing of the original discontinuity.

2. Classification according to the forming process and environment

1) Genetic(primary) discontinuity

Genetic discontinuities are formed during the diagenetic process of rocks and fall into the following three categories:

(1) Sedimentary discontinuity. The bedding plane, stratification, hiatus, and weak sedimentary interlayer are all sedimentary discontinuities. Generally speaking, the sedimentary discontinuity is ductile, and the attitude varies with the change of rock formation. The distribution of marine sedimentary discontinuities is usually stable and clear. In continental rock formation, the discontinuities are often pinch-out or interlocked. The strength of bedding planes, in general, is not necessarily low but can be reduced by interlayer dislocation caused by tectonics or by later weathering. Hiatus includes disconformity and unconformity, which are generally undulating and have paleo-weathering residues, forming a weak interlayer with variable morphology. Sedimentary weak interlayer has low strength and is easily softened by water. Its type and characteristics can be seen in the weak interlayer section.

(2) Igneous discontinuity. The igneous discontinuity is formed during magma intrusion, overflow, and condensation, including flow layers, condensation joints, contact surfaces between

the intrusion and the surrounding rock, and weak contact surfaces formed by intermittent magma overflow. Condensation joints generally have tension characteristics and have an important influence on rock stability and seepage. The contact surface between the intrusion and the surrounding rock sometimes forms a fracture zone or an alteration zone, thus becoming a weak discontinuity. Sometimes, the intrusion and surrounding rock are well-used, so they can not be regarded as discontinuous.

(3) Metamorphic discontinuity. The metamorphic discontinuity can be divided into two types: residual metamorphic discontinuities and recrystallized metamorphic discontinuities. The former is a low-grade metamorphic of sedimentary rocks, and the bedding plane is still preserved, but there are sericite, chlorite, and other scaly minerals densely distributed and oriented arranged in the bedding plane. The recrystallized discontinuities are mainly schistosity and gneissic schistosity. Due to the enrichment and highly oriented arrangement of schistose or columnar minerals, the recrystallized discontinuities often control rock mass properties. Mica schist, chlorite schist, and talc schist in metamorphic rocks often form relatively weak interlayers due to the characteristics of schistosity development, lithological weakness, and easy weathering.

2) Epigenetic (secondary) discontinuity

(1) Discontinuities formed by internal dynamics. Including joints, cleavage, faults, shear zone of interlayer, etc., also called tectonic discontinuity. Except for those who have been cemented, most of these discontinuities are cracked. Among them, the fault and shear zone caused by the interlayer dislocation is large-scale and filled with varying thickness and property of the filler. Some have been in argillization, and its engineering geological properties are very poor. Joints, cleavage, etc., are generally not filled or have a thin thickness of the filling. They mainly affect the integrity and mechanical properties of the rock mass.

(2) Discontinuities formed by external dynamics. This discontinuity is mainly formed by weathering, unloading, and human activities. Its common characteristic is that it is only distributed within the surface or tens of meters below it. Weathering fissures are generally disordered, discontinuous, and mostly filled with clast or mud. Weathering, sometimes also developed along the genetic or tectonic discontinuities, can form weathering interlayer, weathering trough or weathering sac, etc. The unloading joint is formed by the vertical unloading and horizontal stress release caused by the denudation or artificial excavation of the rock mass, such as developing the along slope fissures at the slope of the river and nearly horizontal fissures at the bottom of the valley. Epigenetic filling of the weak interlayer is mainly formed by flowing water or gravity transported clay material filling in existing cracks. The distribution range of blasting fissures is limited, and their distribution density usually decreases rapidly with the

increase of distance from the blasting point.

These descriptions of discontinuity categories are established in engineering practice, and the potential properties can be anticipated from their types. For example, faults are major structures containing weak infillings such as crushed rock and clay gouge, whereas joints have lengths much shorter than faults and joint infillings are often thin and cohesive or absent. However, standard geological names alone rarely give sufficient detailed information for design purposes on the properties of a discontinuity, especially for foundations where particulars of such characteristics as the infilling thickness can significantly influence settlement. For this reason, geological descriptions help understand the general conditions at a site, but further specific geotechnical studies are almost always required before proceeding to the final design.

8.2.2 Suggested methods for the quantitative description of discontinuities

Sets of discontinuities usually occur in orthogonal groups in response to the stress field that has deformed the rock. It is shown in the photograph in Figure 8-2, where the rock mass is cut into polygonal columns by several sets of joints in different directions. Figure 8-3 shows the rock mass characteristics in diagrammatic form. This section describes each of these ten parameters and discusses their influence on foundation performance. Complete mapping and measurement procedures are described in the ISRM publication Suggested Methods for the Quantitative Description of Discontinuities in Rock Masses (ISRM, 1981).

Figure 8-2 Columnar jointing rock mass

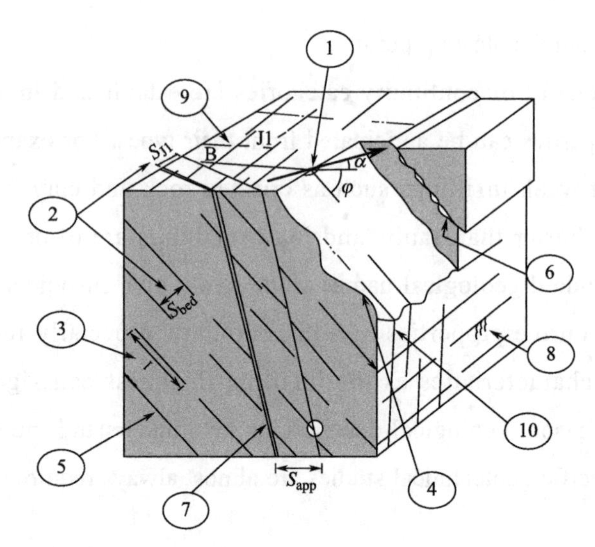

Figure 8-3 Characteristics of discontinuities in rock masses (The following is a description of ①～⑩)

1. orientation

The orientation of discontinuities is expressed as the dip and dip direction (or strike) of the surface.

2. Spacing

Discontinuity spacing can be mapped in rock faces and drill core, with the true spacing being calculated from the apparent spacing for discontinuities inclined to the face, as shown in Figure 8-4. Measurement of discontinuity spacing of each set of discontinuities will define the size and shape of blocks and indicate stability modes such as toppling failure. The spacing is also related to the rock mass strength because, in closely fractured rock, the individual discontinuities will more readily join together to form a continuous zone of weakness.

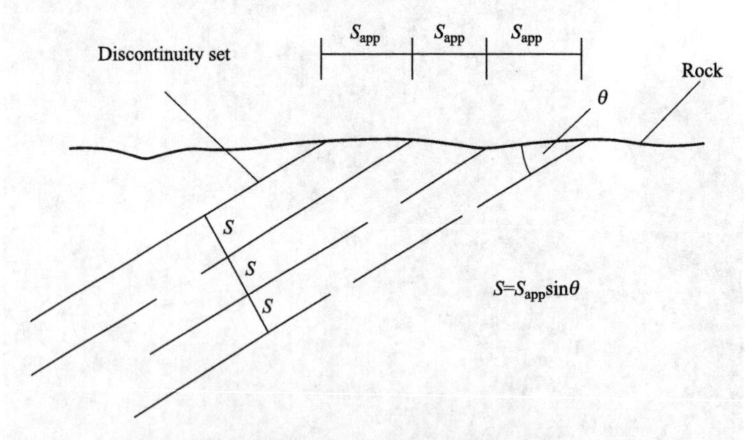

Figure 8-4 True and Apparent Spacing

3. Persistence

Persistence is a measure of continuous length or area of discontinuity. Table 8-1 provides terms that can be used to express persistence values. This parameter defines the size of blocks and the length of potential sliding surfaces, so the mapping should concentrate on measuring the persistence of the set of discontinuities that will have the most significant influence on stability. The persistence of discontinuities is one of the essential rock mass parameters and one of the most difficult to measure. This is because usually, only a small part of the discontinuity is visible in the face, and in the case of drill core, no information on persistence is available.

Table 8-1 Persistence dimension

Persistence	Dimensions (m)
Very low persistence	<1
Low persistence	1~3
Medium persistence	3~10
High persistence	10~20
Very high persistence	>30

4. Roughness

The roughness of a discontinuity surface is often an essential component of the shear strength, especially where the discontinuity is undisplaced and interlocked. Roughness becomes less important where the discontinuity is infilled or displaced, and interlock is lost. Roughness should be measured in the field on exposed surfaces with lengths of at least 2m if possible and in the anticipated sliding direction.

5. Wall strength

The strength of the rock forming the walls of discontinuities will influence the shear strength of rough discontinuities because high stresses are generated at local contact points during shearing. If the rock strength is low relative to the magnitude of these stresses, the asperities will be sheared off, resulting in a loss of the roughness component of the friction angle. The rock strength is quantified by the Joint Compressive Strength (JCS) term. It is often adequate to estimate the compressive strength from the simple field tests or if the core or lump samples are available by carrying out point load tests. The Schmidt hammer test is also a method of estimating the compressive strength of the rock at discontinuity surfaces.

6. Aperture

Aperture is the perpendicular distance separating the adjacent rock walls of an open discontinuity, in which the intervening space is air or water-filled. Aperture is thereby distinguished from the width of a filled discontinuity. It is crucial in predicting the likely behavior of the rock mass, such as deformation under stress changes and permeability, to understand why open discontinuities develop. Possible causes include washing out of infillings, solution of the rock forming the walls of a discontinuity, shear displacement of rough discontinuities, tension features at the head of landslides, and relaxation of steep valley walls following glacial retreat or erosion. The aperture may be measured in outcrops or tunnels provided that care is taken to discount blast-induced open fractures, in drill core if recovery is excellent, and in boreholes using a borehole camera if the walls of the hole are clean.

7. Filling

The filling is the term for material separating the adjacent walls of discontinuities, such as calcite or fault gouge; the perpendicular distance between the adjoining rock walls is termed the width of the filled discontinuity. A complete description of filling material required to predict the discontinuity behavior includes the following: mineralogy, particle size, over-consolidation ratio, water content/permeability, wall roughness, width, and fracturing/crushing of the wall rock. If the infilling is likely to influence the foundation's performance, samples of the material (undisturbed if possible) should be collected, or an in-situ test may be carried out.

8. Seepage

Observations of seepage location provide information on aperture and persistence because the groundwater flow is confined almost entirely in the discontinuity. These observations will also indicate the position of the water table or water tables in the case of rock masses containing alternating layers of low and high permeability rock such as shale and sandstone.

9. Number of sets

The number of sets of discontinuities that intersect one another will influence the extent to which the rock mass can deform without failure of the intact rock. As the number of discontinuity sets increases and the block size diminishes, the greater the block's opportunity to rotate, translate and crush under applied loads.

10. Block size

The block size and shape are determined by the discontinuity spacing and persistence and the number of sets. The block size can be estimated by selecting several typical blocks and measuring their average dimensions, which are then recorded using the terms in Table 8-2. Block shapes include massive, blocky, tabular, columnar, irregular, and crushed.

Table 8-2 Block dimensions

Description	J_v(joints/m^3)
Very large blocks	<1.0
Large blocks	1~3
Medium-sized blocks	3~10
Small blocks	10~30
Very small blocks	>30

8.2.3 Structure type of rock mass

To summarize the rock mass's deformation failure mechanism and the need to evaluate the rock mass's stability, the structure type of the rock mass can be divided according to the degree of jointing of the rock mass. In 2016, the MOHURD (Ministry of Housing and Urban-Rural Development) of my country issued the latest national standard, "*Code for hydropower engineering geological investigation*" (GB 50287—2016), which divided the rock mass structure into five major categories and 13 sub-categories. The essential characteristics are shown in Table 8-3.

Table 8-3 Structure type of rock mass

Type	Subclass	Rock structure characteristics
Massive structure	Monolithic structure	The rock mass is complete and massive, with undeveloped discontinuities whose spacings are greater than 100cm
	Blocky structure	The rock mass is moderately intact and massive, with slightly developed discontinuities whose spacings are generally greater than 50~100cm
	Sub-block structure	The rock mass is moderately intact, in a sub-block shape, with moderately developed discontinuities whose spacings are generally 30~50cm
Layered structure	Huge thick layered structure	The rock mass is intact, with huge thick layers, the discontinuities are not developed, and the spacings are greater than 100cm
	Thick layered structure	The rock mass is moderately intact, with thick layers and slightly developed discontinuities whose spacings are generally 50~100cm
	Medium thick layered structure	The rock mass is moderately intact, with medium-thick layers and moderately developed discontinuities whose spacings are generally 30~50cm
	Interlayered structure	The rock mass is moderately intact or poor in integrity, inter-layered, with well-developed or developed discontinuities whose spacings are generally 10~30cm
	Thin layered structure	The rock mass is poor integrity, thin-layered, with well-developed discontinuities whose spacings are generally less than 10cm
Mosaic structure	Mosaic structure	The integrity of the rock mass is poor, the rock blocks are tightly interlocked, with the moderately developed to very well developed discontinuities whose spacings are generally 10~30cm

continued

Type	Subclass	Rock structure characteristics
Cataclastic structure	Block crack structure	The rock mass is poor integrity, with cuttings and argillaceous material filling between the rock blocks. The interlock is medium and tight-relaxed. The discontinuities are moderately developed to very well developed, whose spacing are generally 10~30cm
	Cataclastic structure	The rock mass is moderately broken, with cuttings and argillaceous material filling between the rock blocks, the mosaic is moderately loose-relaxed, the discontinuities are very developed, whose spacing are generally less than 10cm
Loose structure	Fragmentary structure	The rock mass is broken, with rock fragments or argillaceous material filling between the bloks, and the interlock is loose
	Detrital structure	The rock mass is extremely broken, the rock fragments or argillaceous material is interposed with the rock block, and the mosaic is loose

8.3 Mechanical characteristics of rock mass

The most common method of studying the mechanical properties of rocks is by axial compression of a circular cylinder whose length is two to three times its diameter. If the lateral surface of the rock is traction-free, the configuration is referred to as uniaxial compression or unconfined compression (Figure 8-5a). In this case, the resulting state of stress in the rock is $\{\sigma_1 > 0, \sigma_2=\sigma_3=0\}$. If tractions are applied to the lateral surfaces, the experiment is referred to as one of confined compression. For tests done on a circular cylinder, the stresses involved in the two orthogonal directions perpendicular to the cylinder axis are necessarily equal (Figure 8-5b), and the resulting state of stress in the rock is $\sigma_1 > \sigma_2 = \sigma_3 > 0$. This state is traditionally called "triaxial," even though two of the principal stresses are equal. The more general state of stress, in which $\sigma_1 > \sigma_2 > \sigma_3 > 0$, can be achieved with cubical specimens and is known either as "polyaxial" or "true triaxial" (Figure 8-5c).

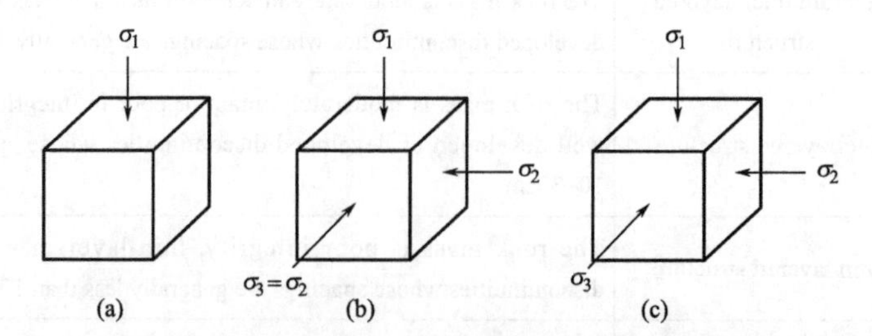

Figure 8-5 Cubic specimen under (a) uniaxial stress, (b) traditional triaxial stress in which the two lateral confining stresses are equal, and (c) true-triaxial stress, in which all three principal stresses are possibly different

Here, considering the behavior of natural rock, we will begin with the simplest form of loading, i.e., uniaxial compression. In its simplest form, the uniaxial compression test is conducted by taking a right cylinder of intact rock, loading it along its axis, and recording the displacement produced as the force increases. In Figure 8-6, we present a typical record of such a test. Note that the force and the displacement have been scaled respectively to stress (by dividing by the original cross-sectional area of the specimen) and to strain (by dividing by the original length). In the curve shown in the figure, the various aspects of the mechanical behavior of intact rock tested under these conditions can now be identified.

At the very beginning of loading, the curve has an initial portion that is concave upwards (the opposite of typical soil behavior) for two reasons:

(1) the lack of perfect specimen preparation, manifested by the ends of the cylinder being non-parallel.

(2) the closing of microcracks within the intact rock.

After this initial zone, there is a portion of essentially linear behavior, more or less analogous to the ideal elastic rock.

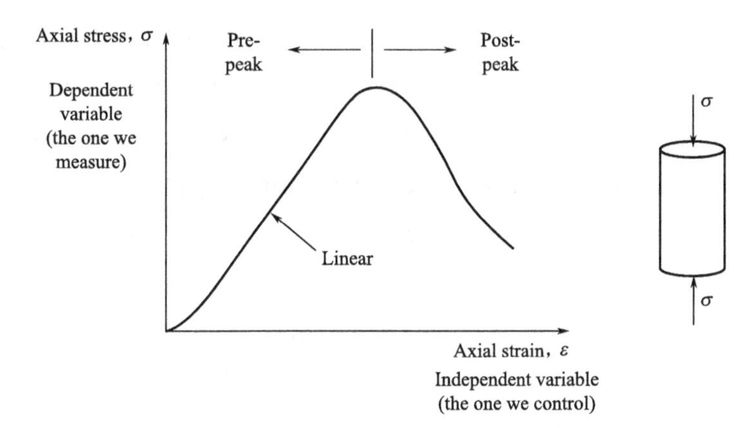

Figure 8-6 The complete stress-strain curve

Remembering that Young's modulus, E, is defined as the ratio of stress to strain, it can be determined in two ways: either by taking the slope of the stress-strain curve at a given point; or by taking the slope of a line connecting two points on this linear portion of the curve (Figure 8-7). The two slopes are the tangent modulus and the secant modulus. The tangent modulus is conventionally taken as the gradient of the σ-ε curve at a stress level corresponding to 50% of the peak stress; the secant modulus may be determined anywhere over the entire linear portion. Naturally, both are approximations to the actual behavior but are functional and adequate for simple elastic applications. However, with increasing numerical and computing capabilities, we can represent the complete stress-strain curve more accurately as a piece-wise linear function if required.

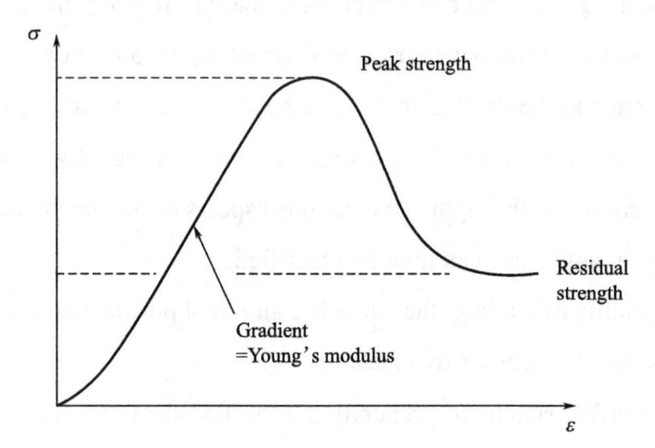

Figure 8-7 The complete stress-strain curve illustrating various mechanical parameters.

A rock under uniaxial compressive stress will not only deform in the direction of the load, but it will also deform in each of the two directions perpendicular to the load. Figure 8-8 shows the strains measured on a cylindrical specimen by Hojem et al. (1975) during confined uniaxial compression of an argillaceous quartzite, with the lateral confining stress, held constant at σ_2=σ_3=6.9MPa. As before, the convention used is that positive normal strains correspond to decreases in the linear dimensions of the specimen. The axial stress vs. axial strain curve exhibits most of the described features, including a pronounced brittle regime.

The strain in the two other directions (i.e., the radial strain, ε_2 =ε_3) is negative, which means that the specimen bulges outward as it is compressed. Within the elastic regime, the magnitude of the radial strain increases nearly in proportion to the axial strain. The negative of the ratio of the transverse strain to the axial strain, $\varepsilon_2/\varepsilon_1$, is known as Poisson's ratio and is denoted by ν . For a linear elastic material, this parameter is independent of stress and is generally in the range 0~0.5.

In the ductile regime, the transverse strains begin to grow (in magnitude) at a much faster rate than does the axial strain. In terms of incremental strains, this behavior could be said to correspond to a value of Poisson's ratio that exceeds unity. As the volumetric strain, $\Delta V/V$, is equal to the sum of the strains in the axial and the two lateral directions, the volumetric strain begins to decrease with an increase in the axial stress. This first occurs at J in Figure 8-8.

Eventually, the lateral strains become sufficiently negative that the total volumetric strain becomes negative; this occurs at K in Figure 8-8. The phenomenon by which the volume of the rock decreases under the action of an additional compressive stress is known as dilatancy. Dilatancy can be ascribed to the formation and extension of open microcracks whose axes are oriented parallel to the direction of the maximum principal stress. By testing specimens in the form of thick-walled hollow tubes, Cook (1970) showed that dilatancy occurs pervasively throughout the entire volume of the rock, and is not a superficial phenomenon localized at the outer boundary.

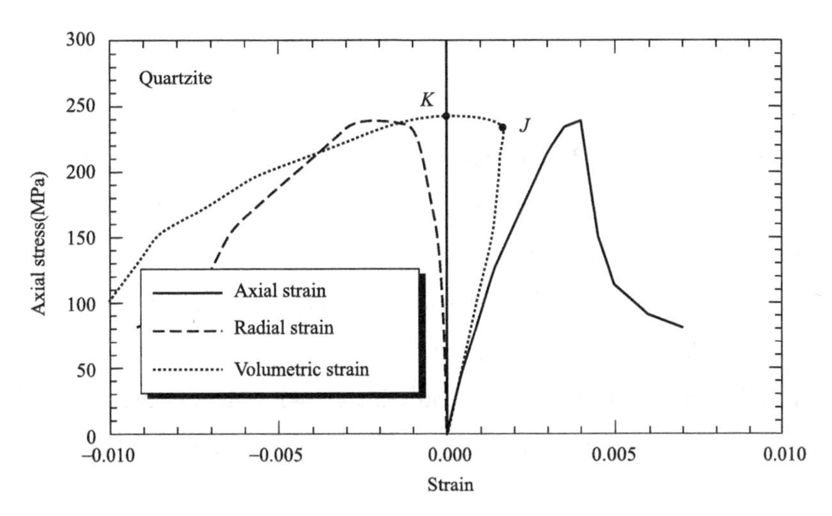

Figure 8-8 Axial strain, m radial strain, and volumetric strain as a function of axial stress, for a cylindrical sample of an argillaceous quartzite, tested under a confining stress of 6.9MPa. (Hojem et al., 1975)

8.4 Weathering of rock

8.4.1 Weathering processes

Weathering is the disintegration and decomposition of geological materials, and it includes any physical or chemical modifications of the materials' characteristics and properties. The final products of rock weathering processes are soils; these may remain as residual soils in their original location on the source rock, or they may be transported as sediment. Transported soils (alluvial, aeolian, or glacial) can be lithified to form new rocks or remain as soils. The contact between soil and rock may be clearly defined or gradual, characteristic of residual soils. The extent to which rock is either intact or as a rock mass is weathered significantly influences its physical and mechanical properties.

Weathered rock materials can be broadly defined as the transition between rock and soil; they present a wide range of geotechnical properties and characteristics that blend with soils and rocks, depending on the degree of weathering (Hojem, Cook, & Heins, 1975). The terms "soft" or "weak" rock are sometimes used to infer weathered materials, although not all soft rocks (e.g., mudstones, siltstones, marls) are a product of only weathering.

When materials with a significant degree of weathering are classified according to their mechanical behavior, the problem is whether they should be considered rocks or soils; in the first case, their properties are undervalued, while in the second case, they are overvalued. Some authors prefer to distinguish soils from rocks by their level of compaction or cementation, structure, and durability.

Weathering processes are controlled by climatic conditions and variations in temperature, humidity, and rainfall. These factors determine the type and intensity of the physical and chemical transformations that affect rocks at and near ground level. Physical actions produce mechanical fracturing of the rock. The most critical activities controlled by climatic factors, especially temperature and humidity, are:

(1) Ice formation: the volume of water held in pores and cracks increases as the temperature drops and ice forms, leading to the rock fracturing.

(2) Solar radiation or "insolation": in arid climates with marked thermal differences over short periods, can induce fracturing due to the stresses from successive dilation and contraction of the rocks.

(3) Salt formation: salt crystallization in pores or cracks in the rocks produces failure and disintegration as the crystals expand.

(4) Hydration: when clays and sulfates are hydrated, their volume increases, producing considerable deformation, which may cause fracturing in the rock.

(5) Capillarity and thermal expansion: fissured and sheet structure minerals, such as micas and gypsum, allow water to penetrate, and with temperature changes, this may cause structural failure as the water expands.

When water is present, temperature controlled chemical processes take place. These occur more quickly and intensely in wet regions than in dry ones and lead to the formation of new minerals or compounds from existing ones. The most important processes are:

(1) Dissolution: the decomposition of minerals by the action of water, leading ultimately to disintegration of the material. Dissolution is a chemical reaction; the dissolution of calcium carbonate causes cracks, joints and existing voids to open up in carbonate and other soluble rocks.

(2) Hydration: the formation of new minerals or chemical compounds from the absorption of water, e.g., gypsum from anhydrite.

(3) Hydrolysis: the decomposition of a mineral or chemical compound by the action of water. The level of hydrolysis depends on the extent that the element ions attract the water molecules.

(4) oxidation and reduction: the formation of new minerals from a combination of a host mineral with oxygen, either by losing one or more electrons and fixing oxygen (oxidation) or by losing oxygen and selecting electrons (reduction).

In cold climates and mountainous elevations with average rainfall, physical weathering will predominate, basically controlled by ice, but in warm tropical climates with high rain, there will be more chemical activity. Figure 8-9 shows the different types of predominant weathering and their intensity depending on temperature and precipitation. Weathering processes affect both intact rock and the rock mass as a whole.

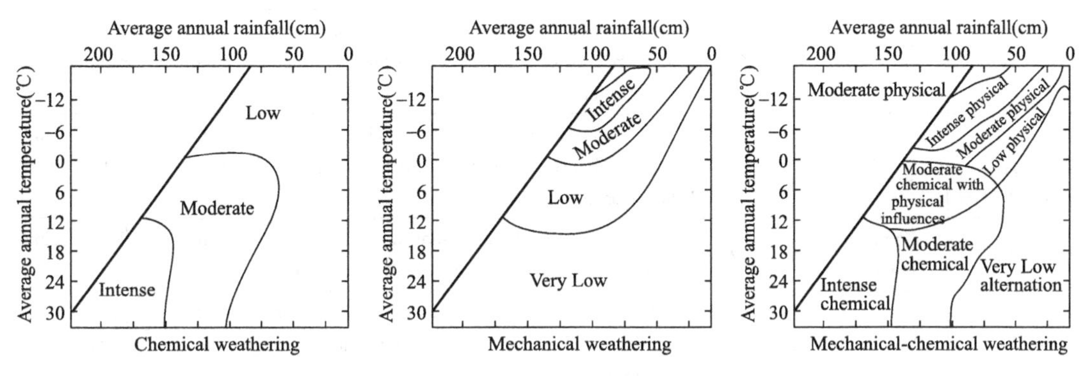

Figure 8-9 Relationship between climate and weathering processes (Embleton and Thornes, 1979)

8.4.2 Degree of weathering

The description of the weathering is important as most construction on or in a rock mass is undertaken at shallow depth within the zone of surface weathering. It should always include comments on the degree and nature of any weathering effects on material or mass scales. It may allow subsequent classification and provide information for separating rock into zones of like character (BSI, 2015). Typical indications of weathering include:

(1) Changes in color.

(2) Reduction in strength.

(3) Changes in fracture state.

(4) Presence, character and extent of weathering products.

1. Degree of weathering by ISRM

Weathering classification may often not be appropriate due to the significant variability in lithologies and other rock mass features, which is valid for engineering purposes. The degree of weathering in a rock mass is evaluated from direct observation of the outcrop and compared with the standard descriptions shown in Table 8-4. A six-stage scale is used: fresh, slightly weathered, moderately weathered, highly weathered, completely weathered, and residual soil. To observe weathering in intact rock, breaking a piece of rock into fragments may occasionally be necessary.

Table 8-4 classification of the degree of weathering in the rock mass (ISRM, 1981)

Degree of weathering	type	Symbol	Description
I	Fresh	F	No visible signs of weathering; perhaps slight discolouration on major discontinuity surfaces
Ⅱ	Slightly weathered	SW	Discolouration indicates weathering of rock material and discontinuity surfaces. All the rock material may be discoloured by weathering and may be somewhat weaker externally than in its fresh condition
Ⅲ	Moderately weathered	MW	Less than half of the rock material is decomposed and/or disintegrated into soil. Fresh or discoloured rock is present either as a continuous framework or as corestones
Ⅳ	Highly weathered	HW	More than half of the rock material is decomposed and/or disintegrated into soil. Fresh or discoloured rock is present either as a discontinuous framework or as corestones
V	Completely weathered	CW	All rock material is decomposed and/or disintegrated into soil. The original mass structure is still largely intact
Ⅵ	Residual soil	RS	All rock material is converted to soil. The mass structure and material fabric are destroyed. There may have been a large change in volume but the soil has not been significantly transported

2. Classification of vertical zonation of weathering crust in China

The surface layer of the lithosphere that is weathered is called the weathered crust. The vertical distance from the surface of the weathered rock to the fresh rock is the thickness of the weathered crust. The rock formation is generally strong on the surface, gradually weakening from the surface down to fresh bedrock. The degree of weathering of the rock is different, and the change of its physical and mechanical properties is also different. Therefore, the performance of adapting to the building is also different, and the corresponding treatment measures are also different.

Consequently, it is very important to zone the weathered rock in the vertical direction properly. At present, the classification signs of rock weathering zones mainly consider the color, rock fragmentation degree, changes in mineral composition, hydraulic properties, physical and mechanical properties, and changes in acoustic wave characteristics. Generally, weathered rocks are divided into 5-stage zones from top to bottom based on this. Namely: fully weathered zone, strong weathered zone, weakly weathered zone, slightly weathered zone, fresh. In 2016, the MOHURD of China issued the latest national standard, "Specifications for Geological Survey of Hydropower Engineering," The classification is shown in Table 8-5. It should be pointed out that this zoning is for keeping a complete weathering profile, and in fact, some bands are often missing.

Table 8-5 classification of the degree of weathering in the rock mass (GB50287-2016)

Weathered zone	Main geological features	The ratio (α) of the longitudinal wave speed of weathered rock to that of fresh rock
Fully weathered zone	1. All rock surfaces are discolored and lose the lustre. 2. The structure of the rock has been completely destroyed. The rock has been disintegrated and decomposed into loose soil or sand, with a large volume change, but has not moved. There are still traces of the original structure. 3. Except for quartz particles, most of the other minerals are weathered and eroded into secondary minerals. 4. The rock feel soft and are dented when hammered. Minerals can be crushed by hand and be dug with a spade	$\alpha<0.4$
Strong weathered zone	1. Most of the rock surfaces are discolored, only some rock blocks maintain the original color. 2. Most of the structure of the rock has been destroyed. Only a tiny part of the rock has been decomposed or disintegrated into the soil, and most parts of the rock are discontinuous skeletons or broken stones. Weathered fissures are developed and sometimes contain a large amount of secondary mud. 3. Except for quartz, feldspar, mica and iron-magnesium minerals have been weathered and eroded. 4. The rock sounds dumb and most parts of it become crisp and fragile when hammered. The rock can be dug with a pickaxe, but the hard parts need to be blasted	$0.4\leqslant\alpha<0.6$
Weakly weathered zone	1. Most of the rock surface or fracture surface is discolored, but the fracture surface still maintains the fresh rock color. 2. The original structure of the rock is clear and complete, but the weathered fissures are developed, and the fissure walls are highly weathered. 3. The iron-magnesium minerals along the fractures are oxidized and corroded, and the feldspar becomes turbid and fuzzy. 4. The rock sounds dumb when hammered and needs to be blasted when excavated	$0.6\leqslant\alpha<0.8$
Slightly weathered zone	1. The rock surface or fracture surface is slight discolored. 2. The original structure of the rock has no change and maintains the original and complete structure. 3. Most of the fissures are closed or filled with calcareous films, and there is weathering and erosion only along the large fissures, or rust film infiltration. 4. The rock sounds crisp when hammered and needs to be blasted when excavated	$0.8\leqslant\alpha<1.0$
Fresh	1. The rock surface keeps the fresh color, and only large fisssures are occasionally discolored. 2. The fissure is tight and welded, and only a few fissure surfaces are stained with rust film or slightly eroded. 3.The rock sounds crisp when hammered and needs to be blasted when excavated	$\alpha=1.0$

8.5 In-situ Rock stress measurement

The unperturbed rock mass contains non-null stress components due to the weight of overlying materials, confinement, and history of stress. Near the surface in mountainous areas, in situ stress can be close to zero at some points or close to rock strength at other issues. In the former case, rocks may fall from the surface and underground excavations because joints are open and weak; in the latter case, the disturbance of the stress field by tunneling or perhaps even surface excavation may trigger the violent release of stored energy.

8.5.1 Influence of the Initial Stresses

Estimating the order of magnitude of stresses and their directions is often possible, but one can never be sure of the margin of error without backup measurements. Such measurements are pretty common in mining practice, but since stress measurements tend to be expensive, they are not routine for civil engineering applications. However, there are several civil engineering situations when knowledge of the state of stress can be helpful, or lack of knowledge might prove so costly that a significant stress measurement program is warranted. For example, when choosing the orientation for a cavern, one hopes to avoid aligning the long dimension perpendicular to the major principal stress. If the initial stress is very high, the shape will have to be selected to minimize stress concentrations vastly. Knowledge of rock stresses also aids in the layout of complex underground works.

The magnitude, direction, and distribution of in-situ stress are related to earthquakes, affect the engineering site's regional stability, and directly impact the design and construction of engineering buildings. For example, in the low-stress area, the rock mass is loose, leaking, and the weathering zone is deep; in the high-in-situ stress area, excavation and unloading will cause the deformation and destruction of the rock mass, fragmentation, and even rockburst. But sometimes, high in-situ stress can also play a beneficial role in engineering. The key is to fully understand the distribution and change law of in-situ stress and understand the influence on the deformation and failure of the rock mass.

In engineering, the level of in-situ stress is not divided by its absolute value but refers to the comparison of horizontal in-situ stress and vertical in-situ stress. At present, both at home and abroad use the ratio of rock strength R_c to the maximum horizontal principal stress σ_{max} to distinguish the level of in-situ stress. For example, the French Tunnel Association, the Japanese Applied Geological Association, and the Donets Basin Coal Mine of the former Soviet Union stipulate the high-stress area, the medium-stress area, and the low-stress area. China's national standard GB/T 50218—94 proposes that the strength-to-stress ratio<4 is an extremely high stress area, and when the strength-to-stress ratio is equal to 4~7, it is a high-stress area. This ratio is retained in the latest national standard GB/T 50218—2014, but canceled the high and extremely high stress statement. Regarding low ground stress, it generally refers to an area where the horizontal ground stress is less than the horizontal stress formed by its weight (Figure 8-10).

Figure 8-10 When the maximum principal stress of a large underground cavern is parallel to the axis of the cavern or the angle of intersection is small, the side walls of the cavern are prone to spalling damage

8.5.2 Estimating the Initial Stresses

1. Vertical stress

The normal vertical stress can generally be assumed to be equal to the weight of the overlying rock, which is 0.027MPa/m on average. Close to the horizontal ground, the principal stress directions are vertical and horizontal. It is commonly assumed that they are also vertical and horizontal in depth (Figure 8-11a). But it is only an assumption to reduce the number of unknowns. This assumption finds reinforcement in Anderson's observations that normal and reverse faults often dip at 60° and 30° , respectively (Jaeger and Cook, 1976). The simplifying assumption that the principal stresses are vertical and horizontal has been widely adopted in practice. Of course, this breaks down at shallow depths beneath the hilly terrain because the ground surface, lacking normal and shear stresses, constantly form a trajectory of principal stress. Below a valley slope, principal stress is normal to the slope and equals zero, while the other two principal stresses are on the slope plane (Figure 8-11b). These stresses also approach zero when the slope of the rock is convex upwards but becomes more significant when the slope is concave upwards. Under the sharp notch of a V-shaped valley, the in situ stresses can be close or at the strength of the rock.

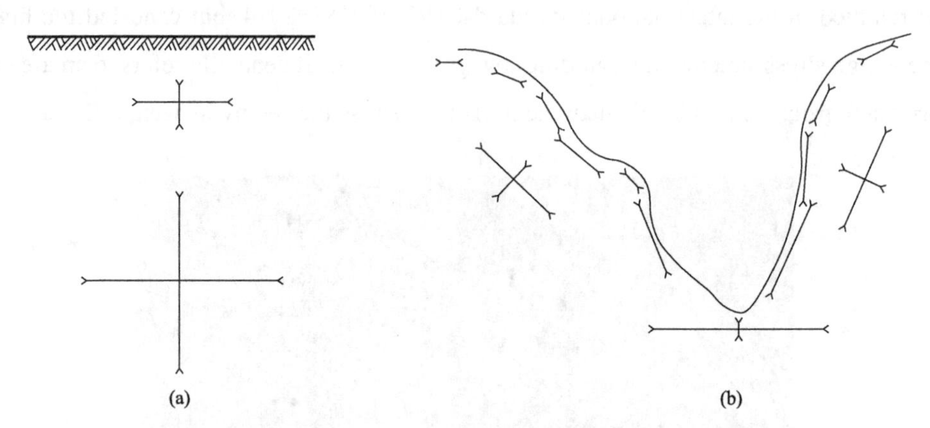

Figure 8-11 The influence of topography on initial stresses

(a) principal stress vectors under a horizontal ground; (b) principal stress vectors under a valley

Over any significant horizontal surface within the ground, the average vertical stress must equilibrate the downward force of the weight of the overlying rock. Hence the rule stated previously:

$$\overline{\sigma_z} = \gamma Z \tag{8-1}$$

Where $\overline{\sigma_z}$ is the average total vertical stress at depth Z in rock with unit weight. This rule has been supported by numerous measurements (Figure 8-16a) and is one of the reliable formulas of stress in situ. However, it can be violated over limited horizontal distances owing to the effects of geological structure. Figure 8-12, for example, shows how the vertical stress might vary

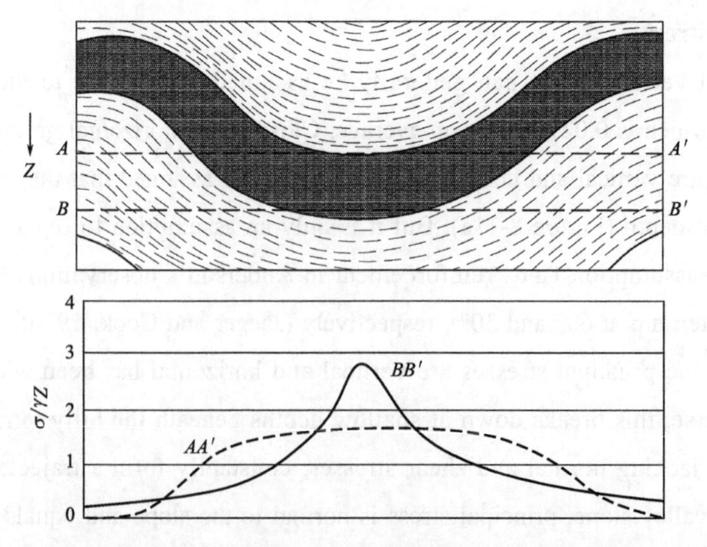

Figure 8-12 The influence of folds in heterogeneous, layered rock on vertical stresses (Goodman,1989)

along horizontal planes cutting through a succession of rigid and compliant beds folded into synclines and anticlines. Along line AA' the stress runs from perhaps 60% greater than γZ under the syncline to zero just beneath the anticline, the more rigid layer serving as a protective canopy and directing the flow of force down the limbs of the fold.

A tunnel driven along line BB' could expect to pass from relatively lower rock in the compliant shales to highly stressed rock as it crossed into more rigid sandstone in passing under the trough of the syncline. If there is a low-strength sheared zone along the contact, produced by slipping between the layers during folding, the vertical stress could be expected to jump in crossing the contact. Since geological structure can alter the vertical stresses and the direction of principal stresses, it is wise to investigate geological effects through analysis in critical applications wherever geological heterogeneities can be expected to deflect the lines of force away from the vertical. Figure 8-13 shows the valley (Jinping Power Station located) stress analysis, performed using the DEM (Distinct Element method), in a region with stream erosion formed a sharply notched topography.

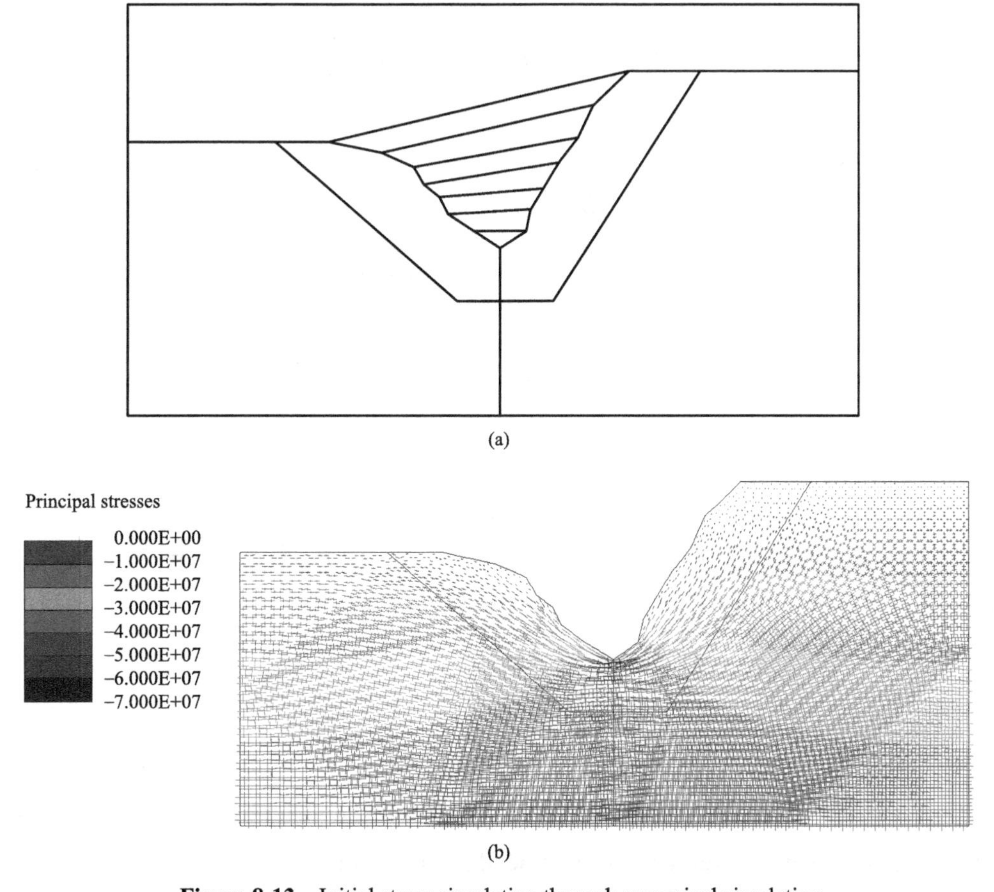

Figure 8-13 Initial stress simulation through numerical simulation

(a) Model before valley formation;(b) Principal stress distribution after valley formation

2. Horizontal stress

Regarding to the magnitude of the horizontal stresses, it is convenient to discuss the ratio of horizontal to vertical stresses. Let

$$K=\frac{\sigma_h}{\sigma_v} \tag{8-2}$$

In a region of recent sedimentation, such as the Mississippi Delta, the theory of elasticity can be invoked to predict that K will be equal to $v/(1-v)$. This expression derives from the symmetry of one-dimensional loading of an elastic material over a continuous plane surface, which infers a condition of no horizontal strain; such a formula has no validity in a rock mass that has experienced cycles of loading and unloading. Consider an element of rock at depth Z_0 with the initial value of $K = K_0$, which is then subjected to unloading by removal of ΔZ thickness of overburden (Figure 8-14). Due to the unloading of $\gamma\Delta Z$ vertical stress, the horizontal stress is reduced by $\gamma\Delta Zv/(1-v)$. Therefore, after erosion of a thickness of rock equal to ΔZ, the horizontal stress at depth $Z = Z_0-\Delta Z$ will become equal to $K_0\ \Delta Z_0 - \gamma\Delta Zv/(1-v)$, and

$$K(Z)=K_0+\left[\left(K_0-\frac{v}{1-v}\right)\Delta Z\right]\frac{1}{Z} \tag{8-3}$$

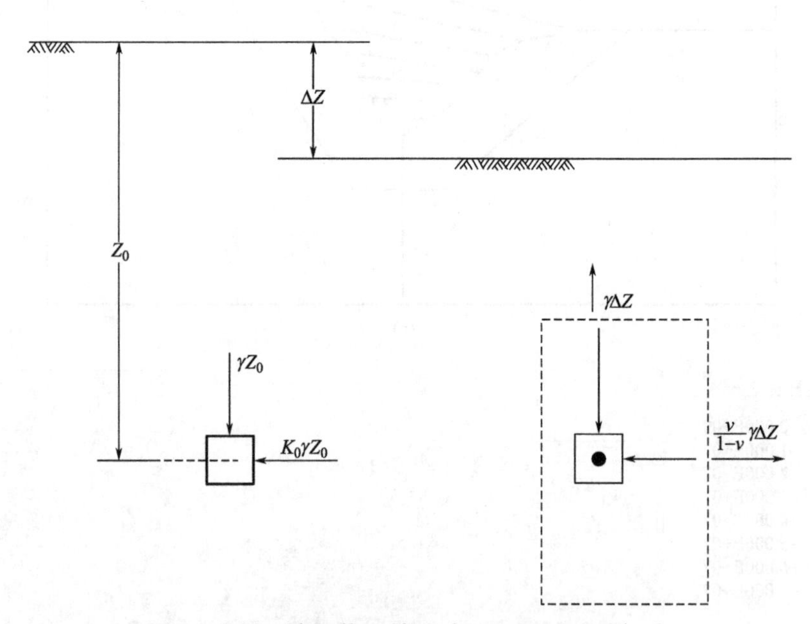

Figure 8-14 The effect of erosion on stresses at depth

Thus, erosion of overlying rock will tend to increase the value of K, the horizontal stress becoming greater than the vertical stress at depths less than a specific value. Other arguments can generate the hyperbolic relationship for $K(Z)$ predicted by Eq. (8-3). While the vertical stress is known to equal γZ, the horizontal stress could lie anywhere in the range of values between the two extremes $K_a\sigma_v$ and $K_P\sigma_v$ shown in Figure 8-15. K_a corresponds to conditions for normal faulting (Figure 8-15b), in which the vertical stress is the major principal stress and failure is by horizontal extension. Assuming Coulomb's law

$$K_a = \mathrm{ctan}^2\left(45+\frac{\phi}{2}\right)-\left[\left(\frac{q_u}{\gamma}\right)\mathrm{ctan}^2\left(45+\frac{\phi}{2}\right)\right]\cdot\frac{1}{Z} \tag{8-4}$$

K_P corresponds to conditions for reverse faulting (Figure 8-15c), in which the vertical stress is the minor principal stress and failure is by horizontal compression, giving

$$K_p = \tan^2\left(45+\frac{\phi}{2}\right)+\frac{q_u}{\gamma}\cdot\frac{1}{Z} \tag{8-5}$$

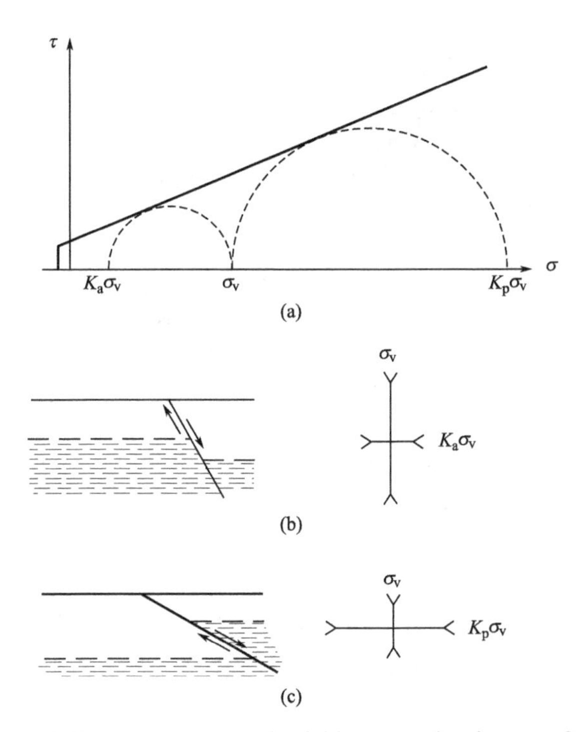

Figure 8-15 Stresses required to initiate normal and reverse faults

If there is no existing fault, we observe that the range of possible values of K such that $K_a \leqslant K \leqslant K_p$ is quite vast. However, near a preexisting fault, q_u can be assumed equal to zero, and the range of K is considerably reduced. Although tension is possible, it has rarely been measured and is to be considered an unusual situation.

Brown and Hoek (1978) examined many published values of in situ stress (Figure 8-16b) and independently discerned a hyperbolic relation for the limits of $K(Z)$, as

$$0.3+\frac{100}{Z}<\overline{K}<0.5+\frac{1500}{Z} \tag{8-6}$$

where Z is the depth in meters and K is the ratio of average horizontal stress to vertical stress. The range in extreme values of K given by this empirical criterion is considerably less than the range K_a to K_P given by Eq. (8-4) and Eq. (8-5) when q_u is not equal to zero, partly because average horizontal stress is being considered. In contrast, the previous criteria refer to maximum and minimum values of horizontal stress. In any event, all the equations for $K(Z)$ presented and the measured data are consistently found to be inverse with Z. Thus, even without measurements, one can estimate, within broad limits, the variation of horizontal stress with depth. While the magnitude of the horizontal stress might be estimated only approximately, it is often possible to offer reasonable estimates for the directions of the horizontal stresses.

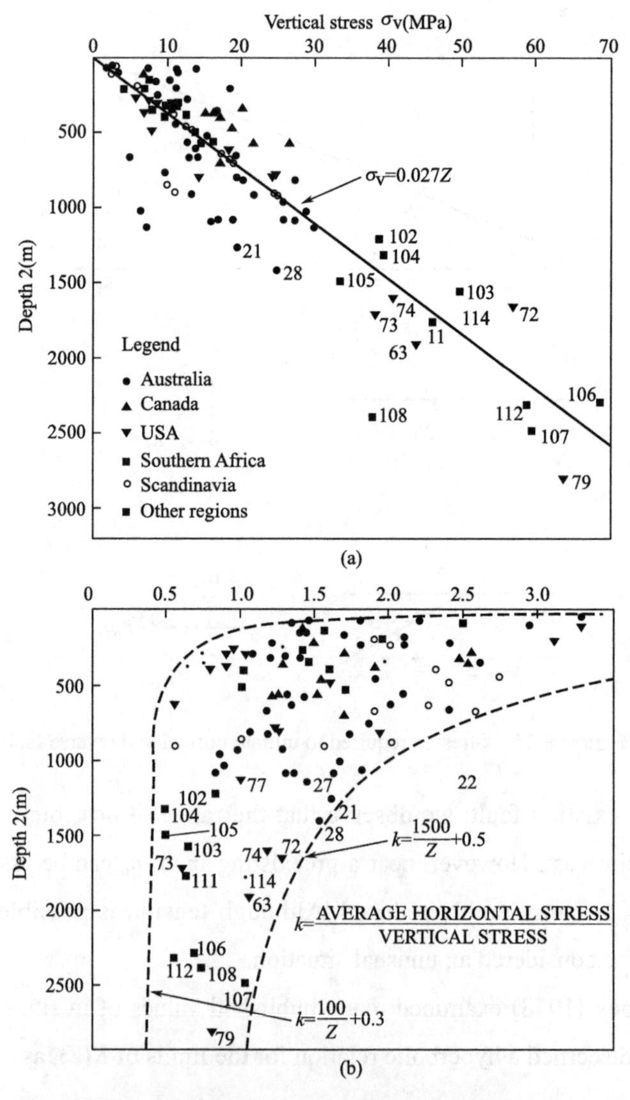

Figure 8-16 Results of stress measurements

(a) vertical stresses; (b) average horizontal stresses (form Brown & Hoek, 1978)

3. Horizontal stress direction

If the present state of stress is a remnant of that which caused visible geological structure, it will be possible to infer the directions of stresses from geological observations. Figure 8-17 shows the relationship between principal stress directions and different types of structures. The state of stress that causes a normal fault has σ_1 vertical, and σ_3 horizontal pointed perpendicularly to the fault trace as seen in the plan. In the case of reverse faulting, the stresses that caused the rupture have σ_3 vertical, while σ_1 is horizontal and directed perpendicular to the fault trace. Axial planes of folds also define the plane of greatest principal stress. Strike-slip faults are created by a state of stress in which σ_1 is horizontal and inclined about 30° with the fault trace, clockwise or counterclockwise as dictated by the sense of motion on the fault. These directions of horizontal stresses are not those of crustal blocks caught and squeezed between pairs of parallel faults; in such blocks, the primary stress state of the crust that is linked directly to the primary rupture surfaces will have superimposed on it the effects of the strain from accumulated fault motions.

Another approach to determining stress directions comes from the occurrence of rock breakage on the walls of wells and boreholes, which tends to create opposed zones of enlargement, termed "breakouts." These features can be seen in caliper logs, photographs, and televiewer logs of boreholes and be aligned from hole to hole in a region. Haimson and Herrick (1985) reported experimental results confirming that breakouts occur along the ends of a borehole diameter aligned with the least horizontal stress.

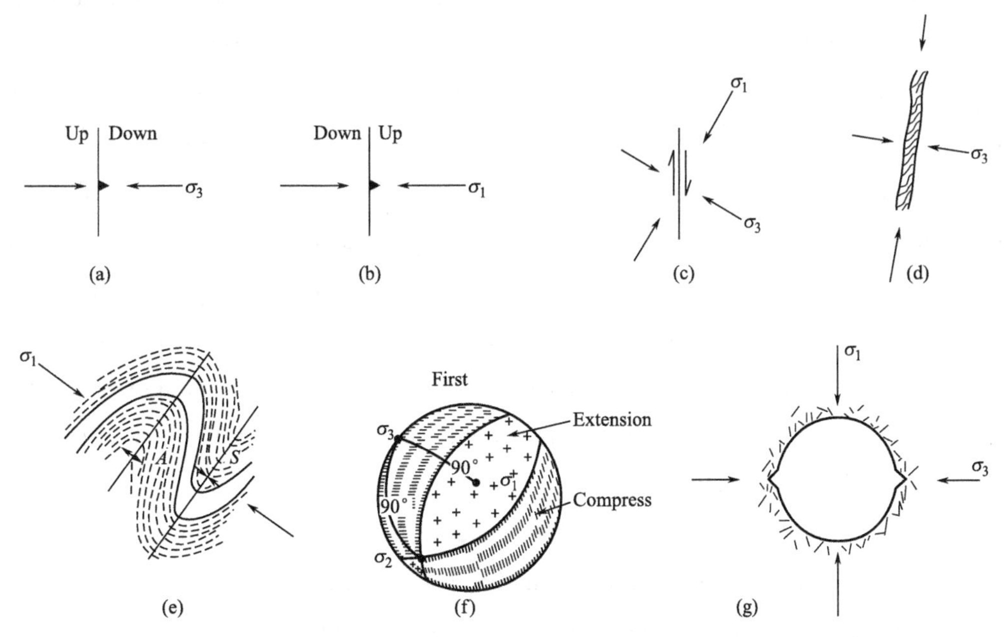

Figure 8-17 Directions of stress(Goodman,1989)

8.5.3 Techniques for Measurement of In-Situ Stresses

Stresses in situ can be measured in boreholes, on outcrops, and in the walls of underground galleries and back-calculated from displacements measured underground. Three of the best-known and most used techniques are hydraulic fracturing, the flat jack method, and overcoring. As will be seen, they are complementary to each other, each offering different advantages and disadvantages. All stress measurement techniques perturb the rock to create a response that can then be measured and analyzed, using a theoretical model to estimate part of the in-situ stress tensor. In the hydraulic fracturing technique, the rock is cracked by pumping water into a borehole; the known tensile strength of the rock and the inferred concentration of stress at the wellbore are processed to yield the initial stresses in the plane perpendicular to the borehole. In the flat jack test, the rock is partly unloaded by cutting a slot and then reloaded; the in-situ stress normal to the slot is related to the pressure required to null the displacement resulting from slot cutting.

In the overcoring test, the rock is entirely unloaded by drilling out a large core sample. In contrast, radial displacements or surface strains of the rock are monitored in a central, parallel borehole. Analysis using an unloaded thick-walled cylinder model yields stress displacement is measured. Precisions are seldom great, and the results are usually considered satisfactory if they are internally consistent and yield values believed to be correct to within about 0.3MPa. The main problem of all stress measurement techniques is that the measurement must be conducted in a region that has been disturbed in gaining access for the measurement; this paradox is handled by accounting for the effect of the disturbance in the analytical technique shown below.

The hydraulic fracturing method makes it possible to estimate the stresses in the rock at considerable depth using boreholes. Water is pumped into a section of the borehole isolated by packers. As the water pressure increases, the initial compressive stresses on the walls of the borehole are reduced and, at some points, become tensile. When the stress reaches $-T_0$, a crack is formed; the down-hole water pressure at this point is P_{c1} (Figure 8-18a). If pumping is continued, the crack will extend, and eventually, the pressure down the hole will fall to a steady value P_s, sometimes called "the shut-in pressure."

To interpret the data from the hydraulic fracturing experiment in terms of initial stresses, we need to determine the orientation of the hydraulically induced fracture.The most significant amount of information coincides with the case of a vertical fracture, and this is the usual result when conducting tests below about 800m. The orientation of a fracture could be observed by

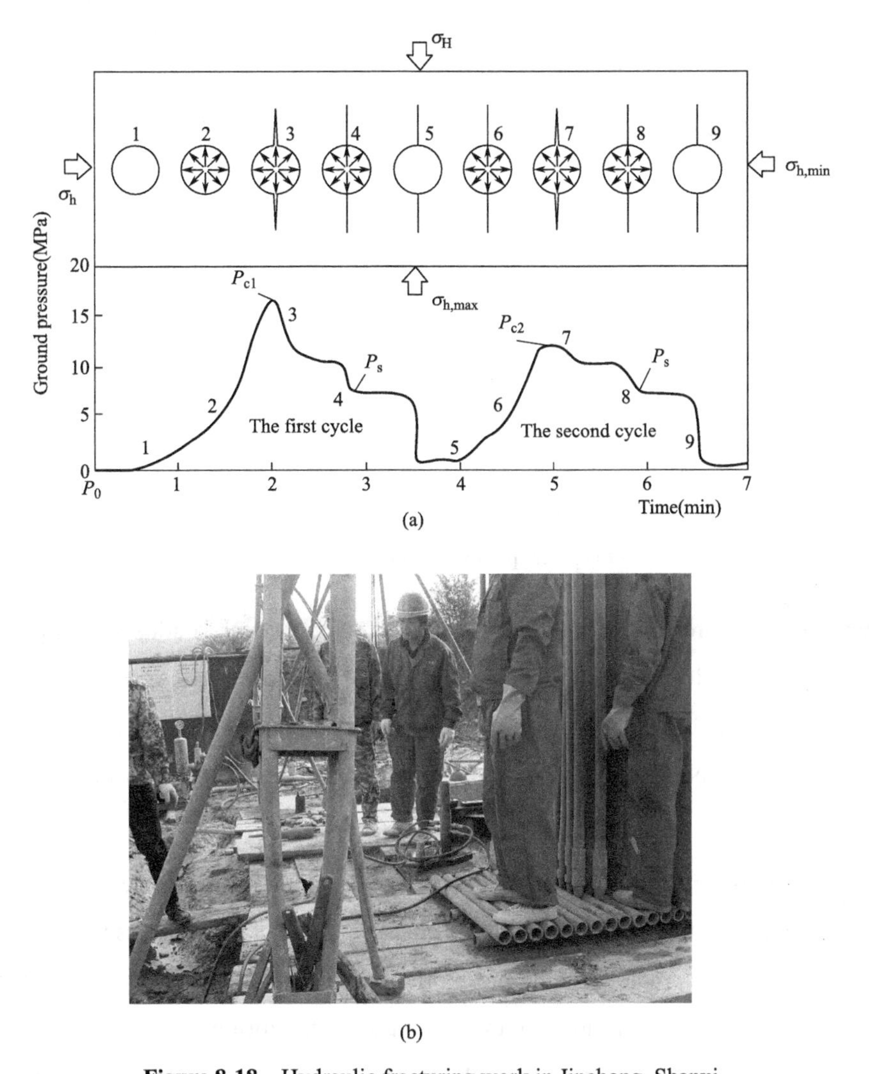

Figure 8-18 Hydraulic fracturing work in Jincheng, Shanxi

(a) Pressure versus time data as water is pumped into the packed-off section; (b) Experiment in progress

using downhole photography or television; however, a crack that closes upon depressuring the hole to admit the camera would be difficult to see in the photograph.

The analysis of the pressure test is simplified if it is assumed that penetration of the water into the pores of the rock has little or no effect on the stresses around the hole. Making such an assumption, it is possible to use the results of the known distribution of stress around a circular hole in a homogeneous, elastic, isotropic rock [Kirsch solution, Eq. (11-1)] to compute the initial stresses at the point of fracture. The tangential stress on the wall of the hole reaches the least magnitude at A and A' (Figure 8-19) where it is

$$\sigma_\theta=3\sigma_{h,min}-\sigma_{h,max} \tag{8-7}$$

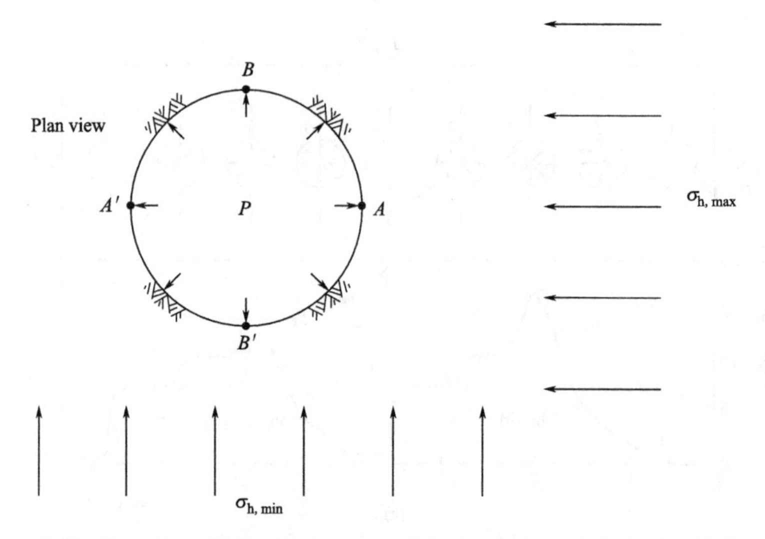

Figure 8-19 Location of critical points around the borehole used for hydraulic fracture

When the water pressure in the borehole is P, a tensile stress is added at all points around the hole equal (algebraically) to $-P$. The conditions for a new vertical tensile crack are that the tensile stress at point A should become equal to the tensile strength $-T_0$. Applying this to the hydraulic fracturing experiment yields as a condition for the creation of a hydraulic fracture

$$3\sigma_{h,min}-\sigma_{h,max}-P=-T_0 \tag{8-8}$$

Once formed, the crack will continue to propagate as long as the pressure is greater than the stress normal to the fracture plane. If the water pressure in the crack was less than or greater than the normal stress on this crack, it would close or open accordingly. In rocks, cracks propagate in the plane perpendicular to σ_3. In the context of hydraulic fracturing with a vertical fracture, this means that the stress normal to the plane of the fracture is equal to the shut-in pressure P_s

$$\sigma_{h,min}=P_s \tag{8-9}$$

Eq. (8-8) and Eq. (8-9) allow the major and minor normal stresses in the plane perpendicular to the borehole to be determined if the tensile strength of the rock is known. If the borehole pressure is dropped and once again raised above the value P_s, the hydraulic fracture will close and then reopen. Let the new peak pressure, smaller than P_{c1}, be called P_{c2}. Replacing T_0, and P_{c1} of Eq. (8-8) with the values 0 and P_{c2}, respectively, and subtracting Equation (8-8) from the resulting equation yields a formula for the tensile strength of the rock around the borehole applicable to the conditions of the experiment

$$T_0=P_{c1}-P_{c2} \tag{8-10}$$

Assuming that the vertical stress equals gravity (γZ) and is a principal stress, the state of stress is now completely known. The experiment yields the major and minor normal stress values and directions in the plane perpendicular to the borehole.

If the rock is porous, water will enter cracks and pores, creating an internal pressure gradient, whereas the theory above presumed a sudden pressure drop across the borehole wall. The effect is to lower the value of P_{c1} and round the peak of Figure 8-18. Haimson (1978) shows how to modify the analysis to solve the principal stresses in this case.

8.6 Rock mass classification

During the probability and preliminary design steps of a project, when very scarce detailed data is ready on the rock mass, and its stress and hydrologic properties, the use of a rock mass classification scheme can be of significant benefit. In its simplest form, it may be used as a checklist to ensure that all relevant information has been taken into account.

It is essential to understand the limitations of rock mass classification schemes (Palmstrom and Broch, 2006). Their use does not (and cannot) replace some of the more elaborate design procedures. However, using these design schemes requires relatively complete information on in situ stresses, rock mass properties, and planned excavation order, none of which may be available early in the project. As this information becomes feasible, the use of rock mass classification schemes should be kept up-to-date and used in conjunction with site-specific analyses.

8.6.1 Engineering rock mass classification

Rock mass classification systems have been advancing for over 100 years since Ritter (1879) tried to formalize an empirical approach to tunnel design, particularly for determining support demands. While the classification schemes are suitable for their original utilization, primarily if used within the bounds of the case records from which they were developed, considerable aattention must be exercised in applying rock mass classifications to other rock engineering problems.

Most multiparameter classification systems have been based on civil engineering case studies, and all elements of the geological engineering feature of the rock mass have been included. In underground hard rock mining, however, particularly at deep levels, rock mass weathering and the influence of water usually are not meaningful and may be neglected. Different classification systems place different emphases on the various parameters. It is advised that at least two ways be used at any site during the early stages of a project.

1. Terzaghi's rock mass classification

The earliest attributing to the use of rock mass classification for the design of tunnel support is in a document by Terzaghi (1946) in which the rock loads, carried by steel sets, are estimated based on a detailed classification (Table 8-6). While no worthwhile goal would be served by including details of Terzaghi's category in this analysis on the design of support, it is fascinating to check the rock mass descriptions contained in his original article because he draws attention to those characteristics that dominate rock mass behavior, particularly in situations where gravity constitutes the dominant driving force. The clear and concise definitions and valuable comments in these descriptions are good examples of the type of engineering geology information that is most useful for technical design.

Table 8-6 Terzaghi's rockmass classification

Class	Type of Rocks	Descriptions
Ⅰ	Hard and intact	The rock is unweathered. It contains neither joints nor hair cracks. If fractured, it breaks across intact rock. After excavation, the rock may have some popping and spalling failures from roof. At high stresses spontaneous and violent spalling of rock slabs may occur from the side or the roof. The unconfined compressive strength is equal to or more than 100MPa
Ⅱ	Hard stratified and schistose	The rock is hard and layered. The layers are usually widely separated. The rock may or may not have planes of weakness. In such rocks, spalling is quite common
Ⅲ	Massive, moderately jointed	A jointed rock, the joints are widely spaced. The joints may or may not be cemented. It may also contain hair cracks but the huge blocks between the joints are intimately interlocked so that vertical walls do not require lateral support. Spalling may occur
Ⅳ	Moderately blocky and seamy	Joints are less spaced. Blocks are about 1m in size. The rock may or may not be hard. The joints may or may not be healed but the interlocking is so intimate that no side pressure is exerted or expected
Ⅴ	Very blocky and seamy	Closely spaced joints. Block size is less than 1m.It consists of almost chemically intact rock fragments which are entirely separated from each other and imperfectly interlocked. Some side pressure of low magnitude is expected. Vertical walls may require supports
Ⅵ	Completely crushed but chemically intact	Comprises chemically intact rock having the character of a crusher-run aggregate. There is no interlocking. Considerable side pressure is expected on tunnel supports. The block size could be few centimeters to 30cm
Ⅶ	Squeezing rock – moderate depth	Squeezing is a mechanical process in which the rock advances into the tunnel opening without perceptible increase in volume. Moderate depth is a relative term and could be from 150 to 1000m
Ⅷ	Squeezing rock – great depth	The depth may be more than 150 m. The maximum recommended tunnel depth is 1000m
Ⅸ	Swelling rock	Swelling is associated with volume change and is due to chemical change of the rock, usually in presence of moisture or water. Some shales absorb moisture from air and swell. Rocks containing swelling minerals such as montmorillonite, illite, kaolinite and others can swell and exert heavy pressure on rock supports

2. Rock quality designation index (*RQD*)

The Rock Quality Designation index (*RQD*) was produced by Deere (Deere et al., 1967) to present a quantitative assessment of rock mass quality from drill core logs. *RQD* is described as the percentage of intact core pieces longer than 100mm (4 inches) in the whole length of the core. The core must be at least NW size (54.7mm or 2.15 inches in diameter) and drilled with a two-tube barrel. The correct procedures for measuring the length of core pieces and the calculation of *RQD* are summarized in Figure 8-20.

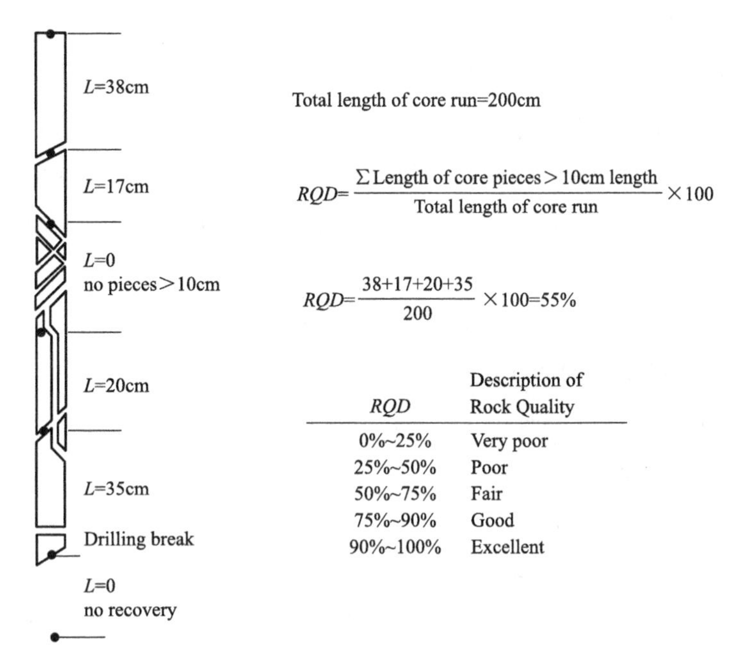

Figure 8-20 Procedure for measurement and calculation of *RQD* (After Deere, 1989)

Palmström (1982) suggested that, when no core is available but discontinuity traces are visible in surface exposures or exploration adits, the *RQD* may be estimated from the number of discontinuities per unit volume. The suggested relationship for clay-free rock masses is:

$$RQD = 115 - 3.3\,J_v \tag{8-11}$$

where J_v is the sum of the number of joints per unit length for all joint (discontinuity) sets known as the volumetric joint count.

The *RQD* is a direction-dependent parameter, and its value can change significantly, depending on the direction of the drill. The use of the volumetric joint count can be pretty helpful in reducing this directional dependence. The *RQD* is designed to represent the quality of the bedrock in situ. When using diamond drill core, care must be taken to ensure that fractures, which have been caused by handling or the drilling process, are identified and ignored when determining the value of *RQD*. When using the Palmström relation for exposure mapping, blast fractures should not be included in the J_v estimation.

Deere's *RQD* was extensively used, particularly in North America, after its presentation.

Cording and Deere (1972), and Deere and Deere (1988) tried to relate *RQD* to Terzaghi's rock load factors and rock bolt requirements in tunnels.

8.6.2 Geomechanics Classification

Bieniawski (1976) announced the specifications of a rock mass classification named the Geomechanics Classification or the Rock Mass Rating (*RMR*) system. Over the years, this method has been successively improved as more case records have been examined. It should be aware that Bieniawski has made significant changes in the ratings assigned to different parameters (Bieniawski, 1989). The following six parameters are used to assign a rock mass using the *RMR* system:

(1) Uniaxial compressive strength of rock material.

(2) Rock Quality Designation (*RQD*).

(3) Spacing of discontinuities.

(4) Condition of discontinuities.

(5) Groundwater conditions.

(6) Orientation of discontinuities.

In utilizing this classification system, the rock mass is divided into many structural regions, and each area is classified separately. The boundaries of the structural regions usually match with a significant structural feature such as a fault or a change in rock type. In some cases, substantial alterations in discontinuity spacing or features within the same rock type may require dividing the rock mass into some small structural regions.

8.6.3 Rock Tunnelling Quality Index Q

Based on an evaluation of a large number of case histories of underground excavations, Barton et al. (1974) of the Norwegian Geotechnical Institute introduced a Tunnelling Quality Index (Q) for the ascertainment of rock mass properties and tunnel support demands(Table 8-7). The numerical value of the index Q varies on a logarithmic scale from 0.001 to a maximum of 1,000 and is defined by:

$$Q=\frac{RQD}{J_n}\times\frac{J_r}{J_a}\times\frac{J_w}{SRF} \tag{8-12}$$

where RQD is the Rock Quality Designation; J_n is the joint set number; J_r is the joint roughness number; J_a is the joint alteration number; J_w is the joint water reduction factor; SRF is the stress reduction factor.

In describing the definition of the parameters used to determine the value of Q, Barton et al. (1974) offer the following explanations:

The first quotient (RQD/J_n), representing the rock mass structure, is a rough test of the block or particle size, with the two extreme values (100/0.5 and 10/20) differing by a factor of 400. If the quotient is defined in units of centimeters, the extreme "particle sizes" of 200 to 0.5cm are seen to be crude but fairly practical approximations. Probably the largest blocks should be several times this size and the smallest fragments less than half the size.

The second quotient (J_r/J_a) expresses the roughness and friction characteristics of the joint walls or filling materials. The quotient is weighted in favor of rough, undamaged joints in direct contact. It is expected that such surfaces will be close to maximum strength, will expand sharply when sheared, and will therefore be particularly conducive to tunnel stability. When the rock joints have fine clayey mineral coatings and fillings, the strength is drastically reduced. However, contact of the rock face after small shear movements can play a critical role in preserving the excavation of the final failure. In the absence of contact with the rock wall, the conditions are very unfavorable to the stability of the tunnel. The "friction angles" are a little beneath the remaining strength values for most clays and are probably down-graded by the fact that these clay strips or fillings may tend to consolidate when shear, at least if normal consolidation or if softening and swelling has happened. The swelling pressure of montmorillonite may also be a portion in this regard.

The third quotient (J_w/SRF) is compose of two stress parameters. SRF is a measure of:

(1) Loosening load in the event of an excavation through shear zones and clay-bearing rock.

(2) Rock stress in competent rock.

(3) Squeezing loads in incompetent plastic rocks.

It can be regarded as a total stress parameter. The parameter J_w is a means of water pressure, which harms the shear strength of joints due to a decrease in effective normal stress. Water may, in addition, cause softening and possible out-wash in the case of clay-filled joints. Combining these two parameters in terms of inter-block effective stress has proved impossible because paradoxically, a high value of effective normal stress may sometimes signify less stable conditions than a low value, despite the higher shear strength. The quotient (J_w/SRF) is a complicated empirical factor describing the active stress.

Table 8-7 Classification of rock mass based on Tunnelling Quality Index Q

Q	>400	100~400	40~100	10~40	4~10	1~4	0.1~1	0.01~0.1	<0.01
Class	Exceptionally good	Extremely good	Very good	good	fair	poor	Very poor	Extremely poor	Exceptionally poor

8.6.4 China's engineering rock mass classification standards

In the early 1990s, Chinese researchers summarized the experience of domestic rock mass classification and collected 460 sets of measured values from 103 domestic projects for statistical analysis. It is finally determined that the uniaxial saturation compressive strength of the rock and the integrity Coefficient of the rock mass, the two parameters are used to evaluate the essential quality of rock mass. In 2014, the MOHUR issued the latest national standard, "Engineering Rock Mass Classification Standard" GB/T 50218—2014.

1. Basic quality classification of rock mass

The basic quality *BQ* of the rock mass is calculated according to the following formula:

$$BQ=100+3R_c+250K_v \tag{8-13}$$

Where: BQ is the basic quality index of the rock mass; R_c is the uniaxial saturated compressive strength of the rock (MPa); K_v is the integrity coefficient of the rock mass, that is, the square of the ratio of the longitudinal velocity of the rock mass to the longitudinal velocity of the rock.

To make the weight reasonable, when $R_c>90K_v+30$, it should be calculated by $R_c=90K_v+30$ and K_v; when $K_v>0.04R_c+0.4$, it should be calculated by the formula of $K_v=0.04R_c+0.4$ and R_c. After obtaining the BQ value according to Eq. (8-13), the basic quality level of the rock mass can be determined according to Table 8-8.

Table 8-8 Basic quality classification of rock mass

Class	Qualitative characteristics of the basic quality of rock mass	Basic quality index of rock mass
I	Hard rock with intact rock mass	>550
II	Hard rock with rock mass of moderately intact; moderately hard rock with intact rock mass	550~451
III	Hard rock with rock mass of moderately broken; moderately hard rock with rock mass of moderately intact; soft rock with rock mass of intact	450~351
IV	Hard rock with rock mass of broken; moderately hard rock with rock mass of moderately broken to broken; softer rock with rock mass of moderately intact to moderately broken; soft rock with rock mass of intact to moderately intact	350~251
V	Soft rock with rock mass of broken; soft rock with rock mass of moderately broken to broken; all extremely soft rocks and all broken rocks	≤250

2. Determination of the quality of engineering rock mass

The detailed grading of engineering rock mass should be based on the basic quality classification of engineering rock mass, combined with the characteristics of different types of engineering, according to the groundwater state, initial stress state, the integrated relationship between the engineering axis, and the main discontinuities and other correction factors, to determine all kinds of engineering rock mass quality indicators.

The influence of the initial stress state of the rock mass on the level of the underground engineering rock mass should be based on the strength-to-stress ratio determined by the corresponding initial stress surrounding rock strength as the corrective control factor. When the initial stress state of the rock mass has a measured stress result, the measured value should be used. When there are no measured results, the evaluation can be made according to the special geological phenomena such as the engineering buried depth, excavation depth, topography, geotectonic movement history, main structural lines, a core cake in the drill hole, and rockburst in the excavation process and so on.

For the special rocks such as expansibility and solubles, the grade of engineering rock mass should be comprehensively determined according to its particular deformation and failure characteristics, the degree of karst development, and its influence on engineering.

Reference

[1] AYDAN Ö. Rock Mechanics and Rock Engineering:Volume 1: Fundamenta-ls of Rock Mechanics[M]. Leiden, CRC Press, 2019.

[2] BARTON N, LIEN R, LUNDE J. Engineering classification of rock masses for the design of tunnel support[J]. Rock Mechanics and Rock Engineering, 1974, 6(4):189-236. https://doi.org/10.1007/BF01239496.

[3] BIENIAWSKI Z T. Rock mass classification in rock engineering[J]. Exploration for rock engineering, proc. of the symp, 1976, 1: 97-106.

[4] BIENIAWSKI Z T. Engineering rock mass classifications: a complete manual for engineers and geologists in mining, civil, and petroleum engineering[M]. New York: John Wiley & Sons, 1989.

[5] BROWN E T, HOEK E. Trends in relationships between measured in-situ stresses and depth[J]. International Journal of Rock Mechanics & Mining Sciences & Geomechanics Abstracts, 1978, 15(4):211-215. https://www.rocscience.com/assets/resources/learning/hoek/Trends-in-Relationship-between-Measured-In-Situ-Stresses-and-Depth-1978.pdf.

[6] BS 5930: 1999. Code of practice for site investigations[S]. Bristish Standard Institution. London, 1999.

[7] BS 5930: 2015. Code of practice for ground investigations[S]. Bristish Standard Institution. London, 2015.

[8] COOK N. An experiment proving that dilatancy is a pervasive volumetric property

of brittle rock loaded to failure[J]. Rock mechanics, 1970, 2(4): 181-188. https://doi.org/10.1007/BF01245573.

[9] CORDING E J, DEERE D U.Rock tunnel supports and field measurements[C]. Proc. North American rapid excav. tunneling conf., Chicago, (eds. K.S. Lane and L.A. Garfield) 1, 601-622. New York: Soc. Min. Engrs, Am. Inst. Min. Metall. PetrolmEngrs, 1972.

[10] DEERE D U, HENDRON A J, PATTON F D, et al. Design of surface and near surface construction in rock[C]. In C. Fairhurst (Ed.), Failure and breakage of rock, proc. 8th U.S. symp. rock mech., 237-302. New York: Soc. Min. Engrs, Am. Inst. Min. Metall. Petrolm Engrs, 1967.

[11] DEERE D U, DEERE D W. The rock quality designation (RQD) index in practice[M]// Rock classification systems for engineering purposes. ASTM International, 1988.

[12] DEERE D U, DEERE D W. Rock Quality Designation (RQD) after Twenty Years[R]. DEERE (DON U) CONSULTANT GAINESVILLE FL, 1989. https://www.nrc.gov/docs/ml0037/ML003749192.pdf.

[13] DE Vallejo L G, FERRER M. Geological engineering[M]. CRC Press, 2011.

[14] DUNCAN C W. Foundations on rock[M]. London: E&FN SPON, 2005.

[15] GOODMAN R E. Engineering Geology-Rock in Engineering Construction[M]. New York: John Wiley & Sons, 1993.

[16] GOODMAN R E. Introduction to rock mechanics (Vol. 2)[M]. New York: John Wiley & Sons,1989.

[17] HARRISON J, HUDSON J, POPESCU M. Engineering Rock Mechanics: Part 2. Illustrative Worked Examples[J]. Applied Mechanics Reviews, 2002, 55(2).

[18] HAIMSON B C. The hydrofracturing stress measuring method and recent field results[J]. International Journal of Rock Mechanics & Mining Sciences & Geomechanics Abstracts, 1978, 15(4):167-178.

[19] HAIMSON B C, HERRICK C G. In situ stress evaluation from borehole breakouts. Experimental studies[C]//US symposium on rock mechanics. 26. 1985: 1207-1218.

[20] HOJEM J P M, COOK N G W, HEINS C. A stiff, two meganewton testing machine for measuring the 'work-softening' behaviour of brittle materials[J], South Afr. Mech. Eng. 1975, 25: 250-70.

[21] HOEK E. Practical rock engineering(2000 edition)[J], 2000. https://www.rocscience.com/assets/resources/learning/hoek/Practical-Rock-Engineering-Full-Text.pdf.

[22] HUDSON J A, HARRISON J P. Engineering rock mechanics: an introduction to the principles[M]. London: Elsevier, 2000.

[23] ISRM (International Society for Rock Mechanics). Basic geological description of rock masses[J]. Int. J. Rock Mech. Min. Sci. & Geomech. Abstr., 1981, 18: 85-110.

[24] JAEGER J C, COOK N. Fundamentals of Rock Mechanics[M]. 2nd ed. London: Chapman & Hall, 1976.

[25] JAEGER J C, COOK N, ZIMMERMAN R W. Fundamentals of Rock Mechanics[M]. 4th ed. Oxford: Blackwell, 2007.

[26] MOHURD (Ministry of Housing and Urban-Rural Development), PRC. Engineering Rock Mass Classification Standard: GB/T 50218-2014[S]. Beijing: China Planning Press, 2014. http://www.mohurd.gov.cn/wjfb/201508/t20150829_224347.html.

[27] MOHURD (Ministry of Housing and Urban-Rural Development), PRC. Code for hydropower engineering geological investigation: GB50287-2016[S]. Beijing: China Planning Press, 2016. http://www.mohurd.gov.cn/wjfb/201702/t20170224_230737.html.

[28] PALMSTROM A, BROCH E. Use and misuse of rock mass classification systems with particular reference to the Q-system[J]. Tunnelling and underground space technology, 2006, 21(6): 575-593.https://doi.org/10.1016/j.tust.2005.10.005.

[29] RITTER W. Die Statik der Tunnelgewölbe[M]. Berlin: Springer, 1879.

[30] TERZAGHI K. Rock defects and loads on tunnel supports[J]. Rock Tunneling with Steel Supports, 1946.

[31] ZHU W, ZHAO J. Stability analysis and modelling of underground excavations in fractured rocks[M]. Boston: Elsevier, 2003.

Exercises

Review Questions

(1) Describe the characteristics of the rock mass and the geological factors affecting the stability of the rock mass.

(2) Describe the concept of discontinuity and the structural characteristics of the rock mass.

(3) How to classify the types of discontinuities?

(4) Describe the basis of rock mass structure type classification and its engineering significance.

(5) Describe the composition and characteristics of natural stress in the rock mass.

(6) Describe the characteristics of geostress fields in valley area.

(7) What is the engineering significance of in-situ stress research?

(8) Describe the concept and grading standards of RQD.

(9) Describe the calculation method of rock mass basic quality BQ.

(10) The hydraulic fracturing technique has measured the stress in a granitic rock mass (Harrison & Hudson, 2000). Two tests were conducted in a vertical borehole: one test at a depth of 500m and the other test at a depth of 1000m. The results were as follows:

Depth (m)	Breakdown pressure, P_B (MPa)	Shut-in pressure, P_S (MPa)
500	14.0	8.0
1000	24.5	16.0

Given that the tensile strength, σ_t, of the rock is 10MPa, estimate and list the values of σ_1, σ_2 and σ_3 at the two depths. State all of the assumptions you have to make in order to produce these estimates. Are any of them doubtful? State whether the two sets of results are consistent with each other, and justify your reasons for the statement. Are the results in agreement with trends exhibited by collated worldwide data?

9
Engineering Geological Study of Dam Foundations

9.1 Introduction

Compared to soils, most rocks are solid and stiff, and carrying a structural load down to the rock basement usually assures a suitable bearing. However, enormous loads of skyscrapers or bridge piers can cause pressures approaching the bearing capacity of even strong rocks. Homes, warehouses, and other lightweight structures rarely create large loads, even weak rocks, but may require geological surveys in cavernous or mined-out substrata or areas of expansive rocks. Bridges require foundations to be constructed through water and soil to bedrock and place piers on steep valley sides where rock slope stability analysis becomes part of the foundation engineering work (Figure 9-1a).

The dam is one of the most critical development infrastructures in any country. They are used for many different purposes, including irrigation, water supply, flood control, electricity production, storage of mine wastes, etc. The dam is the most critical hydraulic structure in a water conservancy complex. It intercepts water flow, raises the water level, and bears immense water pressure and other loads. The dam body transmits water pressure and other loads and its weight to the foundation or the rock masses on both sides of the dam, so the pressure on the rock mass is vast. In addition, water can also infiltrate into the rock mass, soften, muddy, and dissolve certain rock formations and produce uplift pressure that is not conducive to stability. Therefore, dam construction has high requirements for the stability of the foundation rock mass. It is also true that dams can create relatively large inclined loads at their base and valley side abutments. Concrete arch dams transfer some of the reservoir and structural load to the abutment rock (Figure 9-1b). In contrast, concrete gravity and concrete buttress dams primarily direct the foundation rock load. Earth and rock-fill dams create minor, usually tolerable stresses and deformations in rock foundations.

The relationship between dam safety and geology has been one of the most important

(a)

(b)

Figure 9-1 Foundations of a bridge and a dam in very steep terrain

(a) The U. S. Bureau of Reclamation built the Glen Canyon Bridge across a steep canyon in Navajo sandstone; (b) Glen Canyon arch dam

research topics in rock mechanics and geological engineering since the 1960s. Many of the most severe accidents have been due to foundation problems. The failures of the Malpasset Dam (France) in 1959, the Vajont Dam (Italy) in 1963, and the Teton Dam (USA) in 1975 are among the best known. Although the problems were different in each case, all three failures were related to geological conditions. In Malpasset, the structure of the gneiss and schist rock mass, dipping downstream, was conducive to the stability of the dam and its foundations as a whole. Still, the geological surveys for the project did not identify or at least did not take sufficiently into consideration a fault transversal to the schistosity on the left slope. When the reservoir was filled with water, the resulting pore pressures created the required conditions for the failure of the foundation defined by these discontinuities. The foundation failure of the 66.5 m high dam caused the dam's collapse and immediate emptying of the reservoir. The resulting flooding caused 421 deaths in the town of Fréjus.

At present, significant advances in engineering geology and rock mechanics mean that geological surveys for dam projects can detect possible foundations and slope stability problems, analyze how these could affect the safety of the dams, and allow the necessary corrective measures to be taken (de Vallejo & Ferrer, 2011).

9.2 Compression deformation and bearing capacity of dam foundation rock mass

The design of a foundation requires that the bearing pressure and bond (adhesion) allowable in each geological unit be established for the base and sides of the foundation member. The values selected must have a margin of safety against loss of load-carrying capacity (bearing "failure") and must work without large deflections. These values are usually taken from building codes in routine work, which provide safe conservative pressures and reflect regional experience(Goodman, 1989).

9.2.1 Deformability of dam foundation rock

Deformability means the capacity of a rock to strain under applied loads or in response to unloads on excavation. The primary forms of rock mass instability that lead to the failure of the dam foundation are compression deformation and sliding failure. For gravity dams, compression deformation is mainly caused by the settlement of the dam foundation. In contrast, for arch dams, in addition to the settlement and deformation of the dam foundation, there are also near-horizontal deformations caused by the thrust direction of the arch end. The deformation modulus is very high for a challenging and contact rock mass, and the compression deformation is small. When the deformation is uniform, it has no evident impact on the safety and stability of the dam. However, when uneven subsidence occurs, or the rock mass on one bank deforms considerably, tensile stress may be generated in the dam body, which may cause cracks and even destroy the entire dam body. In particular, the arch dam is particularly sensitive to the uneven deformation of the rock masses on both sides of the bank, so the requirements are rigorous. The geological factors that cause uneven deformation are mainly the following three aspects.

(1) The lithology is different in soft and hard rock, and the deformation modulus is also different. As shown in Figure 9-2, a dam seated astride varying rock types having dissimilar deformability properties. As a result, large unequal deflections are caused, which leads to shear and diagonal tension stresses in the dam body. Under normal circumstances, clay shale, mudstone, strongly weathered rocks, and loose sediments, especially silt, cohesive soil layers with significant water content, etc., are rocks that are prone to have substantial subsidence and

deformation.

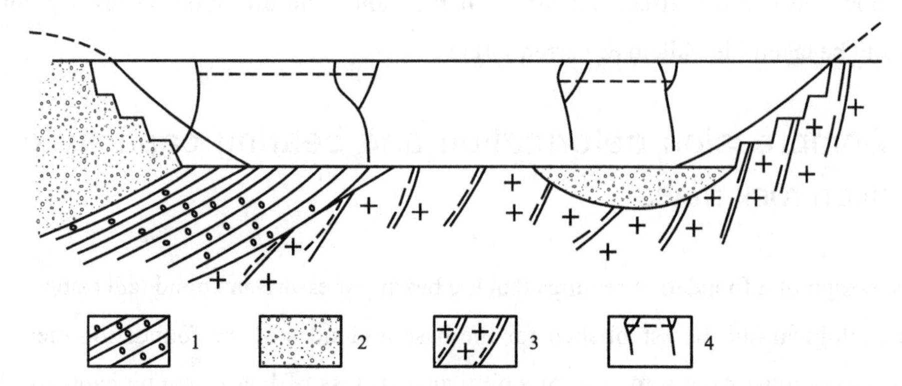

Figure 9-2 Cross-section of a concrete dam with variable deformability in the foundation rock

1—Gravel-bearing claystone; 2—Gravel; 3—Granite gneiss; 4—Subsidence and cracks

(2) When there are large fault fracture zone, dense fracture zone, unloading fracture zone and other soft structural planes in the rock mass of dam foundation or bank, especially when the opening fracture is developed and the fracture plane is roughly perpendicular to the direction of pressure, it is easy to produce large subsidence deformation.

(3) There are dissolution caves or submerged hollowing in the rock body, resulting in collapse and uneven deformation.

The occurrence and distribution of the above-mentioned weak rock formations and weak structural planes also have a significant impact on the deformation of the rock mass. For example, when weak rock formations are distributed on the surface, large subsidence and deformation are prone to occur. When they are distributed near the dam toe (Figure 9-3), it is easy to cause the dam body to tilt and overturn downstream; and when they are distributed near the dam heel, it is easy to cause the rock mass to break.

9.2.2 Dam foundation rock mass bearing capacity

The bearing capacity of the dam foundation rock mass refers to the maximum load pressure that the foundation can withstand to ensure the safety and stability of the building, so it is also called the allowable bearing capacity. It includes not only not allowing the damage caused by excessive subsidence and deformation but also not allowing the damage caused by the fracture or shear slip of the foundation rock mass. So it is a comprehensive indicator. It is mainly used in the initial stage of design or small-scale projects where the geological conditions are relatively simple. Large, medium-sized, or important hydraulic structures must calculate the settlement deformation and anti-sliding stability according to the deformation test or shear test index.

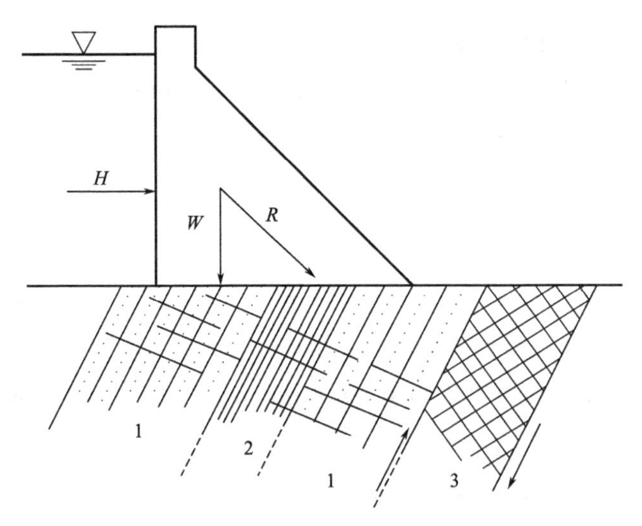

Figure 9-3 Transverse section of dam foundation with uneven lithology

1—sandstone; 2—shale; 3—fault rock

The determination of the bearing capacity of rock foundation mainly includes three methods: empirical analogy, reduction based on compressive strength, and field load test.

1. Empirical analogy

The empirical analogy method is an empirical method for determining the bearing capacity of a dam foundation. It is selected based on the experience data of the completed projects, project samples, and geological conditions. Some domestic codes list the value of the bearing capacity of the rock foundation. For example, the basic values of the bearing capacity of bedrock listed in the "Engineering Rock Mass Classification Standard"(MOHURD,2014) are as shown in Table 9-1. The bearing capacity is shown in Table 9-2.

Table 9-1 Basic value of dam foundation bearing capacity f_0

Rock mass level	I	II	III	IV	V
f_0 (MPa)	>7.0	7.0~4.0	4.0~2.0	2.0~0.5	<0.5

Table 9-2 Rock allowable bearing capacity table(kPa)

Joints development degree	Joints developed well	Joints developed fairly	Joints developed poorly
Joint spacing (cm) / Rock	2~20	20~40	>40
Hard rock	1500~2000	2000~3000	>3000
Relatively soft rock	800~1000	1000~1500	1500~3000
Soft rock	500~800	700~1000	900~1200
Very soft rock	200~300	300~400	400~500

2. Compressive strength reduction method

Multiplying the uniaxial saturated compressive strength (R_b) of the rock by the reduction factor (ψ) to find the bearing capacity is the most widely used simple method. Its calculation formula:

$$f=\psi R_b \tag{9-1}$$

The "Code for Design of Building Foundation" (MOHURD, 2012) stipulates that the value of the reduction coefficient ψ in the formula shall be determined by local experience based on the integrity of the rock mass and the spacing, width, occurrence, and combination of structural planes. When inexperienced, 0.5 can be used for complete rocks, 0.2 to 0.5 for relatively intact rocks, and 0.1 to 0.2 for rather broken rocks. Table 9-3 introduces the commonly used reduction coefficient value methods in water conservancy and hydropower projects.

Table 9-3 Empirical method for determining allowable bearing capacity of dam foundation

Rock	Undeveloped joints (Spacing>1.0m)	Joints relatively developed (Spacing 1~0.3m)	Joint developed (Spacing 0.3~0.1m)	Joints well developed (Spacing<0.1 m)
Hard and semi-hard rock (R_b>30MPa)	1/7 R_b	(1/7~1/10)R_b	(1/10~1/16) R_b	(1/16~1/20) R_b
Weak rock (R_b<30MPa)	1/5R_b	(1/5~1/7) R_b	(1/7~1/10) R_b	(1/10~1/15) R_b

The compressive strength multiplied by the reduction factor is a relatively rough method, which is only suitable for the initial design stage or medium and small water conservancy projects. The bearing capacity converted by the value of ψ is often conservative. If conditions permit, it can be determined by triaxial compression test or load test calculation.

3. Field load test

This is an in-situ test based on the size and direction of the engineering force that the rock mass actually bears, and it is more in line with the actual situation. The test can measure the elastic modulus, deformation modulus, and Poisson's ratio of the rock mass, which can be used to calculate the subsidence of the dam foundation. This method is accurate and reliable, but the test is more complicated and costly, and it is mostly used in large and medium-sized projects.

9.3 Stability Analysis for Gravity Dams

This work aims to check the stability safety for gravity dams under probable load combinations (Stelle et al., 1983). Since transverse joints intersect conventional gravity dams into independent dam monoliths, the stability analysis is usually conducted on two-dimensional principles. A three-dimensional analysis (e.g., multiple-wedge analysis) may be required concerning particular geometric and loading conditions (e.g., bank-slope dam monolith).

Because the contact surface between the dam body and rock mass is the joint surface of two kinds of materials, the horizontal thrust force of the dam body is significant and restricted by construction conditions. Its shear strength is often low. Therefore, in the design of gravity dams, it is necessary to check the anti-sliding stability along the dam foundation. The results must meet the requirements of the safety degree of anti-sliding stability in the design code (Chen, 2015).

The sliding failure process of the gravity dam revealed by theoretical and experimental studies is very complicated. Most countries use design codes in which the sliding stability analysis is based on a mixed way of semi-theoretic and semi-empirical, presuming that failure surfaces can be any combinations of planes and curves. Very often, in practice, failure surfaces are assumed to be planar for simplicity. It should be noted that for an analysis to be realistic, these assumed failure planes have to be kinematically possible. Conventionally, sliding stability is expressed as a safety factor against sliding (MWR, 2018) using the limit equilibrium method. A sliding failure will occur along the presumed slip surface when the applied shear force exceeds the shear strength.

9.3.1 Failure Modes

Four sliding failure types appear on rock-base (Figure 9-4): surface sliding, shallow sliding, deep sliding, and mixed sliding.

Surface sliding. The surface sliding is the shear slip that occurs along the concrete foundation of the dam and the contact surface of bedrock. It is mainly controlled by the shear strength of the dam and bedrock interface.

Shallow sliding. This type of failure occurs when the shear strength of the rock base of the dam is lower than the strength of the concrete and bedrock contact and lower than the strength of the deep rock mass. Shear damage occurs within the shallow rock mass.

Deep sliding. It is a sliding of the rock mass along the existing discontinuities under the engineering stress condition. There is only a discontinuity in the local bedrock, and the possibility of deep sliding occurs when a particular combination can constitute a dangerous slip.

There are also forms of mixed sliding between several slip modes.

The sliding discontinuity can be divided into two categories: sliding control surface and cutting surface. The boundary conditions of the deep slip are formed by the combination of them with free faces. As shown in Figure 9-5, the deep sliding is caused by cutting the structure

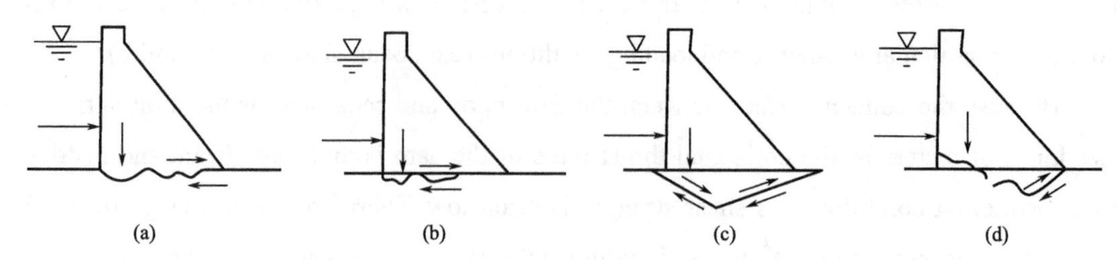

Figure 9-4 Schematic diagram of dam foundation slippage types

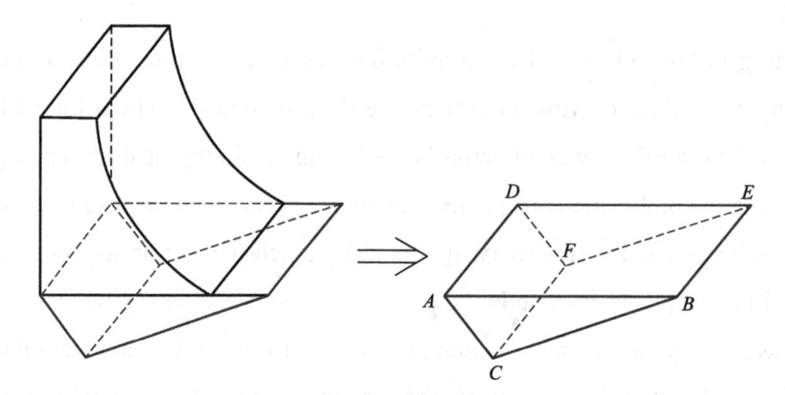

Figure 9-5 The boundary conditions of the deep slip.*ABCDEF* and *ACDF* are cutting surfaces, *CBEF* is sliding surface, *ABED* is free surface

surface around the foundation, forming a sliding body that can slip. The sliding body is composed of a weak structural plane that could be a sliding surface, a cutting surface separated from the surrounding rock mass, and a free surface with free space. The sliding surface, the cutting surface and the free surface constitute the boundary conditions of the sliding of the rock mass.

9.3.2 Stability Analysis Along Dam Base

A dam base is often critical due to its weak concrete-rock bond and more significant horizontal resultant thrust force. The stability against sliding along the base is therefore calibrated obligatory in the design.

1. Friction safety factor

The friction factor K is expressed as a ratio of friction resistance to slip driving on the plane concerned. The resistance is assumed to be purely frictional, and no cohesion can be considered. The plane is horizontal or inclined at a small angle of α (Figure 9-6).

For a horizontal slip plane, the friction factor K is the ratio of the maximum resistance friction to the applied shear force, which can be calculated using:

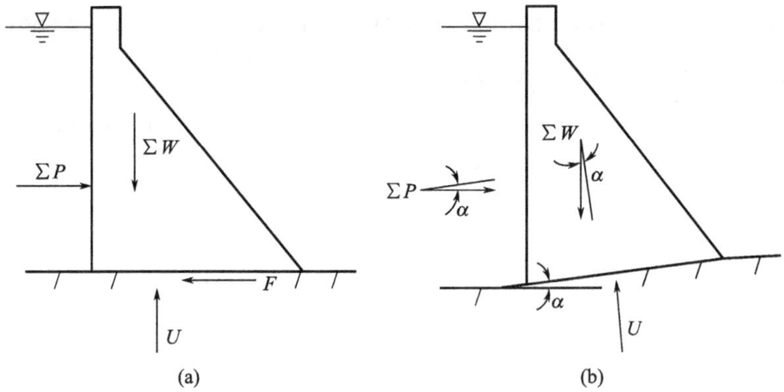

Figure 9-6 Stability against sliding along dam base
(a) Sliding along horizontal dam base; (b) sliding along inclined dam base

$$K=\frac{\text{resistance friction}}{\text{shear}}=\frac{f(\sum W-U)}{\sum P} \tag{9-2}$$

where $\sum W$ is resultant of forces normal to the assumed slip plane (kN/m);$\sum P$ is resultant of forces parallel to the assumed slip plane (kN/m); U is resultant of uplift exerted perpendicular to the slip plane (kN/m); and f is friction coefficient of the assumed slip plane.

If the plane is inclined at a slight angle of α, the previous expression is modified as follows:

$$K=\frac{f(\sum W\cos\alpha-U+\sum P\sin\alpha)}{\sum P\cos\alpha-\sum W\sin\alpha} \tag{9-3}$$

The uplift U is always perpendicular to the slip plane, and the angle α is defined as positive where sliding operates in an uphill sense. The gravity dam base is frequently excavated to give a slight positive inclination of α as to obtain higher K, enlightened by Eq. (9-3).

2. Shear-friction safety factor

The shear-friction safety factor K' is defined as the ratio of total shear resistance mobilized on a slip plane to total slip driving. With this approach, both the cohesion and frictional components of shear strength are accounted for, and

$$K'=\frac{f'(\sum W-U)+c'A}{\sum P} \quad \text{or}$$

$$K'=\frac{f'(\sum W\cos\alpha-U+\sum P\sin\alpha)+c'A}{\sum P\cos\alpha-\sum W\sin\alpha} \tag{9-4}$$

where f' is shear frictional coefficient of the concrete-rock bond plane; c' is cohesion representing the unit shearing strength of concrete-rock bond plane under zero normal stress (kN/m^2); A is the area of contact or slip plane (m^2); and α is the inclination angle of the slip plane defined as positive where the sliding operates in an uphill sense. Although the above formulae are far from perfect yet seemingly correct and straightforward, they have been used in the design of gravity dams for many years. Since the 1980s, the shear-friction safety factor has been prevailing among Chinese engineers. The allowable safety factors against sliding along dam bases are listed in Table 9-4 (Chen, 2015).

Table 9-4 Allowable safety factors against sliding along dam base

Allowable Safety factor	Load combination	Grade of dam		
		1	2	3
K	Basic	1.10	1.05	1.05
	Special(1)(catastrophe flood)	1.05	1.00	1.00
	Special(2)(earthquake)	1.00	1.00	1.00
K'	Basic	3.0		
	Special(1)(catastrophe flood)	2.5		
	Special(2)(earthquake)	2.3		

(1)and(2)represent different scenarios.

It should be indicated that the forces $\sum W$ and $\sum P$ are directly proportional to the square of dam height, whereas the area of contact plane A is directly proportional to the dam height only. Therefore, the cohesion c' contributes less to the higher dams' stability if the shear factor K' in Eq. (9-4) is employed.

It is widely recognized that the shear frictional coefficient f' or the corresponding friction angle φ' is a more stable parameter (Lo et al., 1991). By contrast, cohesion c is not so stable and is influenced by factors such as the roughness of the concrete-rock bond plane, the type and integrity of foundation rock, and the geologic structures. Considerable variation in cohesion may emerge for one specific rock within the confines of the dam site in consequence of local weathering or alteration. Cohesion may also be ignored due to saturation for some incompetent rocks. The parameters f', c', and f may be referred to in Table 9-5 (Code for geological investigation of water conservancy and hydropower projects) in the planning phase.

However, starting from the preliminary design, thorough investigation, extensive in situ, and laboratory tests are highly demanded to confirm these design parameters (Lo et al., 1991). Applied to the middle- and lower dams in the middle-class projects, when there is no possibility for in situ tests, the parameters listed in Table 9-5 may also be used in the design, subject to appropriate verification through laboratory tests.

Table 9-5 Strength parameters of foundation rock mass and contact surface (MOHURD,2008)

Classification of rock	Contact surface between concrete and rock			Rock mass		
	f'	c' (MPa)	f	f'	c'(MPa)	f
I	1.50~1.30	1.50~1.30	0.85~0.75	1.60~1.40	2.50~2.00	0.90~0.80
II	1.30~1.10	1.30~1.10	0.75~0.65	1.40~1.20	2.00~1.50	0.80~0.70
III	1.10~0.90	1.10~0.70	0.65~0.55	1.20~0.80	1.50~0.70	0.70~0.60
IV	0.90~0.70	0.70~0.30	0.55~0.40	0.80~0.55	0.70~0.30	0.60~0.45
V	0.70~0.40	0.30~0.05	0.40~0.30	0.55~0.40	0.30~0.05	0.45~0.35

N.B. 1. f' and c' are the parameters against shearing, and f is the friction coefficient.

2. The parameters listed in the table are only applicable for hard rocks, and the reduction considering softening coefficient is for soft rocks.

9.3.3 Stability Analysis Along Deep-Seated Slip Planes

Founded on a solid and intact rock foundation, the gravity dam designed and built using modern theory and technique is very reliable. The stability against sliding along the dam base is merely necessary to be checked. However, weak geologic structure faces (discontinuities, including joints, seams, faults, etc.) are ubiquitous within rock masses, which might dominate the dam stability in a manner of deep-seated sliding inside the dam foundation. The Austin Dam (the USA, H = 15m) is a typical lesson of failure due to the water softened shale within the foundation (Wise, 2005).

The stability issue resulting from discontinuities in the dam foundation has been widely studied worldwide since the 1960s. According to statistics data of nearly one hundred China's dams under construction or design, about one-third are affected by serious adverse geologic structure faces which lead to consequences such as construction delay or even shut down, important design revisions or even change of dam site and reduction of dam height, limitation of operation reservoir level or reinforcement after completion.

The study on the deep-seated sliding proceeds in the following steps:

(1) The defects within the foundation are investigated, and the slip surfaces (boundaries) are specified.

(2) The strength parameters of the slip surfaces are evaluated.

(3) The safety calibration method and corresponding allowable safety margin are chosen.

(4) The countermeasures for improving stability are selected and justified.

The safety against deep-seated sliding in dam foundation is calibrated mainly using the limit equilibrium method (LEM) in China. For significant and complicated dam projects, the finite element method (FEM) and geomechanical model test are required for comprehensive analysis and evaluation.

1. Stability analysis along single slip plane

For single slip plane problem shown in Figure 9-7, the safety factor K' can be given by

$$K'=\frac{f'_B(\sum W\cos\alpha-U+\sum P\sin\alpha)+c'_B A}{\sum P\cos\alpha-\sum W\sin\alpha} \tag{9-5}$$

where f'_B is friction coefficient of the discontinuity; c'_B is the cohesion of the discontinuity; and A is the area of the discontinuity. The angle α is defined as positive if the sliding operates in uphill sense.

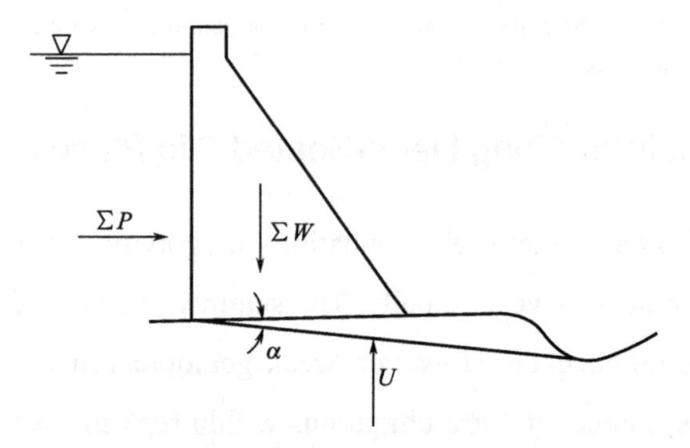

Figure 9-7 Stability analysis along single slip plane

2. Empirical strength parameters

The Chinese engineering geology sector in charge of water resources and hydropower project recommends empirical strength parameters for the discontinuities listed in Table 9-6 (GB 50487—2008). These parameters are summarized from the in situ large-scale tests and laboratory middle-scale tests, which were carried out for more than 30 hydropower projects (Chen, 2015).

Concerning the allowable safety factor for the deep-seated sliding stability, there are no detailed specifications in the corresponding Chinese design codes due to the complexity and uncertainty in the foundation rock properties. The allowable safety factor adopted for the gravity dam design in the Three Gorges Project (Hubei, China, H=185m) is K'=3.0 for basic load combinations, K'=2.3~2.5 for special load combinations; the allowable safety factor for the gravity dam design in the Xiangjiaba Hydropower Project (Sichuan, China, H=161m) is K'=3.0~3.5 for basic load combinations, K'=2.0~3.0 for special load combination (1) (catastrophe flood), K'=2.3~2.5 for special load combination (2) (earthquake).

The criteria mentioned above on shear-friction safety factors are relatively conservative for the lower-to-medium gravity dams or barrages affected by seams daylighting downstream

(Figure 9-8). Under such circumstances, the Chinese engineer usually employs the analysis methods and criteria for lower head hydraulic structures on a soft foundation: the cohesion c' is generally ignored, and the friction factor K is applied for the stability calibration. In engineering practice, the choice safety factor usually is higher than that from the code (Table 9-4), for example: $K = 1.2$ for the Baishan Gravity Arch Dam (Jilin, China, $H = 149.5$m); $K = 1.3$ for the Dahua gravity dam (Guangxi, China, $H = 78.5$m); $K = 1.1$~1.2 for the Tongjiezi Gravity Dam (Sichuan, China, $H = 82$m).

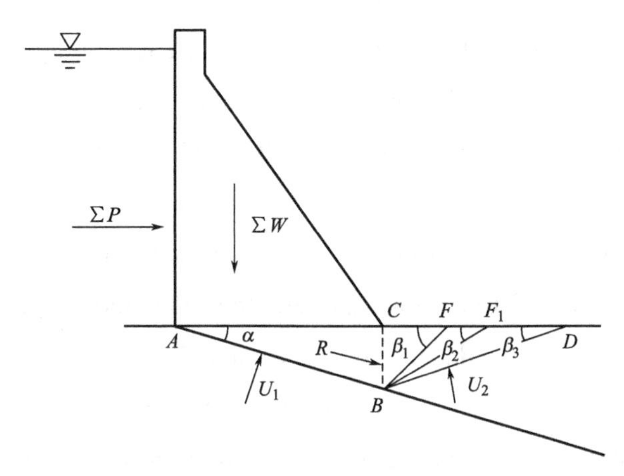

Figure 9-8 Potential slip plane within the resisting wedge

Table 9-6 Strength parameters of discontinuity (GB 50487—2008)

Type of discontinuity		f'	c'(MPa)	f
Well-cemented		0.90~0.70	0.30~0.20	0.70~0.55
Unfilled		0.70~0.55	0.20~0.10	0.55~0.45
Weak seam	Rock-laden debris filled	0.55~0.45	0.10~0.08	0.45~0.35
	Argillite-laden debris filled	0.45~0.35	0.08~0.05	0.35~0.28
	Debris-laden argillite filled	0.35~0.25	0.05~0.02	0.28~0.22
	Argillite filled	0.25~0.18	0.01~0.005	0.22~0.18

N.B. 1. The f' and c' are shear parameters and f is the friction parameter.

2. The parameters listed in the table are appropriate only to the discontinuities within the hard rock.

3. For the discontinuities within the soft rock, the parameters listed in the table are subject to reduction.

4. For the well-cemented and unfilled discontinuities, the strength parameters are selected among the upper and lower bounds according to the roughness of discontinuities.

9.4 Seepage deformation of dam foundation

After the reservoir is impounded, the dam foundation will cause the movement, structural deformation, and even destruction of the dam body under the pressure seepage flow, called seepage deformation. Severe seepage deformation not only affects engineering benefits but also endangers the stability of the dam.

9.4.1 The manifestation of seepage deformation of dam foundation

1. Piping / Erosion

The tiny particles filled between the large particles of the sand foundation move with the seepage flow under the action of the hydrodynamic pressure formed by the seepage water flow, and often produce "sand boiling" phenomenon downstream of the dam foundation, resembling piping, also known as "mechanical erosion." In the rock foundation area, if the cracks are filled with soluble salt (such as gypsum, calcite, etc.), or the crack filling is cemented by soluble salt, due to the chemical dissolution of groundwater and the action of water pressure, "chemical erosion" is often formed. For example, the foundation of a dam in a reservoir in Sichuan Province had large caves due to chemical erosion. In southern my country, red beds (red sandstone, conglomerate, shale, etc., containing soluble materials) often appear karst landforms, all of which are related to chemical erosion. In addition, burrowing animals (such as voles, badgers, earthworms, ants, etc.) sometimes damage the soil structure. If channels are formed inside and outside the embankment, they can also create piping, or "biological erosion," such as the Yellow River in China.

2. Flowing soil/ Pop-off

Flowing soil generally occurs in zones dominated by cohesive soils. Because the soil is relatively dense, when particles have a specific cohesive force, the fine particles are not easily taken away by the seepage water under the flowing water pressure. Still, they can move or uplift as a whole. This phenomenon is called flowing soil/pop-off. Sometimes, the dam foundation is composed of a binary structured layer, especially in a zone where the upper layer is cohesive soil and the lower layer is sandy soil. In that case, fluid soil may be generated when the hydrodynamic pressure of the seepage flow in the lower layer exceeds the weight of the overlying soil. This kind of seepage deformation often causes pieces of soil damage, water, or sand turning in the outflow zone of the downstream dam foot. If without treatment, it will directly threaten the safety of the dam.

9.4.2 Cause analysis of seepage deformation of dam foundation

According to a large number of experimental data, it is proved that the cause of the above-mentioned various erosion and flow-soil phenomena is essentially the result of the action of the hydrodynamic pressure of the seepage water on the rock (soil) body. When the hydrodynamic pressure is greater than the resistive force of the rock (soil) body, seepage deformation will occur. Here, the physical and chemical properties of rock (soil) body itself, such as the grain size, the nature of the cracks, the structure, the degree of compactness, the cementation, and the water permeability are the fundamental internal factors; the hydrodynamic pressure is the external factor. External causes work through external causes, and neither is indispensable.

1. Analysis of rock (soil) structure factors

Examples of seepage deformation on the dam foundation surface show that the structure of the rock (soil) body, especially the inhomogeneity of the particle composition, is the main reason for the formation of erosion and flowing soil. Piping easily occurs when the average diameter of the most coarse and fine particles in the earth is significant. Theoretically, when the pore diameter between uniformly spherical particles is 1/6 of the particle diameter, any particles smaller than this pore diameter can pass through, as shown in Figure 9-9. However, under natural conditions, there is water around the fine particles, and the shape of the particles can not be ideal spherical. Studies show that only when $D_{big}/d_{small} > 20$, can create piping.

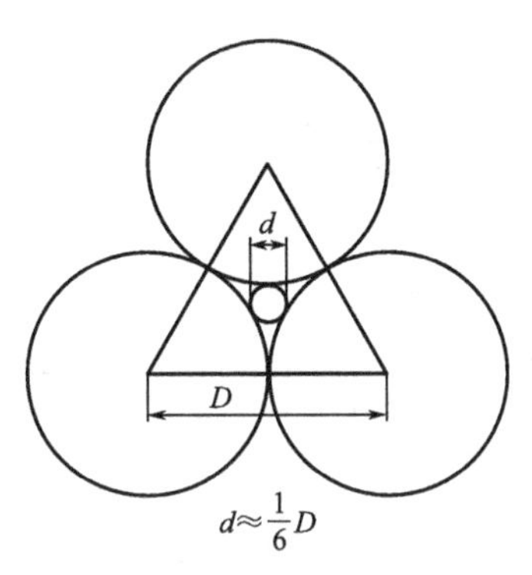

Figure 9-9 The relationship between uniform spherical particle diameter and pore diameter

The influence of the coefficient of heterogeneity (uneven) η on soil particle size distribution is significant. For example, the soil with $\eta < 10$ is prone to seepage deformation mainly in the form of flowing soil; the soil with $\eta > 20$ is inclined to seepage deformation in the main form of piping; $10 < \eta < 20$, may produce runny soil, and may also make piping.

If the soil layer has a binary or a multi-layer structure, it should be analyzed in detail according to the burial conditions of the soil layer. In the case of a binary structure,when the cohesive soil is on top and the sandy soil is on the bottom, and the cohesive soil is thick and complete, seepage deformation is not easy to occur. But when the cohesive soil is thin or incomplete, it is easy to produce fluid soil uplift downstream of the dam, and successively have lower soil piping. Suppose there is a pinch-out layer, lens, and other soil layers, and the cohesive soil layer from upstream to downstream gradually thinning. In that case, the sand and gravel layer below it gradually thickens, and the increase of the cross-section will weaken the seepage pressure downstream. On the contrary, if the sand gravel layer pinches downstream, the seepage pressure will increase significantly, and these places are prone to flowing soil or piping.

2. Hydrodynamic pressure

Hydrodynamic pressure is the pressure exerted by the seepage flow on a unit volume of soil/rock, and its magnitude is mainly related to the hydraulic gradient of the seepage flow and the unit weight of the water, namely

$$P_d = \gamma_\omega I \tag{9-6}$$

Where P_d is hydrodynamic pressure (kN/m^3); γ_ω is unit weight of water (kN/m^3); I is hydraulic gradient of seepage flow.

Because $I = \frac{\Delta H}{L}$ (ΔH is the head difference, L is the seepage distance), the larger the ΔH or the smaller the L, the greater the value of I and the greater the possibility of seepage deformation (Figure 9-10).

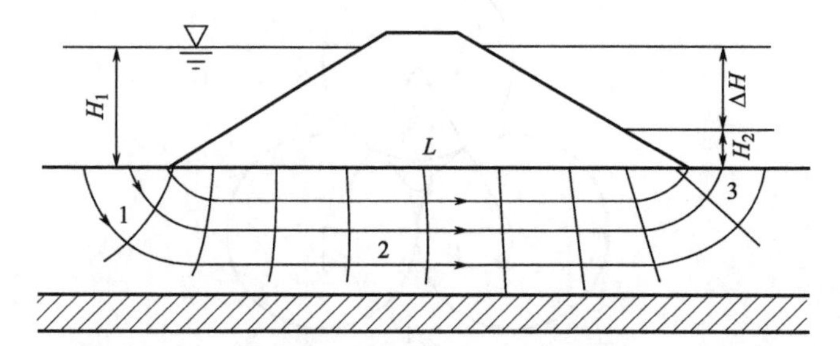

Figure 9-10 Schematic diagram of dam foundation seepage

1—The upstream seepage direction of the dam foundation is downward; 2—the seepage direction of the dam foundation is horizontal; 3—the downstream seepage direction of the dam foundation is downward; H_1—upstream water surface; H_2—downstream water surface; L—seepage distance; ΔH=water level difference ($\Delta H = H_1 - H_2$)

The seepage flow flows from up to down in front of the dam (mainly at the foot of the dam slope). The permeable water flows from upstream to downstream at the dam foundation and is approximately horizontal under the dam. The direction of hydrodynamic pressure is consistent with the direction of water flow, which is also roughly flat. In this case, if the resistance of soil particles to hydrodynamic pressure is less than hydrodynamic pressure, they will move downstream along the hydrodynamic pressure direction. At the toe of the downstream slope of the dam, it is the escape section of seepage water flow, and the direction of flow and dynamic water pressure is from bottom to top, so seepage deformation is most likely to occur here.

9.5 Dam foundation treatment

The naturally occurring rock mass is the product of natural history, and it has experienced various geological invasions and changes for a long time. Therefore, the geological conditions of any dam site are not entirely in line with the ideal architectural design requirements, and there will be some bad geological problems. However, for various bad geological conditions, they can be dealt with under normal circumstances as long as they are evident in advance. The requirements for safety and stability can be guaranteed. The treatment of bad geological problems can be divided into foundation clearing, rock mass reinforcement, and seepage prevention and drainage.

9.5.1 Foundation clearing

Foundation clearing is the excavation of the loose, weak, weathered, broken rock layers or shallow, weak interlayers on the surface of the dam foundation (Figure 9-11). Place the dam on a relatively fresh and intact rock mass. The depth of dam foundation excavation and the determination of the foundation surface elevation are essential issues in design and construction. It will significantly impact the investment, construction period, safety, and stability of the

Figure 9-11 Cleaning of the foundation surface by pressure water and compressed air for the Orlik Dam in Bohemia (Záruba & Mencl,1976)

entire project. Because the excavation of foundation pits of large hydropower projects often reaches a depth of tens of meters, the excavation can reach millions of m^3, and the excavated part will have to be backfilled in the future. In addition, there are still problems such as foundation pit drainage, foundation pit slope stability, and in-situ stress. For example, in the Ertan Hydropower Station on the Yalong River, the dam foundation is invaded by Permian basalt and syenite. Four comparison schemes have been proposed for the foundation surface through particular research:

(1) Option Ⅰ is to use fresh and slightly weathered rock mass as the dam foundation, with an average embedded depth of 64.6m;

(2) Option Ⅱ is a weakly weathered lower part with an average embedded depth of 53.7m;

(3) Scheme Ⅲ is the weakly weathered middle section, with an average embedded depth of 46.1m;

(4) Option Ⅳ is based on making full use of the weakly weathered middle section, and a small part is placed on the weakly weathered upper rock mass, with an average embedded depth of 38.6m.

After comparison, Option Ⅳ is 7.5m less than Option Ⅲ (preliminary design plan), which reduces the excavation of 800000m^3 of stone and 370000m^3 of concrete. It saves 61.17 million Yuan of investment (calculated according to the 1986 quota), and the construction period also can be shortened by 11 months. If compared with Option I, the savings are much more significant.

In recent years, many people have conducted many special studies and demonstrations on the elevation of the dam foundation surface. In actual work, the dam foundation excavation depth is determined based on the degree of weathering or the quality of the rock mass. Under normal circumstances, high dams can be built in the lower part of the hard rock in the slightly or weakly weathered zone. After sufficient demonstration and treatment, they can also be created (or partly)in the middle of the weak weather zone. The middle dam can be built in the middle or part of the upper part of the weakly weathered zone. The higher terrain on both sides of the strait can be appropriately relaxed. But thorough research and treatment should be done when unfavorable special conditions such as weak structural surfaces or special dam requirements.

The foundation clearance requirements for earth-rock dams are lower than those for concrete dams. Because only the humus soil, silt, high plastic soft soil, quicksand layer, and other rocks and soil layers with increased compressibility and low shear strength are removed when clearing the foundation. It can be based on a loose sedimentary layer. For rock formations with a faster weathering rate, when the foundation pit is exposed for a long time, a protective layer should be reserved, or other protective measures should be taken. In addition, the foundation surface should be slightly undulating and inclined upstream as much as possible. When excavating near the bank, attention should be paid to whether the slope toe will endanger the stability of the slope after being dug.

9.5.2 Reinforcement of dam foundation rock mass

The rock mass below the foundation surface often has more or fewer cracks, pores, and fault fracture zones. To increase the strength of the rock mass, some reinforcement measures can be taken, which can also reduce the amount of foundation pit excavation. There are usually the following measures.

1. Consolidation grouting

Consolidation grouting is to press a suitable cementitious slurry (mostly cement slurry) into the cracks or pores of the bedrock by drilling holes in the rock foundation so that the broken rock mass is cemented into a whole to increase the strength of bedrock. Almost all concrete dam foundations in China adopt this kind of measure, and even some earth and rockfill dams also use consolidation grouting to reinforce the dam foundation. Under normal circumstances, good results can be achieved, but when there is mud filling in the fissures, a certain pressure must be used to press clean water for flushing.

According to practical experience, grouting holes are generally arranged in a plum blossom shape, with a hole spacing of 1.5~3.0m, depending on the effective range of grout diffusion. The hole depth is determined according to the requirements of strengthening the rock mass. Shallow hole consolidation grouting is generally 5~8m, and the deepest is no more than 15m. Under exceptional circumstances, such as deep crack distribution, deep hole consolidation grouting can also be carried out. The grouting holes are generally straight. Sometimes, they can also be arranged as inclined holes roughly perpendicular to the prominent cracks or other weak surfaces to improve the effect.

2. Installation of anchors

When there are weak planes in the local bedrock that control the sliding of the rock mass, to enhance the anti-sliding stability of the rock mass, prestressed anchor rods (or steel cables) can be used for reinforcement. The method is to drill through the weak plane, penetrate the hard and intact rock mass, anchor the prestressed steel bar or steel cable, and seal the hole with cement mortar (Figure 9-12). When conditions permit, large-diameter reinforced concrete pipe columns can also be used for anchoring. In 1964, this method was used for the first time at the right abutment of the Meishan Multi-Arch Dam. There were 250 anchor holes, 170 holes on the hillside, and 80 holes on the 13th arch-gravity dam. The whole distance is 2~3m, the hole depth is 25~40m, and the steel cable diameter is 5mm (Figure 9-13). After that, this method also has been used to anchor the rock mass of the dam abutment at the Maotiaohe Third-level Power Station and the Lubuge Hydropower Station.

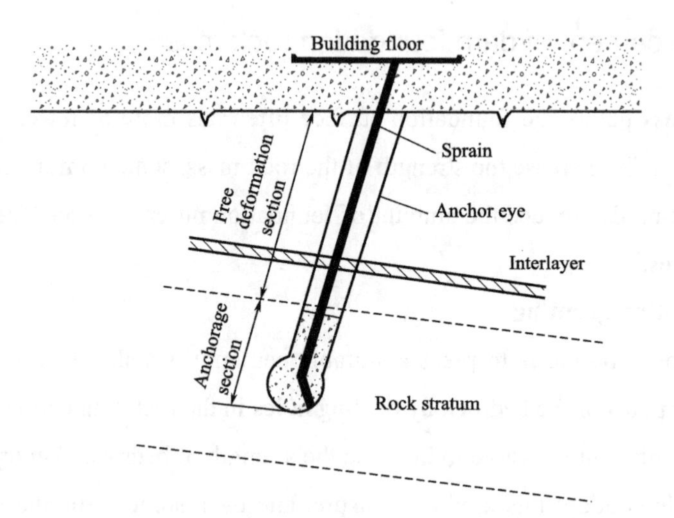

Figure 9-12 Schematic diagram of the anchoring structure

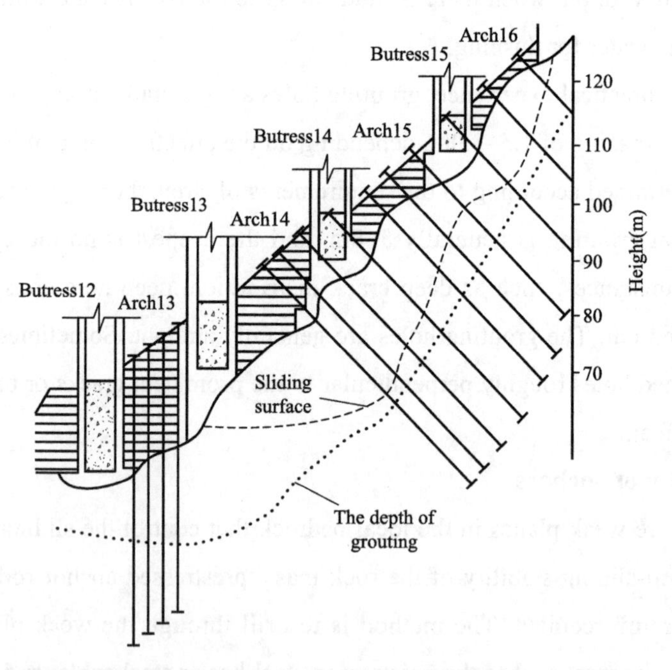

Figure 9-13 Anchorage on the right bank of the Meishan dam

3. Treatment of fracture zones

Special treatment is required when there are large-scale weak fracture zones under the dam foundation, such as fault fracture zones, weak interlayers, argillaceous layers, cystic weathered zones, fissure-intensive zones, etc.

1) Treatment of weak fracture zone with high inclination angle

The main treatment methods for the high inclination and broken zone include concrete plugs, concrete beams, concrete arches, etc. The concrete plug is to excavate the weak and broken area to a certain depth and then backfill the concrete to increase the foundation's strength (Figure 9-14a).

Generally, trenches with inverted trapezoidal cross-sections are dug along the broken zone. The excavation depth should be calculated and determined according to factors such as the foundation rock's stress and the width of the fractured zone. In general, 1.0~1.5 times the width can be taken.

When the lithology of the weak fracture zone is loose and weak, the strength is deficient, and the width is large, the concrete beam or arch structure can be adopted. It can transfer the load to both sides on a hard and intact rock mass (Figure 9-14b). When there are deep trenches in the overburden layer and deep weathered trenches in the river bed, due to the difficulty of deep excavation, the form of beams or arches can be used and then combined with treatment measures such as grouting and horizontal anti-seepage.

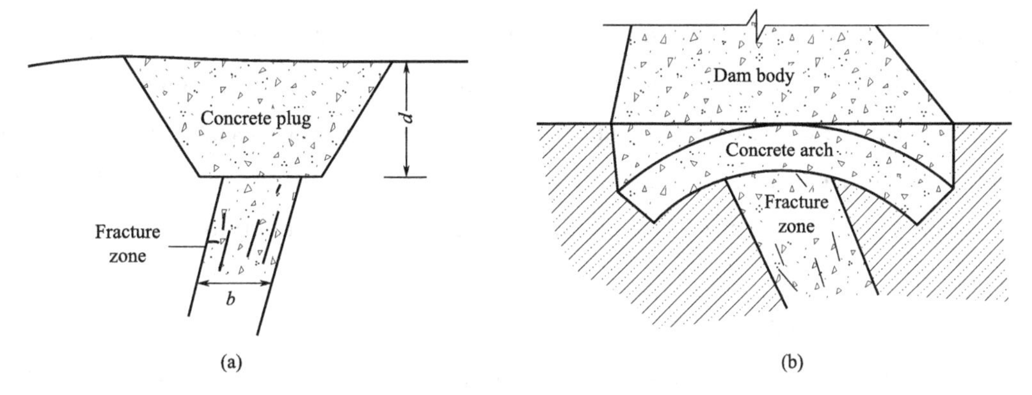

Figure 9-14 Schematic section of concrete plug and arch for dam foundation treatment

(a) Concrete plug; (b) Concrete arch

2) Treatment of weak and weak fracture zone with gentle dip

When the gentle dipping weak fracture zone is buried shallow, it can be wholly excavated and backfilled with concrete (Figure 9-15a). It is the safest and most reliable method. If the burial is deep, it needs to use the hole excavation (flat or inclined hole). The deep excavation can be equipped with a shaft (Figure 9-15b). After excavation and backfilling, consolidation grouting can still be carried out. It can also be partially excavated under stable conditions to reduce the workload. When the weak fracture zone tends to be downstream or upstream, a flat hole can be dug at regular intervals along its direction. The top and bottom of the hole are embedded in a competent and intact rock layer and then backfilled with concrete to form a concrete key (Figure 9-15c) to improve the anti-sliding ability. When it is inclined to the bank, a series inclined well can be dug along with its inclination and backfilled with concrete.

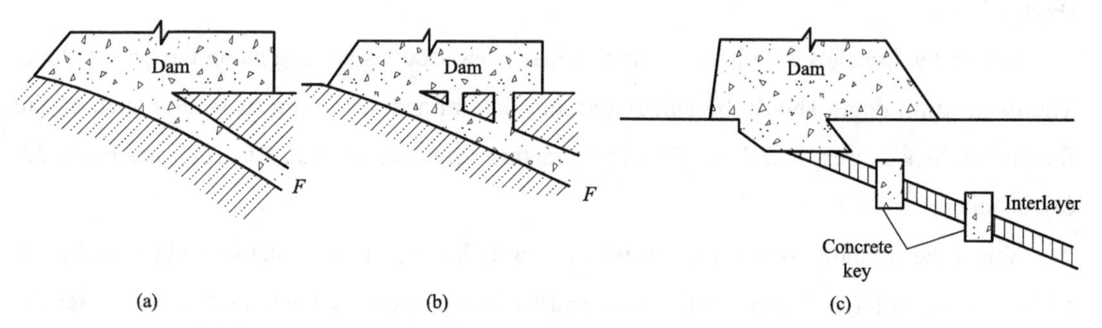

Figure 9-15 Treatment of weak and weak fracture zone with gentle dipping angle

(a) Fully excavated and backfilled with concrete; (b) With shafts; (c) With concrete keys

9.5.3 Seepage prevention and drainage measures

The anti-seepage and drainage measures of the dam foundation are crucial. It is an essential means to prevent the seepage deformation of the foundation and reduce the uplift pressure. The general principle is that anti-seepage measures, such as grouting curtains, should be set on the water-facing surface of the dam or its upstream parts to minimize the seepage flow of the dam foundation. Drainage measures, such as drainage wells, holes, etc., are set in the dam foundation downstream of the water surface (behind the seepage prevention curtain) to reduce the seepage pressure.

1. Curtain grouting

Curtain grouting is a grouting project in which grout is poured into the cracks and pores of rock mass or soil layer to form continuous water blocking curtain to reduce seepage flow and pressure. Dam foundation curtain grouting is usually arranged near the upstream of the dam foundation surface (Figure 9-16), which is the most commonly used grouting project with high technological requirements. The top of the curtain is connected to the dam body. The bottom is deep into the relatively impervious rock layer to prevent or reduce groundwater penetration in the dam foundation. The drainage system located downstream can also reduce the uplift pressure of the seepage flow on the dam foundation. Since the 20th century, curtain grouting has been the primary method of anti-seepage treatment of hydraulic structures and plays an essential role in ensuring the safe operation of hydraulic structures.

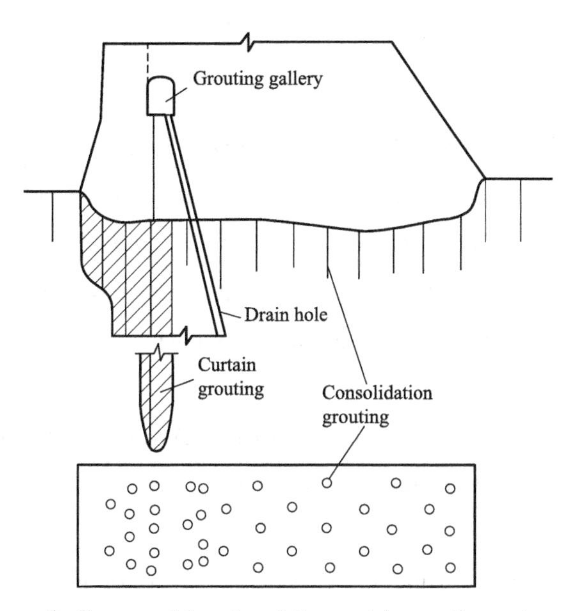

Figure 9-16 Schematic diagram of dam foundation curtain grouting and consolidation grouting

2. Drainage measures

Although anti-seepage curtains and other anti-seepage measures are installed in the rock mass of the dam foundation, there will still be a small amount of water flowing around or through the curtain. Drainage holes are set in the dam foundation downstream of the curtain To reduce the seepage pressure. Usually, 2 to 3 rows of drainage holes, drainage pipes, corridors, or water collection wells can be set to drain the water out of the dam body (Figure 9-17).

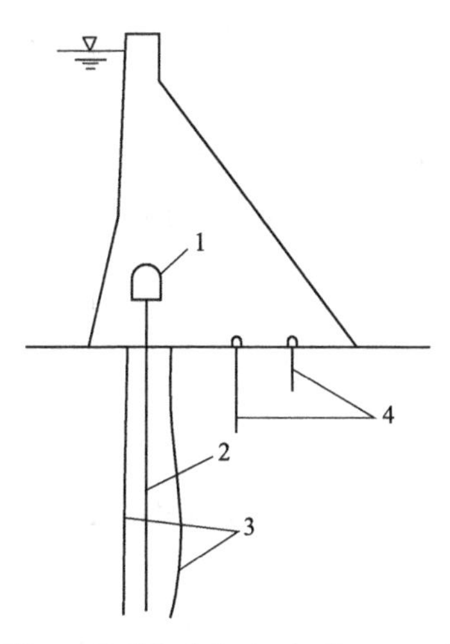

Figure 9-17 Schematic diagram of impermeable curtain

1—grouting gallery; 2—curtain grouting drilling; 3—slurry diffusion range; 4—drain holes and drainage galleries

In addition to the dam foundation treatment measures mentioned above, some structural measures can also be taken in the design, such as increasing the section area of the dam body, expanding the foundation, setting up the supporting wall, adding a gravity pier on the abutment of the dam, deepening the tooth wall, and installing impervious bedding connected with the inclined wall for earth and rockfill dam, etc. These contents will be further discussed in specialized courses.

Reference

[1] CHEN S. Hydraulic structures[M]. Berlin: Springer, 2015.

[2] DE VALLEJO L G, FERRER M. Geological engineering[M]. Leiden: CRC Press, 2011.

[3] GOODMAN R E. Introduction to rock mechanics[M]. 2nd ed. New York: Wiley, 1989.

[4] LO K Y, OGAWA T, LUKAJIC B, et al. Measurements of strength parameters of concrete-rock contact at the dam-foundation interface[J]. Geotechnical Testing Journal, 1991, 14(4): 383-394.

[5] MOHURD (Ministry of Housing and Urban-Rural Development), PRC.Code for Design of Building Foundation Foundation: GB 50007—2011[S]. Beijing: China Building Industry Press, 2012.

[6] MOHURD (Ministry of Housing and Urban-Rural Development), PRC. Engineering Rock Mass Classification Standard: GB/T 50218—2014[S], 2014.

[7] MOHURD (Ministry of Housing and Urban-Rural Development), PRC. Code for geological investigation of water conservancy and hydropower projects: GB 50487-2008[S], 2008.

[8] MWR(Ministry of Water Resources of the People's Republic of China). Design specification for concrete gravity dams: SL 319-2018[S], 2018.

[9] STELLE W W, RUBIN D I, BUHAC H J. Stability of concrete dam: Case history[J]. Journal of Energy Engineering, 1983, 109(3): 165-180.

[10] WISE E C. The day Austin died[J]. Penn Lines, 2005, 40(9): 8–11.

[11] ZÁRUBA Q, MENCL V. Engineering geology[M]. Amsterdam: Elsevier, 1976.

Exercises

Review Questions

(1) What are the main geological factors causing uneven deformation of dam foundation rock mass?

(2) What are the main methods for determining the bearing capacity of rock foundations?

(3) Try to describe the types and characteristics of sliding failure of dam foundation rock mass.

(4) How to analyze the boundary conditions of sliding failure of dam foundation rock mass? What conditions are most likely to form slip surfaces?

(5) What are the sliding resistance factors to be considered in the failure of dam foundation rock mass?

(6) How to analyze the boundary conditions of rock mass sliding in abutment?

(7) What is the reason for calculating and excluding the c value in calculating the anti-sliding stability of the dam foundation? What principles should be considered in evaluating the concrete dam's test results?

(8) What are the measures to improve the stability of dam foundation rock mass?

10 Engineering Geological Study of Rock Slopes

10.1 Introduction

Slope refers to the surface of the earth's crust with a lateral surface, usually including natural slope and artificial slope, two types. Slopes undergo continuous development due to various processes that shape them. Of significant importance for all engineering works are earth movements caused by natural forces or human interference, depending on the type and number of factors and their interaction (Zaruba & Mencl, 1976).

A variety of engineering activities require the excavation of rock. In civil engineering, projects include transportation systems such as highways, railways, dams for power production and water supply, and industrial and urban development. In mining, open pits account for a significant portion of the world's mineral production. The dimensions of open pits range from areas of a few hectares and depths of less than 100m, for some high-grade mineral deposits and quarries in urban areas, to areas of hundreds of hectares and depths as great as 800m, for low-grade ore deposits. The overall slope angles for these pits range from near vertical for shallow pits in good quality rock to flatter than 30° for those in very poor-quality rock (Wyllie & Mah, 2004). Figure 10-1 shows the Palabora open pit in South Africa that is 830m deep and an overall slope angle of 45° ~50° ; this is one of the steepest and deepest pits in the world (Stewart et al., 2000). The upper part of the pit is accessed via a dual ramp system, which reduces to a single ramp in the lower part of the pit.

In recent years, the process of industrialization in China has developed rapidly. The shallow ore on the surface has been exhausted, and the mining has evolved to the depths. The mining of the underground ore layer will form a goaf. The appearance of the goaf destroys the initial stress distribution of the mine, redistributes the internal stress, affects the stability of the slope rock mass, and triggers a series of engineering geological disasters, such as surface subsidence, collapse, and landslide, etc. Once a geological catastrophe occurs, it will significantly threaten people's lives and property safety and affect the mining of underground ore bodies. In 1980, the Yanchi River Phosphate Mine in Yuan'an, Hubei destroyed and buried the entire mining area after the mountain collapsed, causing 284 deaths and heavy losses(Li et al., 2016). The collapse graph of the Yanchi River is shown in Figure 10-2. In 2009, in Wulong, Chongqing, a landslide occurred in the Jiwei Mountain due to underground mining activities, which buried 12 houses in the mining area. Seventy-four people died in the accident, and eight people were injured(Yin, 2010).

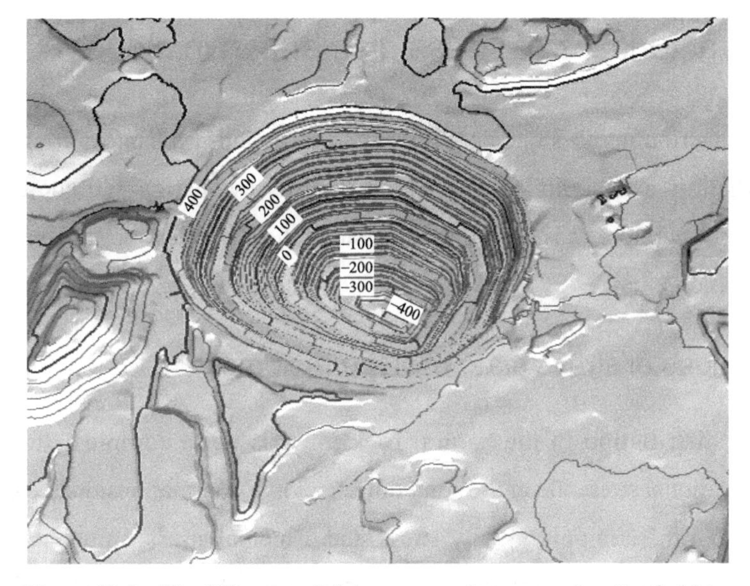

Figure 10-1 The 830m deep Palabora open-pit copper mine, South Africa

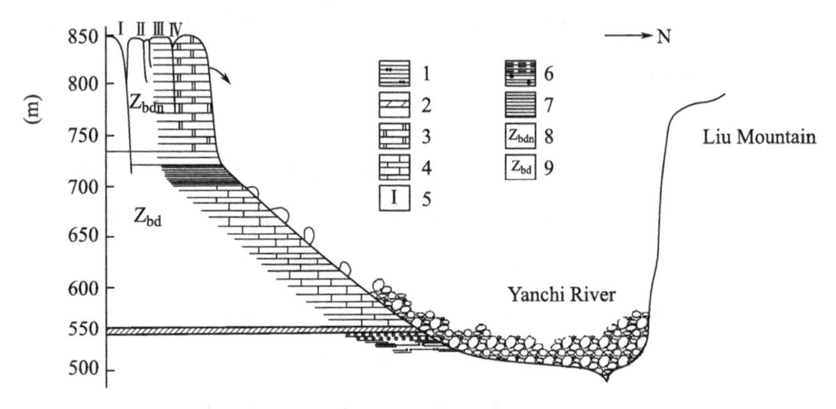

Figure 10-2 The collapse diagram of the Yanchi River

1—Gray-black silty shale; 2—Phosphate rock layer; 3—Thick massive dolostone; 4—Thin to medium thick dolomite; 5—Fracture number; 6—Dolomitic mudstone and sandy shale; 7—Thin to medium thick lamellar dolomite; 8—Upper Sinian Dengying Formation; 9—(upper Sinian) Doushantuo Formation

In addition to these artificial excavations, the stability of natural rock slopes may also be of concern in mountainous terrain. For example, highways and railways in river valleys may be situated below such slopes or cut into the toe, detrimental to stability. One of the factors that may influence the stability of natural rock slopes is the regional tectonic setting. Safety factors may only be slightly greater than one where there is rapid uplift of the landmass and corresponding down-cutting of the streams, together with earthquakes that slacken and displace the slope. Such situations exist in seismically active areas such as the Pacific Rim, the Himalayas, and central Asia (Wyllie & Mah, 2004).

10.2 Characteristics of stress distribution in slope

The stress distribution in slope determines the form and mechanism of slope deformation and failure, which is also significant for slope stability evaluation and good prevention and control measures. Therefore, it is crucial to understand the characteristics of stress distribution in slope after formation.

10.2.1 Changes of stress state of slope rock mass

The stress distribution in the natural rock mass is quite complex. In addition to the ubiquitous gravitational stress, there are sometimes tectonic, thermal, residual, etc. It is generally believed that when there is only gravity stress, the principal stress (initial stress) in rock and soil is straight and horizontal before the slope is formed. That is, vertical stress is the maximum principal stress, and horizontal stress is the minimum principal stress. According to the numerical simulation analysis and photoelastic test, the stress field of the slope has the following characteristics before failure:

(1) Apparent deflection occurred in the trajectory of principal stress in a slope (Figure 10-3). The maximum principal stress σ_1 near the slope surface is almost parallel to the slope, while the minimum principal stress σ_3 is nearly perpendicular. The shear stress in the lower part of the slope is almost horizontal, and its general trend is to increase from the inside to the outside. It gets more prominent when the shear stress is closer to the foot of the slope. In the interior of the slope, the stress field gradually restores to its original state.

(2) Stress concentration occurs in the slope body. An apparent stress concentrated zone is formed near the slope toe. It can be found that the steeper the slope, the more pronounced the concentration. The stress concentration zone of the slope foot is related to the stress difference between the maximum principal stress σ_1 and the minimum principal stress σ_3. Meanwhile,

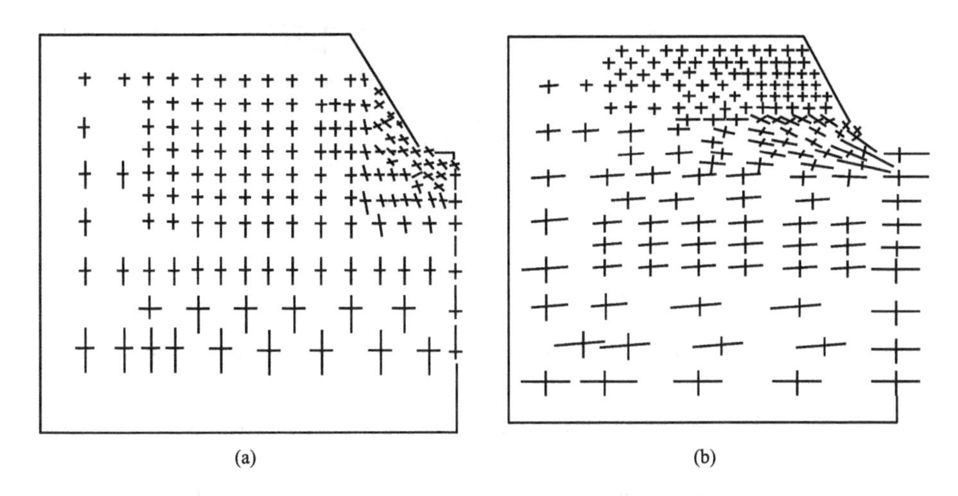

Figure 10-3 Principal stress trajectories solved by numerical simulation. Note the size of the line segment represents the magnitude of the stress (Coates, 1981)

(a) Horiontal field stress is one-third the vertical stress; (b) Horiontal field stress is three times of the vertical stress

the maximum shear stress concentration formed a domain with evident shear strain. Tension zones are formed in some parts of the surface or the top of the slope due to the tensile stress created by the release of horizontal stress (Figure 10-4). Since a jointed rock mass is weak in tension, it is interesting to examine the tensile regions within a slope. It was found that the toe tensile stress was increasing by an increase in the slope angle. A tensile zone may occur at the ground surface behind the crest of the slope. This is dependent mainly on lateral latent stress. With increasing slope angles and lateral latent stress, the tensile zone increases in area and tends towards the slope crest and further down towards the toe (Figure 10-6). This agrees with the findings of Duncan and Goodman (1966).

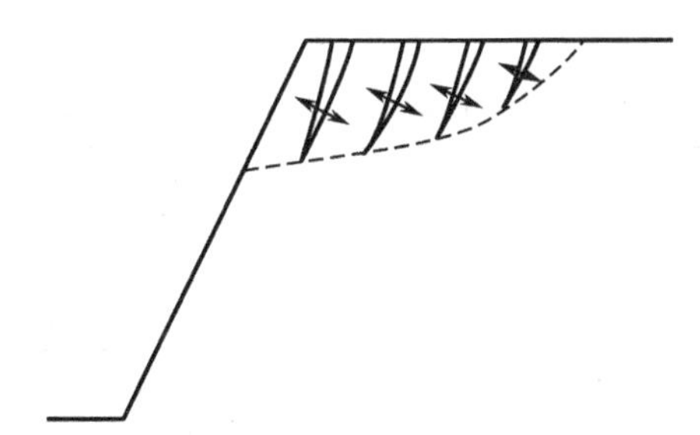

Figure 10-4 The tension domain near the crest of a slope

(3) Due to stress deflection, the path line of maximum shear stress in the slope also changed from the original straight line to the arc shape concaving to the slope surface (Figure 10-5).

(4) Because the slope is almost free of the lateral pressure, it changes into a bidirectional stress state. It gradually transforms into a three-way in the interior of the slope body.

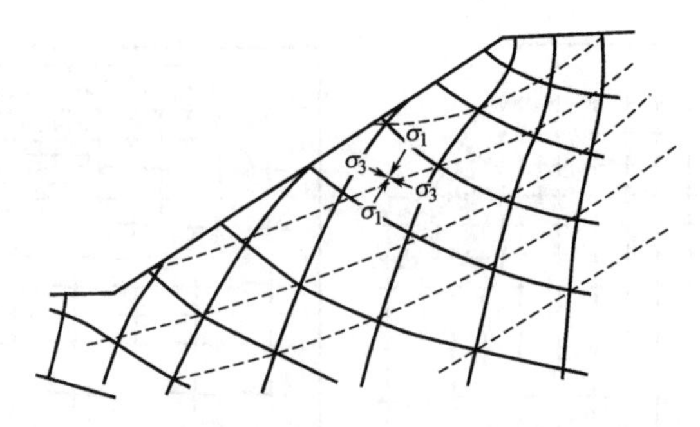

Figure 10-5 Schematic diagram of the relationship between shear stress and major principal stress trajectory in a slope

Solid line—principal stress trajectory; Dotted line—shear stress trajectory

10.2.2 The main factors influencing the stress distribution of slope rock mass

1. Influence of the initial stress

In some areas, the initial stress is intense, especially in the zone with strong tectonic movement. There is often large horizontal ground stress in the rock mass. In the uplift zone, deep valleys usually occur, forming high and steep valley slopes. Horizontal initial stress of the slope in the valley near the free face often creates stress concentration, significantly impacting the rock mass stress distribution. It is mainly manifested in aggravating the stress differentiation, enhancing the stress concentration near the slope foot, increasing the tension zone of the surface and top of the slope, and intensifying the degree and scope of slope deformation and failure. The table (Table 10-1) summarizes dimensionless values of the maximum shear stress at the toe in a vertical section through a slope, determined in the two and three-dimensional study.

Table 10-1 Maximum shear stresses at the toe of the slope(Stacey,1973)

Two-dimensional investigation		Three-dimensional investigation			
Lateral latent stress $\sigma_L/\rho gH$	Maximum shear stress at the slope toe $\tau_{max}/\rho gH$	lateral latent stress $\sigma_L/\rho gH$	Maximum shear stress at the slope toe		
			Non-circular pit		Circular pit
			Section in the short axis	Section in the long axis	
0	0.60	0	0.24	0.23	0.29
1	3.75	1	1.97	1.64	1.74
3	11.77	3	5.29	4.50	4.70

The larger the initial horizontal stress of rock mass is, the greater the influence will be. Figure 10-6 shows that the lateral horizontal stress is $\sigma_L=3\gamma H$, resulting in a tension zone at 30°.

When the rock mass has no lateral horizontal stress(σ_L =0), the 45° slope does not nearly have a tension zone, and the 60° slope begins to have a tension zone. The comparison shows that the horizontal initial stress of rock mass has a considerable influence on slope stability.

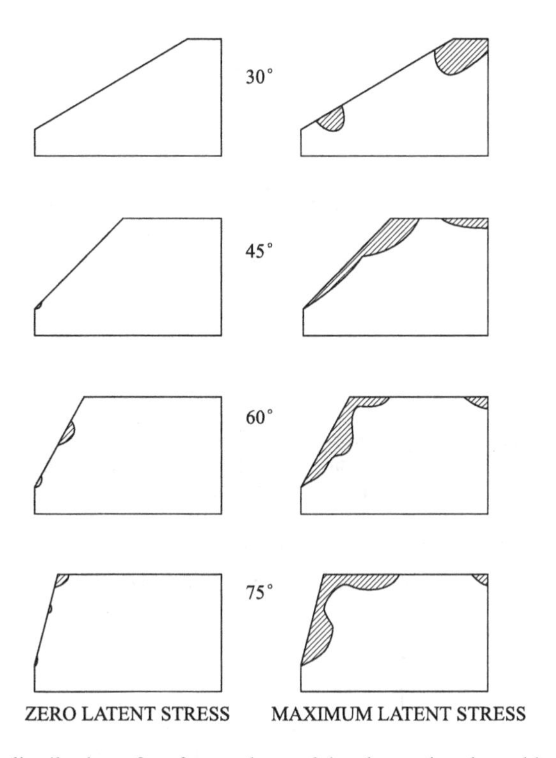

Figure 10-6 The tension distribution of surface and top of the slope when lateral horizontal stress is σ_L=0 and $\sigma_L=3\gamma H$(Stacey,1970)

2. Influence of the slope shape

Slope shape includes slope angle, slope bottom width, and slope form, which significantly influence the stress distribution of slope rock mass. Slope angle will affect the concentration of stress at the toe of the slope and the tension zone at the top of the slope (Figure 10-6). The more concentrated the slope foot stress, the higher the maximum shear stress. The relationship between slope angle and maximum shear stress (at the toe of slope) is shown in Figure 10-7. Curves are offered for both zero pit floor width, and residual toe stress conditions under lateral latent boundary stresses σ_L=0 and $3\rho gH$.

The width of the valley bottom (or pit bottom) is closely related to the mutual extrusion of the slopes on both sides of the valley. In Figure 10-8, the pit floor width was increased from w=0~6H. The maximum shear stress at the toe of the slope is plotted versus the floor width for a slope of =30° , σ_L =3gH. It is found that the curve decreases steeply to a minimum near w=0.8h, after that remaining constant for all practical purposes.

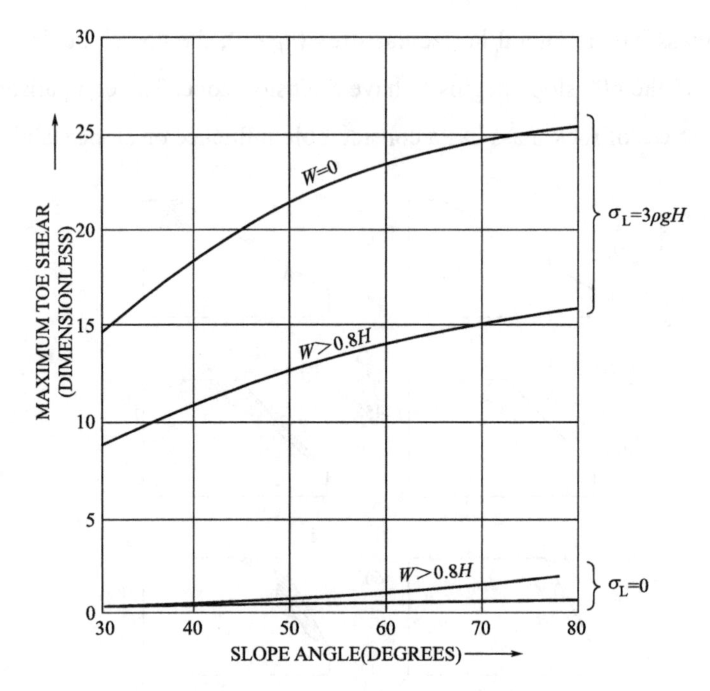

Figure 10-7 Effect of slope angle on maximum shear stress at the toe (Stacey, 1970)

W—valley bottom width; *H*—slope height

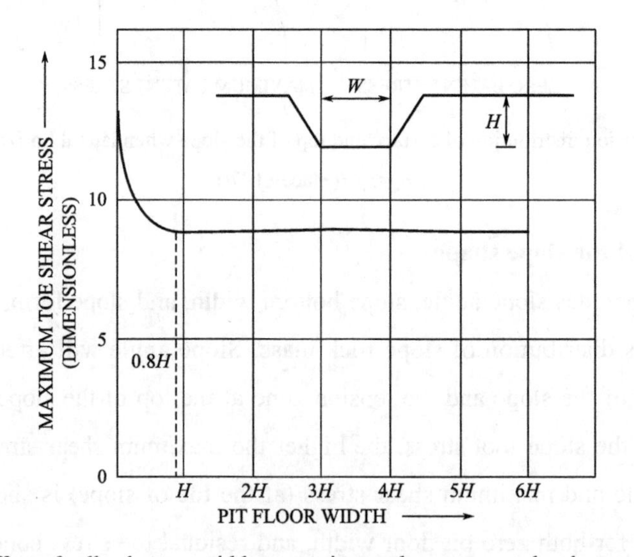

Figure10-8 Effect of valley bottom width on maximum shear stress at the slope toe (Stacey, 1970)

The slope can be divided into a concave shape, convex shape, and straight shape according to the shape of the slope surface. The concave slope is commonly found in the steep wall at the back edge of the landslide, the head of a gully, the source of branch valley, the turning place of the open pit, etc. Both sides support this kind of slope in the pitch direction. The stress concentration phenomenon at the slope toe will be reduced, and the maximum shear stress value will therefore be significantly reduced. The smaller the radius of curvature, the more significant the tendency of shear stress to slow down. Convex slope, on the contrary, even suffers tension in the strike direction, which is not conducive to stability.

3. Influence of the discontinuity

Rock mass structure has an evident influence on the stress characteristics of a slope. The inhomogeneity and discontinuity of rock mass will cause stress concentration and discontinuous distribution. The stress concentration level around the weak surface varies with the strength of the surrounding rock. In rock mass of low strength, such as claystone, tuff, carbonaceous shale, slate, phyllite, and marl, etc., the stress concentration level is also low. In hard rock mass, the stress concentration level is higher along the weak surface, and the stress along the hard rock side is higher at the interface between soft and hard rock mass. The characteristics of stress concentration are closely related to the occurrence of the discontinuity and the direction of the principal compressive stress, as shown in Figure 10-9.

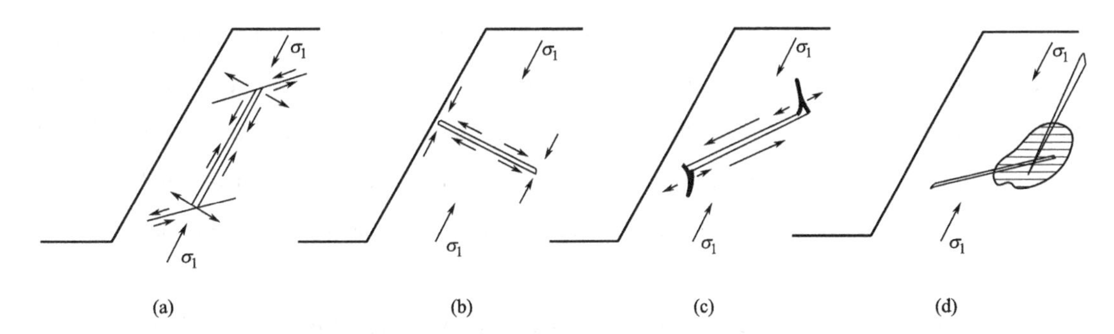

Figure 10-9 Stress concentration on the discontinuity of a slope

When the principal compressive stress is parallel to the discontinuity, tensile and shear stress concentrations occur near the endpoints of the discontinuity, resulting in tensile cracks on both sides of the discontinuity (Figure 10-9a). The slope deformation and failure caused by tensile cracks are pronounced, which is more likely to occur when the discontinuity is weak and muddy.

In the case of Figure 10-9(b), tension stress along the discontinuity will appear, and compressive stress perpendicular to the discontinuity will appear at the endpoints. It is helpful to the compaction of the discontinuity and the stability of the slope.

Figure 10-9(c) shows that the shear stress is mainly concentrated along the discontinuity, and tensile stress occurs at the endpoint or stress block region. Suppose the discontinuity intersects with the principal compressive stress axis at an angle of 30°~40°. In that case, the maximum shear stress and tensile stress will occur, resulting in the shear slide along the discontinuity easily, which is the most detrimental to the stability of the slope. If the discontinuity is weak, the slope will be challenging to stabilize.

At the intersection of discontinuities, stress is blocked, compressive stress and tensile stress are intensely concentrated (Figure 10-9d), forming a moving source of slope failure, which gradually expands into a slip plane under certain conditions, causing slope failure.

Discontinuities in rock mass are generally not single. Still, combining several discontinuities with a different attitude makes the stress distribution in fractured slopes very complex. It is necessary to make a comprehensive analysis and distinguish the prominent discontinuity and secondary discontinuity to understand the relationship between them fully.

10.3 A Guide to Understanding Landslides

10.3.1 What is a Landslide?

Geologists, engineers, and other professionals often rely on sole and slightly different explanations of landslides. This diversity in definitions reflects the complicated nature of the many disciplines associated with landslide appearances (Highland & Bobrowsky, 2008). The landslide is a general term used to describe the downslope movement of soil, rock, and organic materials under gravity's effects and the landform that results from such activity (Figure 10-10).

Various classifications of landslides are associated with the specific mechanics of slope failure and the properties and characteristics of failure types. Several other phrases/terms are used interchangeably with the term "landslide," including mass movement, slope failure, and so on, and such terms are commonly heard applied to all types and sizes of landslides. Understanding the essential parts of a typical landslide is helpful regardless of the exact definition used or the type of landslide under discussion. Figure 10-11 shows the position and the most common terms used to describe the unique parts (Highland & Bobrowsky, 2008; Cruden,1993; Zhang & Dong, 1995).

Accumulation. The volume of the displaced substance, which lies above the initial terrain surface.

Crown. The practically undisplaced material is still in place and adjacent to the highest parts of the prominent scarp.

Depletion. The volume is bounded by the prominent scarp, the depleted mass, and the original ground surface.

Figure 10-10 A landslide buried part of the highway and blocked traffic

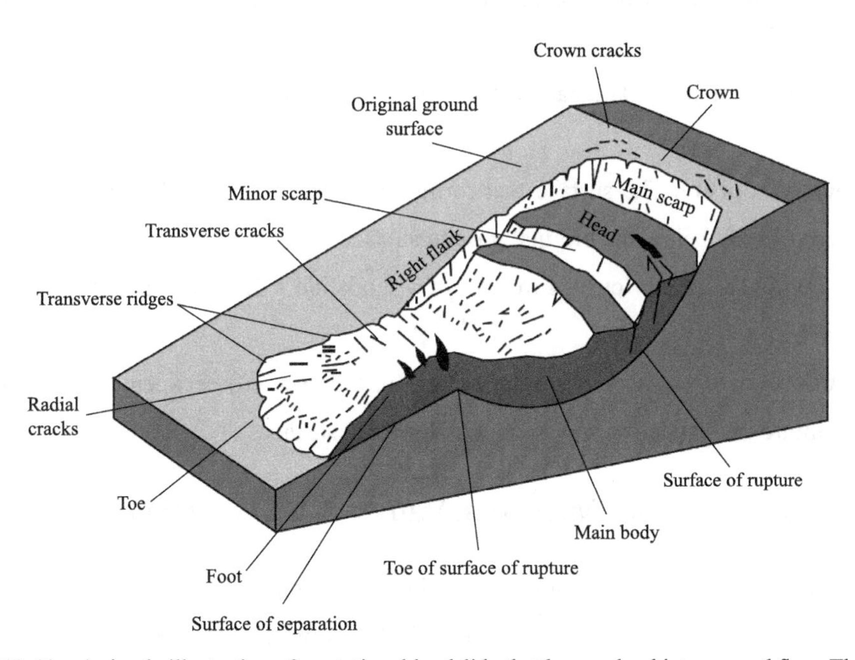

Figure 10-11 A simple illustration of a rotational landslide that has evolved into an earthflow. The image illustrates commonly used labels for landslide parts (Modified from Varnes, 1978)

Depleted mass. The volume of the displaced body, which overlies the rupture surface but underlies the initial terrain surface.

Displaced material. Material displaced from its original location on the slope by removal in the landslide, and it forms both the depleted mass and the accumulation.

Flank. The undisplaced material is adjacent to the sides of the rupture surface. Compass directions are preferable in describing the flanks, but if left and right are used, they refer to the flanks as viewed from the crown.

Foot. The portion of the landslide has moved beyond the toe of the surface of rupture and overlies the original ground surface.

Head. The upper parts of the landslide and the contact between the displaced material and the main scarp.

Main body. The part of the displaced material of the landslide that overlies the surface of rupture between the main scarp and the toe of the surface of rupture.

Main scarp. A steep surface on the undisturbed ground at the upper edge of the landslide, caused by movement of the displaced material away from the undisturbed environment. It is the visible part of the surface of rupture.

Minor scarp. Differential movements within the displaced material produce a steep surface on the displaced material of the landslide.

Original ground surface. The surface of the slope that existed before the landslide took place.

Surface of separation. The part of the original ground surface overlain by the foot of the landslide.

Surface of rupture. The surface that forms (or which has formed) the lower boundary of the displaced material below the original ground surface.

Toe. The lower, usually curved margin of a landslide's displaced material. It is the most distant from the main scarp.

Toe of surface of rupture. The intersection (usually buried) between the lower part of the surface of rupture of a landslide and the original ground surface.

10.3.2 Basic Landslide Types from Varnes & IAEG & USGS

There are many different schemes for classifying slope damage at home and abroad (Zhang et al., 2014). In the 1990s, the Landslide Committee of the International Association for Engineering Geology and Environmental (IAEG) recommended using Varnes's landslide classification as the international standard scheme. Varnes' landslide classification takes the slope's material composition and movement mode into account (Table 10-2).

Table 10-2 Types of landslides, abbreviated version of Varnes' classification of slope movements (Varnes,1978)

TYPE OF MOVEMENT		TYPE OF MATERIAL		
		BEDROCK	ENGINEERING SOILS	
			Predominantly coarse	Predominantly fine
FALLS		Rock fall	Debris fall	Earth fall
TOPPLES		Rock topple	Debris topple	Earth topple
SLIDES	ROTATIONAL	Rock slide	Debris slide	Earth slide
	TRANSLATIONAL			
LATERAL SPREADS		Rock spread	Debris spread	Earth spread
FLOWS		Rock flow (deep creep)	Debris flow (soil creep)	Earth flow
COMPLEX		Combination of two or more principal types of movement		

A landslide is a downslope displacement of rock or soil, or both, occurring on the surface of rupture—either curved (rotational slide) or planar (translational slide) rupture—in which much of the substance often moves as a coherent or semicoherent mass with small internal deformation. It should be noted that, in some cases, landslides may also include other types of movement, either at the commencement of the failure or later, if features change as the displaced material moves downslope.

USGS provided descriptions and illustrations of the various landslides depending on the kind of movement and the material involved (Highland & Bobrowsky, 2008). In brief, material in a landslide mass is either rock or soil (or both); the latter is described as earth if mainly composed of sand-sized or finer particles and debris if composed of coarser fragments. The type of movement describes the internal mechanics of how the landslide mass is displaced. Thus, landslides are expressed using two terms that refer respectively to material and action (that is, rockfall, debris flow, and so forth). Landslides may also form a complex failure encompassing more than one type of movement (rock slide—debris flow).

1. Falls

A fall begins with detachment soil or rock from a steep slope along a surface on which little or no shear displacement has occurred. The material subsequently descends mainly by falling, bouncing, or rolling. Falls are abrupt, downward movements of rock or soil, or both, that detach from steep slopes or cliffs (Figure 10-12). The falling material usually strikes the lower slope at angles less than the angle of fall, causing bouncing. The falling mass may break on impact or begin rolling on steeper slopes or continue until the terrain flattens.

Figure 10-12 The collapse that occurred on the highway in Changdu, Tibet

2. Topple

A topple is recognized as the forward rotation out of a slope of soil or rock mass around a point or axis below the center of gravity of the displaced mass (Figure 10-13). Toppling is sometimes driven by gravity exerted by the weight of material upslope from the displaced mass. Sometimes toppling is due to water or ice in cracks in the mass. Topples can consist of rock, debris (coarse material), or earth materials (fine-grained material). Topples can be complex and composite.

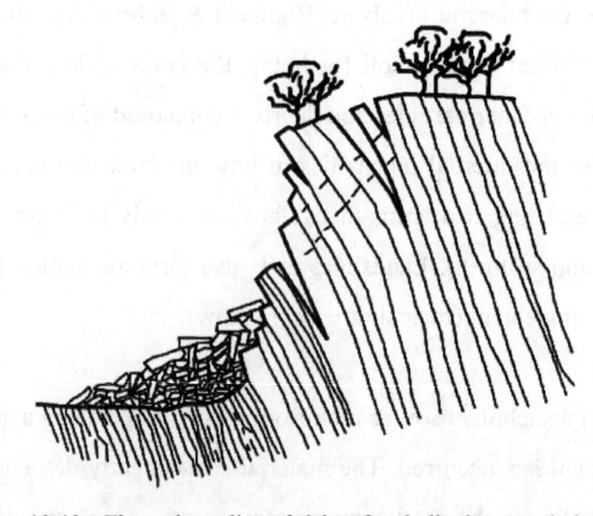

Figure 10-13 Flexural toppling of slabs of rock dipping steeply into face

3. Slides

A slide is a downslope movement of a soil or rock mass occurring on rupture surfaces or relatively thin zones of intense shear strain. Movement does not initially coincide over the whole of what eventually becomes the surface of rupture; the volume of displacing material enlarges from an area of local failure (Figure 10-14).

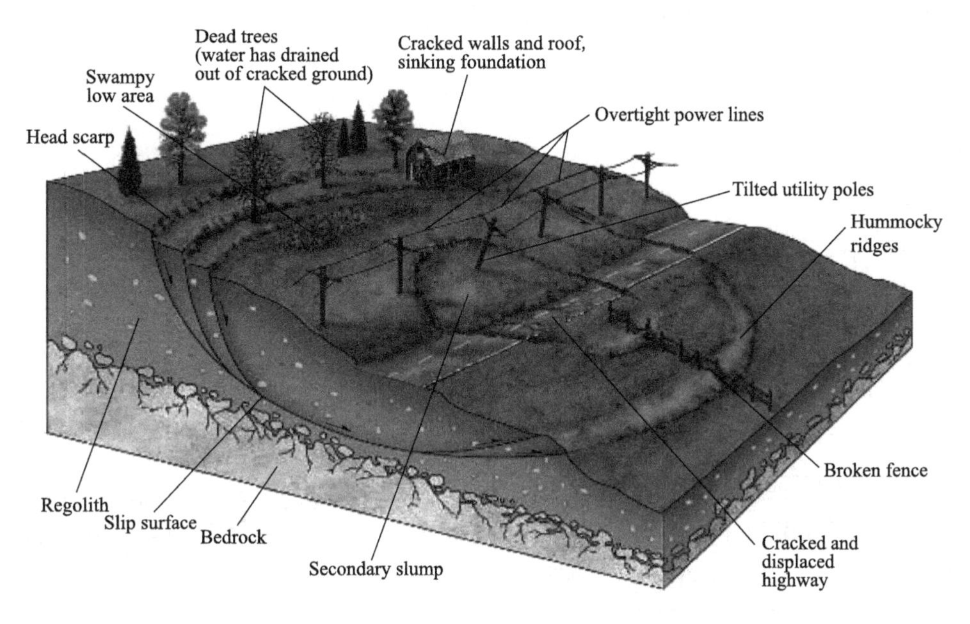

Figure 10-14 Surface features warn that a large slump is beginning to develop

Rotational Landslide: A landslide on which the surface of rupture is curved upward (spoon-shaped), and the slide movement is more or less rotational about an axis parallel to the contour of the slope. The displaced mass may, under certain circumstances, move as a relatively coherent mass along the rupture surface with slight internal deformation. The head of the displaced material may move almost vertically downward, and the upper surface of the displaced material may tilt backward toward the scarp. It is called a slump if the slide is rotational and has several parallel curved planes of movement.

Translational Landslide: The body in a translational landslide moves out, down, and outward along a relatively planar surface with little rotational movement or backward tilting. This type of slide may progress over substantial distances if the rupture surface is sufficiently inclined, unlike rotational slides, which tend to restore the slide equilibrium. The material in the slide may range from loose, unconsolidated soils to wide slabs of rock or both. Translational slides commonly fail along geologic discontinuities such as faults, joints, bedding surfaces, or the contact between rock and soil. In northern environments, the slide may also move along the permafrost layer. The Xintan landslide in the Zigui country of Hubei Province lies on the north bank of the Yangtze River near the little town of Xintan (Figure 10-15). On the 12th of June 1985, a largescale, intense landslide of 30 million m^3 of rock-soil mass occurred. It destroyed Xintan. About 2 million m^3 of rock-soil mass rushed into the Yangtze River, blocking up to half of the river flow. A more than 40m high wave destroyed 64 boats and 13 small steamers along the 10km of the river course near the landslide (Huang et al., 2009).

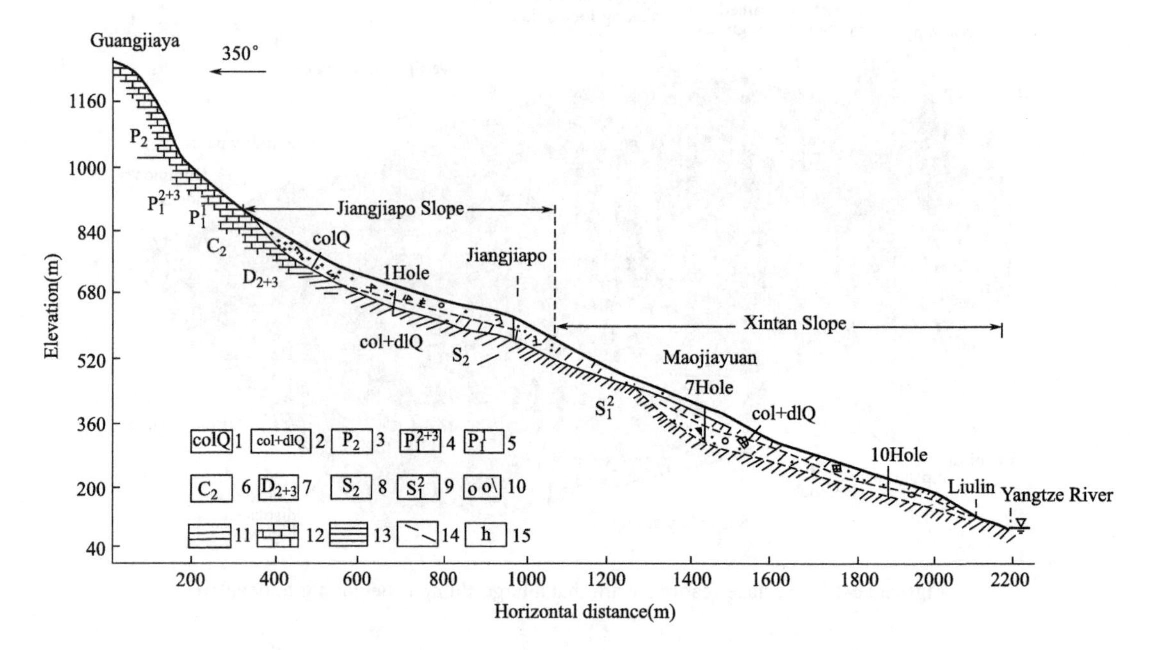

Figure 10-15 The typical section of the Xintan slope

1—Colluviums of the Quaternary system; 2—colluvial slope deposits; 3—Upper Permian system; 4—Sub-Permian series; 5—Maanshan group;6—Huanglong group of Middle Carboniferous; 7—Middle-Upper Devonian series; 8—Shamao group in Middle Silurian series; 9—Luoreluoping group in Sub-Carboniferous series; 10—blocks of limestone; 11—shale; 12—limestone; 13—sandstone; 14—slip-bed inferred; 15—bore hole

4. Spreads

An extension of cohesive soil or rock mass is combined with the fractured mass of cohesive material into the softer underlying material. Spreads may result from liquefaction or flow (and extrusion) of the softer underlying material. Types of spreads include block spreads, liquefaction spreads, and lateral spreads.

The probability of recurring is high in areas that have experienced previous problems. It is the most prevalent in the regions that have an extreme earthquake hazard and liquefiable soils. Lateral spreads are also associated with susceptible marine clays and are a common problem. Figure 10-16 and Figure 10-17 show a schematic and an image of a lateral spread.

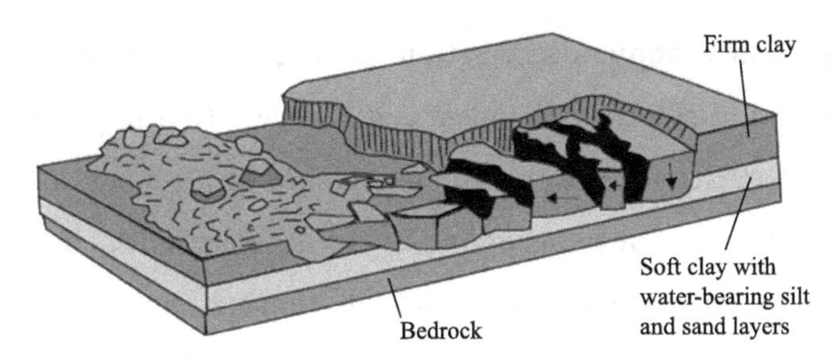

Figure 10-16 Schematic of a lateral spread. A liquefiable layer underlies the surface layer

Figure 10-17 A lateral spread damage to a roadway as a result of a earthquake

5. Flows

A flow is a spatially continuous action in which the shear surfaces are short-lived, closely spaced, and usually not preserved. The element velocities in the displacing mass of a flow resemble those in a viscous fluid. Often, there is a stage of change from slides to flows, depending on the movement's water content, movement, and evolution.

Debris Flows: A form of rapid mass transportation in which loose soil, rock, and sometimes organic material combine with water to create a slurry that flows downslope. They have been informally and inappropriately called "mudslides" due to a large amount of delicate matter present in the flow. Infrequently, as a rotational or translational slide gains velocity and the internal mass loses cohesion or obtains water, it may develop into a debris flow. Dry flows can sometimes occur in cohesionless sand (sand flows). Debris flows can be fatal as they can be remarkably rapid and may occur without any warning.

10.3.3 Landslide classification based on geomechanical modes

Slope deformation failure is a particular type of rock deformation and failure, and it can also be summarized as several similar basic geomechanical modes, namely: creeping (slip)-tensile cracking, slip-compression induced tensile cracking, bending-pull cracking (dumping-pulling cracking), plastic flow-pulling cracking and slipping-bending. There may also be two or more deformation modes in the same slope deformation body, which can be combined in different ways. Similarly, a specific deformation mode can also evolve (transform) into another mode in the course of evolution (Zhang et al., 2014).

The slope mentioned above deformation geomechanics model reveals the inherent mechanical mechanism of slope development and changes. To a large extent determines the possible ways and characteristics of the ultimate failure of the slope rock mass. Therefore, the failure type can be determined according to the deformation mode associated with the mechanical mechanism, such as creep-split landslide, slip-split landslide, bend-split collapse, plastic flow-split landslide and plastic flow-split landslide, etc. Therefore, it can also be called the geomechanical model of slope deformation and failure.

Theory and practice have proved a specific connection between the slope deformation and geomechanical failure model and the rock mass structure. They become the evolution link between various slopes and their ultimate failure. By studying the formation and evolution of multiple models, the purpose of systematic evaluation and prediction of slope stability can be achieved.

10.3.4 Tensile-rupture and sheared-sliding (TRSS) Failure

Under the condition of strong earthquake and the amplification effect of elevation, the stress produced by horizontal seismic wave will be much higher than the tensile strength of rock mass, therefore, the vertical fracture surface nearly parallel to the slope surface is easy to be generated (Figure 10-18). Evidently, under seismic conditions, stress state and response of rock mass differs significantly in many aspects from conventional static gravitational loading condition. Under the gravitational loading condition, the failure mode of the slope is shearing sliding along

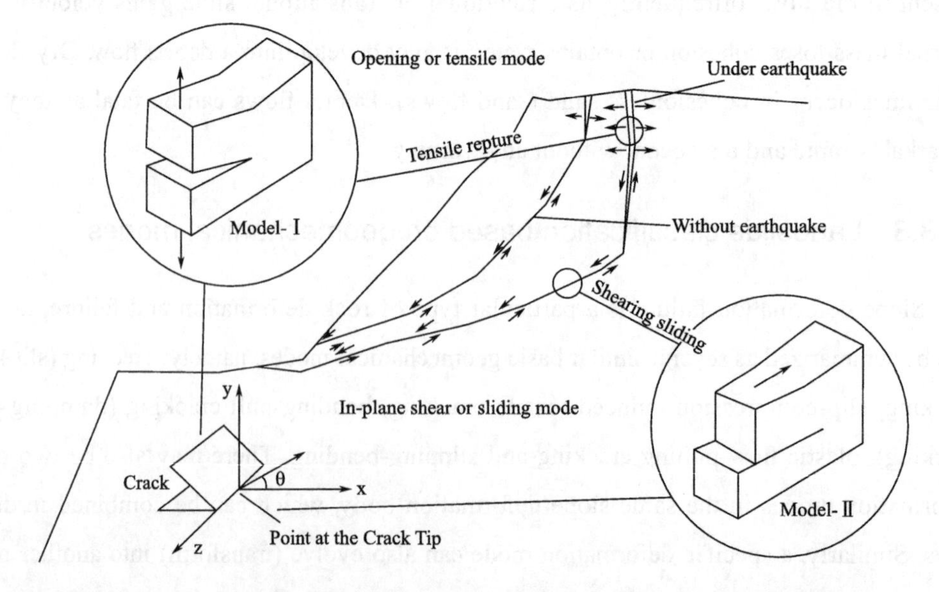

Figure 10-18 The different sliding surfaces under static and earthquake loading conditions

the weak surface, and then the tensile cracks of small-scale occur at the trailing edge. The shear slip surface grows and demonstrates nearly mode of pure shear; the rear fracture is shallower, and the scale is relatively smaller, showing the deformation and failure characteristics of the bottom shear slip. In contrast, under the earthquake conditions, the first failure is the appearance of deep tension cracks, and then shearing slip along the bottom. The rear fracture is deep, steep and rough, and the bottom shear slip surface is relatively short, and it is characterized by shearing-tension, and showing tensile failure of the edge. We can use term “tensile-rupture and sheared-sliding” (TRSS) to describe this kind of destruction phenomenon, that is, under the earthquake condition, the first location of the slope failure is the appearance of tension cracks on the crest, and then shearing sliding along the bottom slip surface; after the tensile fracture and shear fissure are connected, landslide will be formed (Liu, et al.,2020).

Tensile rupture fractures are observed at the top of the slope under earthquake loading condition and sheared sliding is formed after tensile rupture. The stress loading modes and crack coordinates system are also demonstrated.

The Tangjiashan landslide is located on the right bank of Tongkou River in Sichuan Province, China, around 125km from 2008 Wenchuan earthquake epicenter of Yingxiu Town. The landslide is about 2.5km away from Yingxiu—Beichuan part of Longmenshan Central Fault Zone (LCFZ). The Tongkou River snakes through the landslide area (Figure 10-19a). The water level in the Tongkou River during summer was about 664m; the river surface was 100m to 130m in width, and 0.5 to 4m in depth. Before the earthquake, the terrain gradient of the slope was about 40° , the elevation of the slope toe was about 665m and the elevation the slope crest was about 1400m above sea level (Figure 10-20b).

(a)

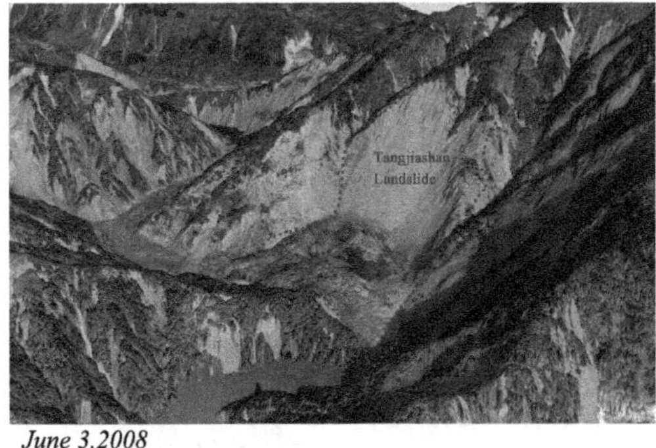

(b)

Figure 10-19 The digital elevation model and remote sensing model of the research zone

(a) before the earthquake (from upstream); (b) after the earthquake (from upstream)

The Wenchuan earthquake (Mw 8.0, 2008) occurred along the LCFZ with an epicenter depth of 19km (shallow earthquake). The seismic intensity in the area of the landslide was XI. Tangjiashan slide was induced by the earthquake almost simultaneously, and it was the most hazardous of the 40 or so valley-blocking landslides triggered by the shock of the world. The main strata of the slope are the lower Cambrian system/($\in_1$), contains siltstone, silicalite, marlite, and mudstone. The rocks are well bedded striking EW and dip moderately to steeply towards the north. The landslide also involved the alluvial deposit of Quaternary sediments (Q_{al}). These strata of the slope are dipping outward (North direction) and forming a bedding slope of steep incline (Figure 10-19a). In contrast, the other side of the valley forms a reverse dip slope.

Combining photograph interpretation by helicopter low altitude flight with the field geological surveying nearby the remnant mountain body behind the scarp after the landslide, it indicates that there are mainly three sets of cracks distributing on the surface of the remnant mountain body after earthquake, which are all tension cracks with the opening width of about 10~50cm. The distribution of three sets of fractures: (1)the set along the strike direction of N40° ~50° E or N70° E~EW with the extended length from 20 to 200m(dip angle is between 60° to 80°), paralleled to the trailing edge of landslide, showing unloading tension failure towards the free face of Tongkou river (Figure 10-20a); (2)the set along the strike direction of N10° W~N10° E, paralleled to the trend of the Big Ditch and the Small Ditch valleys, belonging to unloading tension failure towards the free faces of the Big Ditch and the Small Ditch (Figure 10-20a); (3)the set along the strike direction of N70° W, distributing on the ridge lines, paralleled to the trend of the ridge lines, belonging to the seismic crack in ridge lines of the steep slopes (Figure 10-20a). It can be diagnosed that there existed tension cracks indication of the slope crest before earthquake occurrence, and the seismic loading leads to the development and initiation

Figure 10-20 The slope shape after sliding

(a) Distribution of cracks in the residual mountain after landslide; (b) Engineering geological profile through the Tangjiashan landslide and landslide dam

of tension cracks. Tension cracks expand downward to cause the development of shear cracks and all tracks ran throughout at last. The Tangjiashan landslide is a typical failure mode of TRSS from the field investigation.

After the landslide occurrence, with a volume of 20 million m^3, the debris of landslide mass accumulated and formed the landslide dam with a thickness of 80 to 125m. The landslide that dammed the river covers about 800m along the river bend and about 650m along transverse river direction. About 100 people lost their lives in this disaster. It is supposed to be residents, vehicles, and pedestrians on the roads. If the dam breaks down, the water (240 million m^3) will barrel down into the river. The people in Beichuan County (around 6.5km downstream) will suffer the worst battering if the blockage cannot be excavated timely. The engineer used digging equipment and explosives to dig a drainage channel in the dam to ease water from the dangerous lake and relieve the pressure behind it. On June 7, 2008, the channel was completed and the water can flow downstream.

10.4 The modes of rock slope failure surface

10.4.1 Plane failure

When the sliding surface is flat, it is the simplest case, often seen in bedding landslides controlled by weak surfaces. Assuming that only the weight of the rock mass is considered, the frictional resistance of the lateral cutting surface is not considered, and a unit width perpendicular to the sliding direction is taken for the calculation. The section along the sliding direction is shown in Figure 10-21. *AC* is the sliding surface, its length is *L*, the weight of sliding body *ABC* is *G*, the sliding force is $G\sin\alpha$, the anti-sliding force is $G\cos\alpha\tan\varphi+cL$, ($\varphi$—internal friction angle, c-cohesion), the safety factor *K* can be Calculate as follows:

$$K=\frac{G\cos\alpha\tan\varphi+cL}{G\sin\alpha}=\frac{\tan\varphi}{\tan\alpha}+\frac{cL}{G\sin\alpha} \tag{10-1}$$

If the section *ABC* of the sliding body is a triangle, $G=1/2hL\cos\alpha$, substituting the above formula to simplify:

$$K=\frac{\tan\varphi}{\tan\alpha}+\frac{4c}{\gamma h\sin2\alpha} \tag{10-2}$$

where h is the height of the landslide body; γ is the rock weight; K is the safety factor.

It can be seen from the above formula that the stability safety factor of the slope reduces with the increase of the angle α and the height of the sliding body h; it increases with the rise of the values of φ and c. When most slopes are damaged, they all happen after water seeps into the rock mass. Therefore, the effect of water pressure should be considered in general calculations. In addition, other loads acting on the slope and seismic forces should be considered.

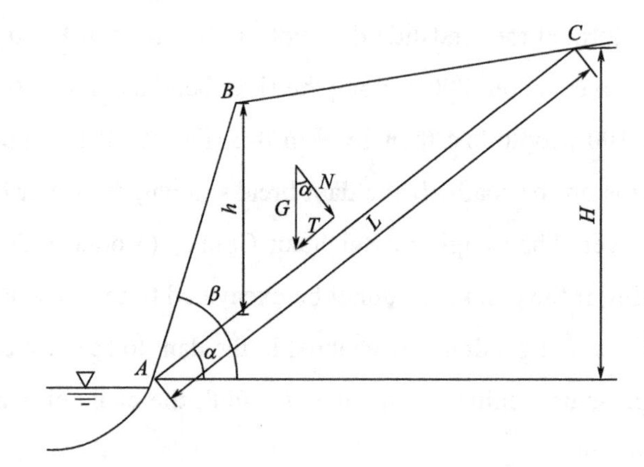

Figure 10-21 Schematic section view of slope stability calculation

10.4.2 Wedge failure

1. Definition of wedge geometry

The wedge in Figure 10-22 is formed by bedding on the left and a conjugate joint set on the right. As in Figure 10-22, the line of intersection daylights in the slope face, and failure occurred. However, sliding occurred almost entirely on the bedding in this wedge, with the joint acting as a release surface. Therefore, the shear strength of the joint has little effect on stability.

The geometry of the wedge for investigating the basic mechanics of sliding is defined in Figure 10-23. Based on this geometry, the general conditions for wedge failure are as follows:

(1) Two planes will always intersect in a line (Figure 10-23a). On the stereographic net, the line of intersection is described by the point where the two great circles of the surfaces intersect, and the orientation of the line is specified by its trend α_i and its plunge ψ_i (Figure 10-23b).

(2) The plunge of the line of junction must be steeper than the average friction angle of the two slide planes and flatter than the dip of the face, that is $\psi_{fi} > \psi_i > \varphi$ (Figure 10-23b and c). The inclination of the slope face ψ_{fi} is estimated in the view at right angles to the line of intersection. Note that the ψ_{fi} would only be equal to ψ_f, the true dip of the slope face if the dip direction of the line of the meeting were the same as the dip direction of the slope face.

Figure 10-22 Typical wedge failure involving sliding on two persistent joints with line of intersection of joints daylighting at toe of rock face (After Hoek, 2000)

(3) The line of intersection must dip in a direction out of the face for sliding to be feasible; the possible range in the trend of the meeting line is between α_i and α_i'(Figure 10-23d).

The Forces parallel to the line and perpendicular to the two sliding faces can be solved to determine the safety factor assuming that the sliding orientation is parallel to the line of intersection of the two sliding planes. This analysis leads to

$$F=\frac{(R_A+R_B)\tan\varphi}{W\sin\psi_i}$$
$$R_A+R_B=\frac{W\cos\psi_i\sin\beta}{\sin\frac{1}{2}\delta} \tag{10-3}$$

The various forces and angles used in these formulae are shown in the individual parts of Figure 10-23. Consolidating these formulae results in

$$F=\frac{\sin\beta}{\sin\frac{1}{2}\delta}\times\frac{\tan\varphi}{\tan\psi_i} \tag{10-4}$$

A direct insight into the fundamental mechanism of wedge instability is achieved by abbreviating the equation to

$$F=K_w\times F_p \tag{10-5}$$

Where F is the safety factor of a wedge supported by friction only; F_p is the safety factor of a plane failure in which the slope face is inclined at, and the failure plane is inclined at; K_w is the wedge factor that depends upon the included angle of the wedge and the angle of tilt of the wedge.

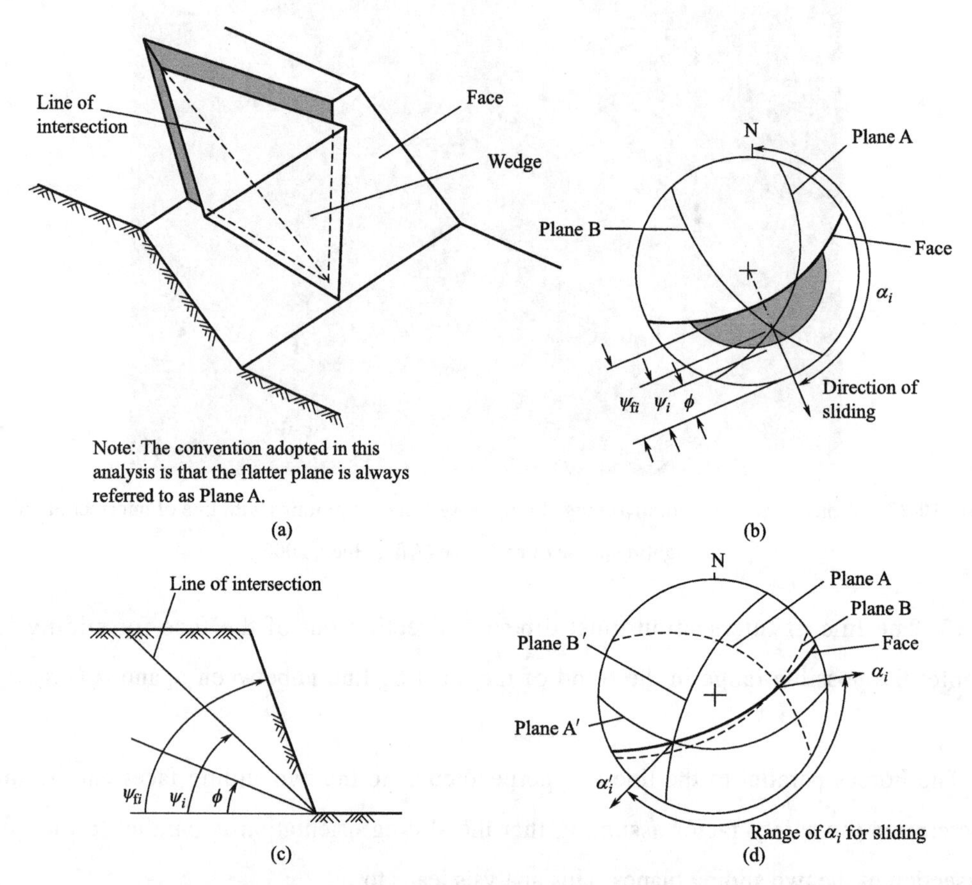

Figure 10-23 Geometric conditions for wedge failure

(a) pictorial view of wedge failure; (b) stereographic showing the orientation of the line of intersection, and the range of the plunge of the line of intersection ψ_i where failure is feasible; (c) view of slope at right angles to the line of intersection; (d) stereonet showing the range in the trend of the line of intersection α_i where wedge failure is feasible

2. 3DEC Results and Discussion

Figure 10-24 shows a wedge failure controlled by two intersecting structural features in the bench of an open pit mine (Itasca,2015). The two structural features defined by:

A: dip = 40° dip direction = 130°

B: dip = 60° dip direction = 220°

The same mechanical properties are assumed for both planes.The mechanical properties selected for this model are: Density ρ = 2000kg/m^3; Joint shear and normal stiffness $K_n = K_s$ =10MPa/m;

Gravity acceleration $g = 10\ m/s^2$. The 3DEC(3 Dimensional Distinct Element Code) model shown in Figure 10-24 consists of four rigid blocks. Three of the blocks are fixed, and only the isolated wedge is allowed to move under gravity. Two of the wedge faces have contact with faces of adjacent blocks. The edge at the back of the wedge has contact with an intersection of two rear blocks. The wedge is first allowed to consolidate under its weight by setting the friction on both planes to a high value and applying adaptive global damping. Then, friction is reduced until failure occurs. A history of vertical velocity of the wedge clearly indicates the simulation step at which instability occurs (Figure 10-25). Using 3DEC, a critical friction angle of $\varphi = 33.12°$ is calculated and compared with the analytical solution $\varphi = 33.36°$.

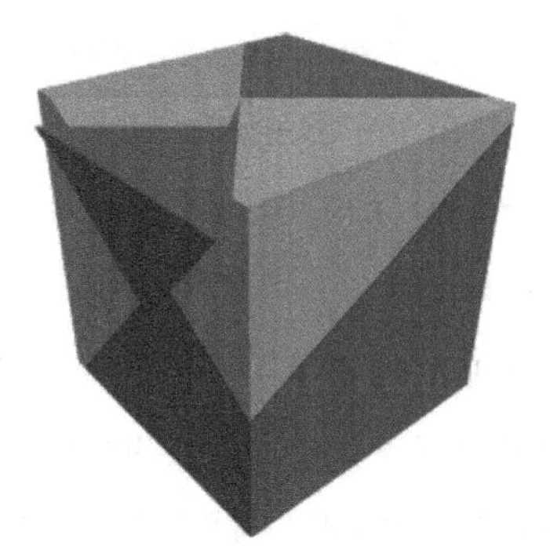

Figure 10-24 Sliding wedge model

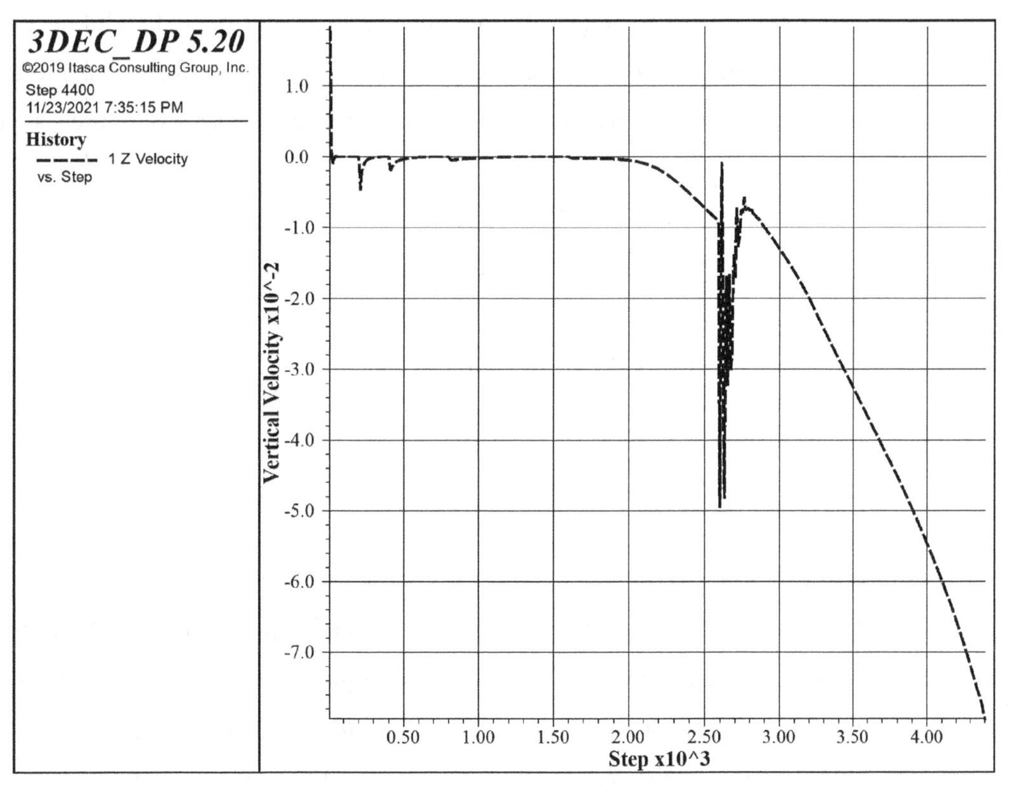

Figure 10-25 History of vertical velocity of the wedge

10.4.3 Circular Failure

Although this textbook is concerned primarily with the stability of rock slopes containing well-defined sets of discontinuities, it is also required to design cuts in weak materials such as deeply weathered or closely fractured rock and rock fills. In such materials, failure occurs along a surface that approaches a circular shape (Figure 10-26), and this section is devoted to a discussion on the stability analysis of these materials.

1. Conditions for circular failure and methods of analysis

In the example of a closely fractured or highly weathered rock, a strongly defined structural model no longer exists, and the slide surface is free to find the line of least resistance through the slope. Perceptions of slope failures in these materials suggest that this sliding surface usually takes the kind of a circle, and most stability methods are based upon this investigation. Figure 10-26 shows a typical circular failure in a highly weathered rock slop. The circumstances under which circular failure will occur arise when the individual particles in a soil or rock mass are minimal compared with the size of the slope. Likewise, soil consisting of sand, silt, and smaller particle sizes will display circular slide surfaces, even in slopes only a few meters in height. Profoundly altered and weathered rocks and rocks with closely spaced, randomly oriented discontinuities will also tend to fail in this mode. It is appropriate to design slopes in these materials to assume that a circular failure process will develop.

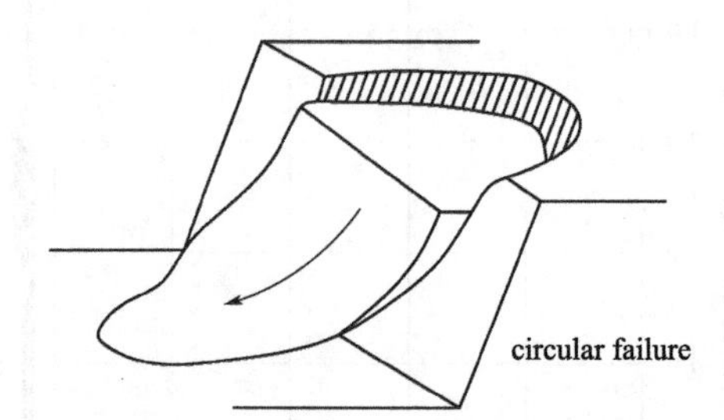

Figure 10-26 Circular failure in highly weathered

2. Stability analysis methods

The stability analysis of circular failures shall follow the limit equilibrium procedure similar to that described above for plane and wedge failures. This procedure compares the available shear strength along the sliding surface with force required to maintain the slope equilibrium. Applying this procedure to circular failures involves the division of the slope into a series of slices that are usually vertical but may be inclined to coincide with certain geological features (Figure 10-27). Each slice is inclined at an angle, and rearranging this equation, we have ψ_b and have an area A. In the simplest case, the forces acting on the base of each slice are the shear resistance S due to the shear strength of the rock (cohesion c; friction angle φ), and forces E (dip angle ψ; height h above base) acting on the sides of the slice. The analysis procedure considers the slice-by-slice equilibrium conditions, and if an equilibrium state is satisfied for each slice, it is also satisfied for the whole sliding mass. The investigations are static indeterminate, and assumptions are needed to compensate for the imbalance between the equations and the unknowns (Wyllie & Mah, 2004).

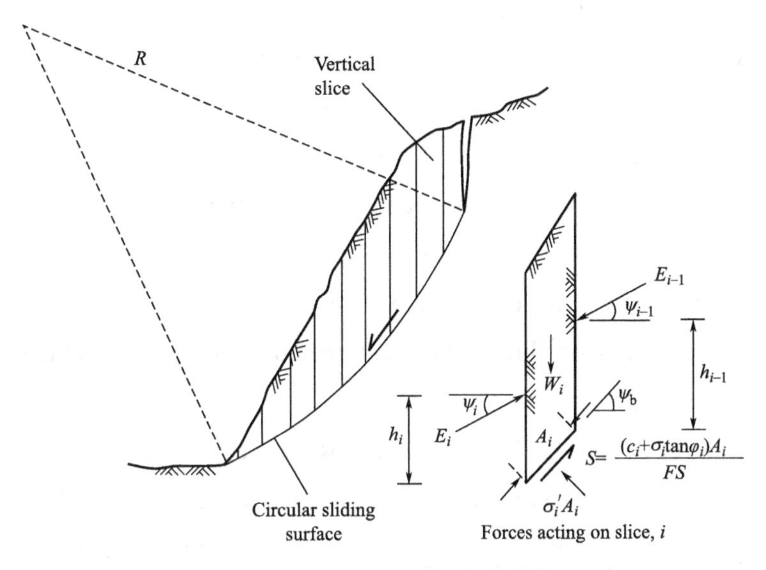

Figure 10-27 Stress analysis of typical sliding surfaces

The various limit equilibrium analysis procedures either make assumptions to balance known and unknown, or they do not satisfy all the equilibrium conditions. For example, the Spencer Method assumes that the inclination of the side forces is the same for every slice. At the same time, the Fellenius and Bishop methods do not satisfy all equilibrium conditions. The factor of safety of the circular failure based on limit equilibrium analysis is defined as

$$FS=\frac{\textit{shear strength available to resist sliding}(c+\sigma\tan\varphi)}{\textit{shear stress required for equilibrium on slip surface}(\tau_e)}$$

and rearranging this equation, we have

$$\tau_e=\frac{c+\sigma\tan\varphi}{FS} \tag{10-6}$$

The solution for the safety factor is to use an iterative process in which an initial estimate is made for FS, which is refined with each iteration. The influence of various normal stress distributions upon the factor of safety has been examined by Frohlich (1955). He found that a lower bound for all safety factors is given by the assumption that the normal stress is concentrated at a single point on the slide surface. Similarly, the upper bound is obtained by assuming that the normal load is concentrated at the two ends of the slide surface.

The shear strength reduction technique (SSRT) has a number of advantages over the method of slices for slope stability analysis. Most importantly, the critical failure surface is found automatically. Application of the technique has been limited in the past due to the long computer run time required. But with the speed of desk computers increasing, the technique is becoming an reasonable alternative to the method of slices, and is being used increasingly in engineering practice (Dawson et al., 1999; Wei et al., 2010). A difficulty with all the limit equilibrium methods is that they are based on the assumption that the failing mass can be divided into slices. This in turn necessitates further assumption relating to side force direction between slices, with consequent implications for equilibrium. The side force assumption is one of the main characteristics distinguishing one limit equilibrium method from another, yet the concept of side forces is entirely artificial.

In the method of SSRT, the *FOS* of a slope is defined here as the factor by which the original shear strength parameters (cohesion and friction angle) must be divided to bring the slope to the point of failure (Griffith & Lane, 1999). The factored shear strength parameters and are therefore given by (Dawson et al., 1999)

$$c_f'=c'/SRF \tag{10-7}$$

$$\varphi_f'=\arctan\left(\frac{\tan\varphi'}{SRF}\right) \tag{10-8}$$

To find the "true" FOS, the value of the strength reduction factor (*SRF*) that will just cause the fail of slope must be systematically searched. When this value is found, *FOS*= *SRF* (Wang, 2011).

The complicated three-dimensional numerical simulation and safety assessment of the flood discharge tunnel slope in the Yangqu hydropower station is implemented by a numerical simulation method.The stability calculation of the slope under the effect of seepage is based on the SSRT, and the calculation results obtained under three different schemes. In the scheme of , the safety factor is 1.26 (Figure 10-28). Comparing and analyzing the potential mode of failure in different schemes, we find that the potential sliding areas are similar, and the areas are mainly located in the slope near the outlet of the tunnel. The average depth of the landslide body is about 25~35m. The landslide body shows the characteristic of shallow sliding.

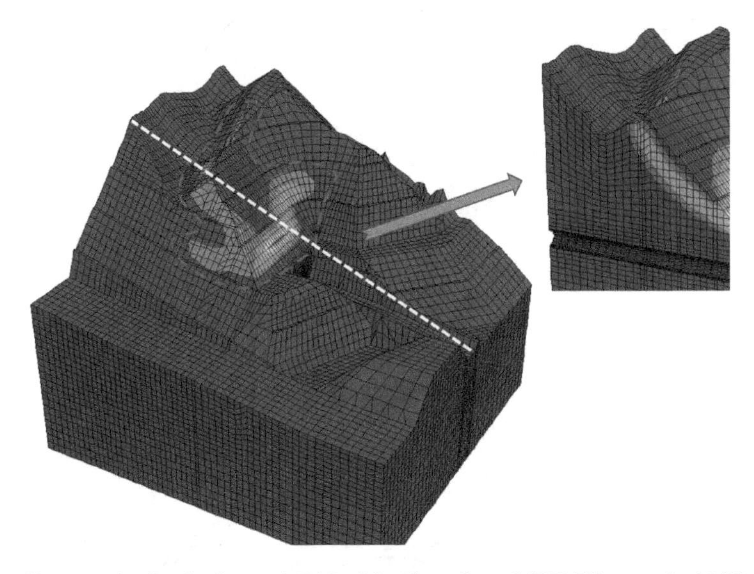

Figure 10-28 Shear strain distribution and *FOS* of the slope from SSRT (Wang et al., 2013)

10.5 Stereographic projection

The stereographic projection simplifies graphical solutions to problems involving the relative orientations of lines and planes in space. In rock mechanics contexts, the stereographic projection is appealing for analyzing the stability of excavations and exploring and characterizing discontinuities in rocks. Many publications in structural geology, crystallography, and rock mechanics show constructions and techniques available using stereographic projection. Valuable references for rock mechanics are Phillips (1972), Hoek and Bray (1977), and Goodman (1976). An important limitation of stereographic projections is that they consider only angular relationships between lines and planes and do not represent the position or size of the feature.

Stereographic projection is about representing planar and linear features in a two-dimensional diagram. The orientation of a plane is represented by imagining the plane to pass through the center of a sphere (Figure 10-29). The line of intersection between the plane and the sphere will then represent a circle, and this circle is formally known as a great circle. Except for the field of crystallography, where upper-hemisphere projection is used, most geologists use the lower part of the hemisphere for stereographic projections, as shown in Figure 10-29. However, most engineering geology textbooks in China use the radiation emitted by the lower pole, that is, upper-hemisphere projection (Figure 10-30). We want to project the plane onto the horizontal plane that runs through the center of the sphere. Hence, this plane will be our projection plane, and it will intersect the sphere along a horizontal circle called the primitive circle.

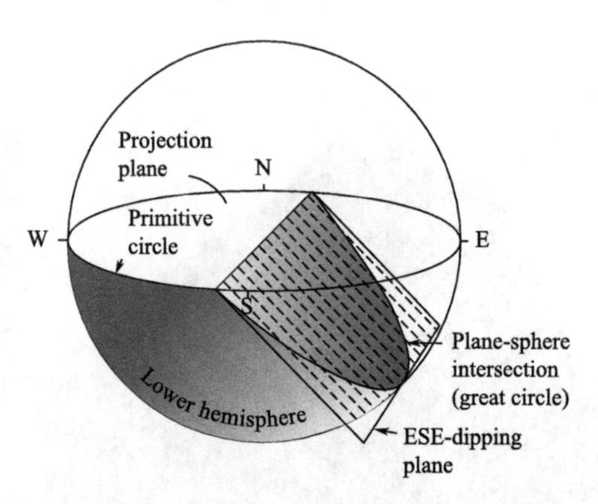

Figure 10-29 Basic concept with lower-hemisphere projection (LG, 2016)

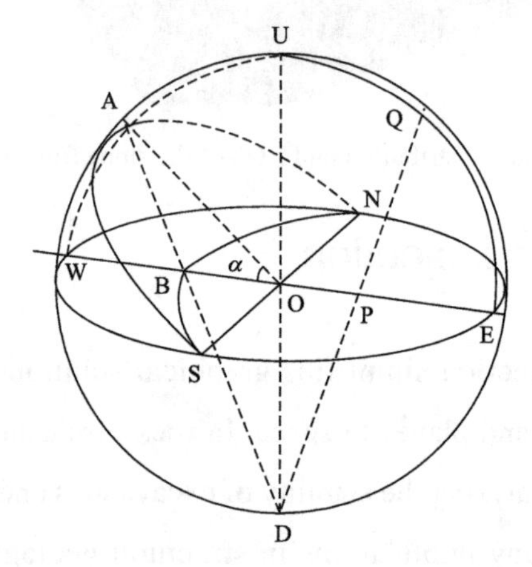

Figure 10-30 Schematic diagram of stereographic projection (upper-hemisphere projection)

Many different types of hemispherical projections are available (Kliche, 2009), and the choice of the proper one depends on the particular problem being analyzed. These projection methods are instrumental in engineering geology studies, such as exploring discontinuous rock blocks underground and on the surface, as in rock slope stability analyses. Structural geologists frequently use hemispherical projections to analyze rock structure interactions and borehole discontinuity data; these projections have also been used to diagnose force vectors (Priest, 1985).

10.5.1 Nets used for stereographic projection

The equal-angle stereonet, is also known as the meridional stereonet or the Wulff stereonet (Figure 10-31). Two types of nets are in general use in geology. The net in Figure 10-31(a) is called a stereographic net; it is also called a Wulff net, after G. V. Wulff, who adapted it to crystallographic use. The net in Figure 10-31(b) is called a Lambert equal-area net, or Schmidt net. In common geologic parlance, both of these nets are referred to as stereonets.

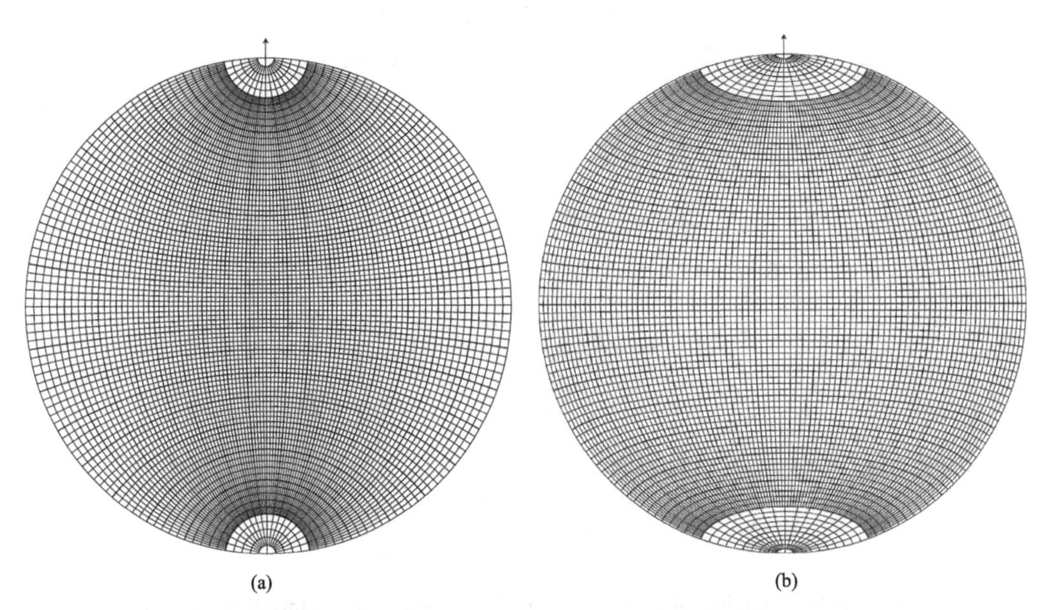

Figure 10-31 Nets used for stereographic projection

(a) Stereographic net or Wulff net; (b) Lambert equal-area net or Schmidt net

The two nets are constructed somewhat differently. On the equal-area net, equal areas on the reference sphere remain same on the projection, which is not the case with the stereographic net. The condition is similar to map projections of the earth; some projections sacrifice accuracy of the area to preserve spatial relationships, while others do the opposite. In preserving the area, the equal-area net does not preserve angular relationships. The construction of the equal-area net does allow the correct determination of angles, and this net may reliably be used even when angular relationships are involved. The relative density of data points is often crucial in structural geology, so most structural geologists use the equal-area net. In crystallography, angular relationships are significant, so crystallographers use the Wulff net. In this book, we will use the equal area net exclusively.

The equal-area net is arranged like a globe of the earth, with north-south lines analogous to meridians of longitude and east-west lines comparable to parallels of latitude. The north-south lines are called great circles, and the east-west lines are called small circles. The net perimeter is called the primitive circle (Figure 10-29); here, "primitive" has the mathematical sense of "fundamental" (Rowland et al.,2013).

10.5.2 Principle of stereographic projection

A line or a plane is imagined to be surrounded by a projection sphere (Figure 10-32). A plane meets the sphere in a trace that is a great circle bisecting the sphere precisely. A line

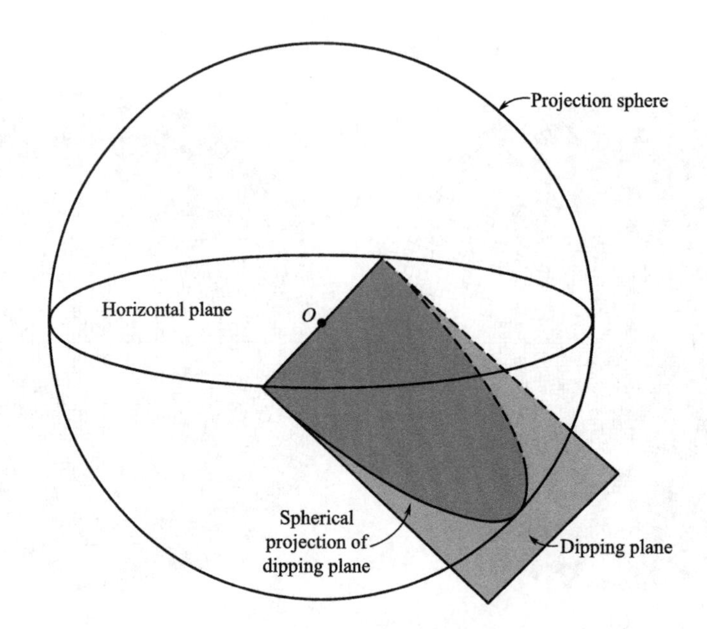

Figure 10-32 Principle of the stereographic projection (lower-hemisphere projection)

crosses the globe in a point. To image features on a sheet of paper, these traces and points are projected from a point at the summit or zenith of the sphere onto the equatorial plane.

As a principle, planes that dip at low angles are represented by great circles having significant curvature and lying closer to the primitive. In contrast, steeply dipping planes are characterized by straighter great circles passing close to the center of the plot. All vertical planes will project as straight lines passing through the center of the stereogram(Waldron,2009; Waldron & Snyder, 2009).

10.5.3 Ploting operations

The following describes the principle of a stereographic projection of the upper hemisphere's point, line, and surface with the emission point as the bottom pole (upper-hemisphere projection)

(1) Projection of points. The emission point of the lower pole is like looking up at any point in the upper hemisphere from the lower pole, and the intersection of the line of sight and the equatorial plane is the projection point. For example, the point M in Figure 10-33(a) is the projection of the point P on the equatorial plane. If point P rotates one circle around the upper and lower poles continuously on the spherical surface, its projection point M also rotates one circle around point O.

(2) Projection of lines. For example, OB in Figure 10-33(b) is a straight line passing through the center of the sphere, and the angle between it and the equator is α, and the projection

of the *OB* line on the equator is *OM*. It can be seen from the figure that the direction of *MO* is consistent with the tendency of the *BO* line. The length of the *OM* line segment changes with the included angle α. The larger the angle α, the shorter the *OM* line; otherwise, the longer. When α=90° , *OM*=0, that is, point *O*. When α=0° , *OM*=*OW*. Therefore, the radius of the great circle can represent the inclination of a line segment in space.

(3) The projection of the surface. For example, *NBSD* in Figure 10-33(b) is an inclined plane passing through the center of the sphere, and its intersection with the sphere is a great circle. Examining the *NBS* from the lower pole, its stereographic projection is *NMSN*, and *NMS* is circular. If the equatorial plane is taken out of the sphere, it is as shown in Figure 10-33(c). It can be seen from the figure:

1) The direction of *NS* represents the direction of the *NBSD* surface.

2) The direction of *MO* represents the inclination of the surface.

3) Like the projection of the line, the length of the *MO* line can reflect the inclination of the surface *NBS*. The scale of the inclination angle is 0° ~90° from *W* to *O*. If there are two intersecting inclined planes, the projection is shown in Figure 10-33(d), and *MO* in the figure is the projection of the intersection line of the two inclined planes.

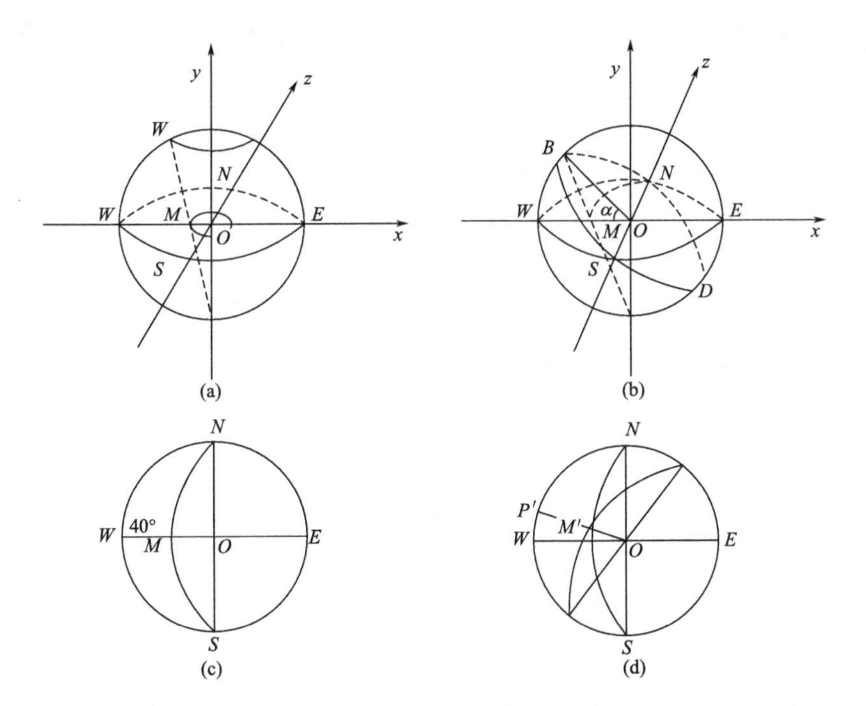

Figure 10-33 Stereographic projection of points, lines and surfaces (upper-hemisphere projection)

10.5.4 Applications of the stereographic projection

Different types of slope failure are associated with different geological structures. The slope designer must recognize the potential stability problems during the early stages of a project (Hoek & Bray,1977). Below we will introduce the use of the projection method to analyze the stability of the slope under the condition of joint cutting.

1. Slopes with a single discontinuity

(1) Reverse slope, the weakness inclines into the slope, and the stereographic projection shows that the slope surface is opposite to the weak plane, as shown in Figure 10-34(a). This slope is not easy to slide along the weak plane, so it is generally stable.

(2) Downward slope, but the inclination angle α of the weakness is smaller than the slope angle β, the stereographic projection is that the weak plane and the slope surface are on the same side, but the slope projection arc is inside the weak plane projection arc (Figure 10-34b). It is weak and facing the void, the slope is separated, and the rock mass is easy to slide along, so it is unstable.

(3) Downward slope, but α is greater than β, and the stereographic projection shows that the slope surface arc is outside the weakness arc (Figure 10-34c). In this case, the possibility of sliding is less. The slope is stable, but the damage will often occur if there are other gently inclined weak planes.

(4) Oblique cross slope, its stereographic projection map is as shown in Figure 10-34(d) (the intersection angle between slope and weakness γ>40°) and Figure 10-34(e) (γ<40°), this kind of slope is moderately stable, and the γ

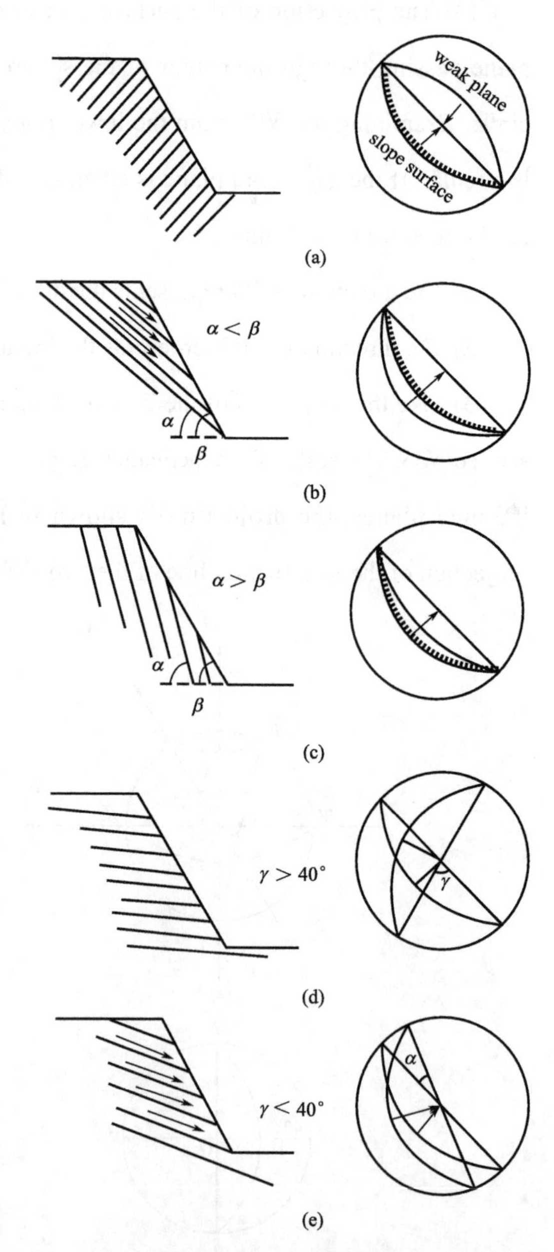

Figure 10-34 Stereographic projection and stability of slopes with a single discontinuity

angle becomes larger, the greater the stability.

2. Slopes containing two discontinuities

1) Analysis of sliding direction

For layered structure slopes or other single-sliding surface slopes, under the condition of pure self-weight, the sliding potential energy along the inclination direction of the sliding surface is the largest. The sliding component force of the self-gravity in the inclination direction of the sliding surface is the largest. Therefore, the inclination direction of the slip surface of a single slip surface slope is its slip direction.

When the slope is cut by two intersecting discontinuities, most of the possible sliding bodies formed are wedge-shaped bodies, and their sliding direction under the action of gravity. It is generally controlled by the inclination of the intersection of the two discontinuities, but there are also exceptions. Here is a general method for judging the slip direction of this type of slope based on the stereographic projection of the disontinuity. On the stereographic projection map, make the projection of the slope and the two discontinuities J_1 and J_2, draw the dip lines AO and BO, and the combined intersection line IO of the two discontinuities (Figure 10-35). The sliding direction of the slope has the following situations:

(1) When the combined intersection line IO of the two discontinuities is between their dip lines AO and BO, the dip direction of IO is the sliding direction of the sliding body. At this time, both discontinuities are sliding surfaces, as shown in Figure 10-35(a).

(2) When the combined intersection line IO of two discontinuities coincides with the dip line of one discontinuity (Figure 10-35b, the dip line BO of IO and discontinuity J_2 coincides), the dip direction of IO also represents the sliding direction of the sliding body. However, at this time, the discontinuity J_2 is the main sliding surface, and the discontinuity J_1 is the secondary sliding surface.

(3) When the combined intersection line IO of the two discontinuities is located on one side of their dip lines AO and BO, the dip direction in the middle of the three is the sliding direction, as shown in Figure 10-35(c). The dip line AO of discontinuity J_1 is the sliding direction. At this time, the sliding body is with a single sliding surface and only slides along the discontinuity J_1, and J_2 only functions as a lateral cutting surface here.

The above is the analysis of the sliding direction of the slope rock mass under the action of gravity. Regarding the sliding direction of the slope under the action of various external forces, you can refer to the relevant books.

2) Sliding Possibility Analysis

Figure 10-36 shows the analysis of the stability conditions of the double-sliding slope formed by the combined cutting of two structural surfaces, which can be divided into five situations.

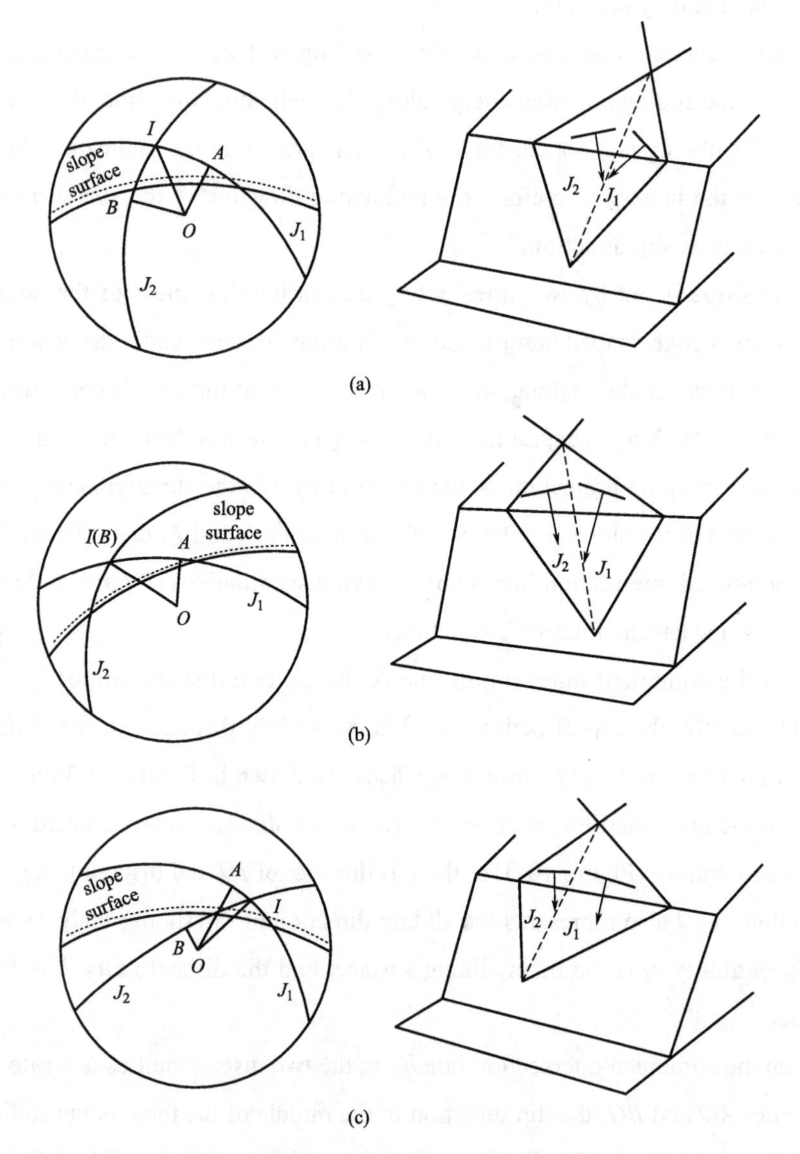

Figure 10-35 Analysis of sliding direction

(1) Unstable condition. As shown in Figure 10-36(a), the intersection point I of the two discontinuities J_1 and J_2 is located between the excavation slope surface S_c and the natural slope surface S_n. The dip angle of the intersection line is smaller than the dip angle of the excavation slope surface but bigger than that of the natural slope surface. If the intersection line IO is exposed on both the bottom and top surface of the slope, the slope is in an unstable state. As shown in the cross-sectional view of Figure 10-36(a), the slashed shaded parts are likely to be

dangerous. However, if the daylight point of the intersection line on the top of the slope is far away from the excavated slope surface, and the intersection line is not exposed on the excavated slope surface but inserts into the slope below, the stability condition is relatively good.

(2) Less stable condition. As shown in Figure 10-36(b), the intersection point I of the projection of the two discontinuities J_1 and J_2 is located outside the projection of the natural slope surface S_n. It shows that although the combined intersection line is gentler than the excavation slope surface S_c, it has no daylight point on the surface of the slope. Therefore, when there is no longitudinal (slope strike) cutting surface on the surface of the slope, the slope can be in a stable state. However, if there is a longitudinal cutting surface, the slope is prone to slippage.

(3) Basically stability condition, as shown in Figure 10-36(c), the intersection point I of the projection of the two discontinuities J_1 and J_2 is located on the projection arc of the excavation slope S_c, indicating that the inclination of the combined intersection line IO of the two discontinuities is nearly equal to that of the excavation slope. The slope is basically stable. The slope angle of the excavation should be the stable slope angle inferred from the stability analysis of the slope.

(4) Stable condition. As shown in Figure 10-36(d), the intersection point I of the two discontinuities J_1 and J_2 is located inside the projection arc of the excavation slope surface S_c, so the inclination angle of the combined intersection line IO of the two structural surfaces is steeper than the inclination angle of the excavation slope surface. The slope is in a more stable state.

(5) The most stable condition. As shown in Figure 10-36(e), the intersection point I is located in the semicircle opposite to the projection of the excavation slope surface S_c, indicating that the combined intersection line IO of the two facets inclines into the slope, and the slope is in its most stable state.

For the sake of simplicity, Figure 10-36 shows the particular case where the dip direction of the intersection line of the two discontinuities is consistent with the dip direction of the slope surface. When the dip direction of the intersection line of the discontinuities is different from that of the slope surface, the analysis and judgment of the slope stability condition are the same. In other words, in the stereo projection graph with the discontinuities and the slope surface, the preliminary judgment of the slope stability can be made according to the position of the intersection of the discontinuities projection. It is very convenient for preliminarily judging the stable conditions of various combinations of discontinuities under the condition of cutting multiple sets.

The above discussed the qualitative analysis of the stability of the slope rock mass cut by the discontinuity. To determine the slope stability quantitatively, the calculation should be carried out based on the strength of the weakness. A slope composed of more than two groups of discontinuities can be researched by real proportion projection.

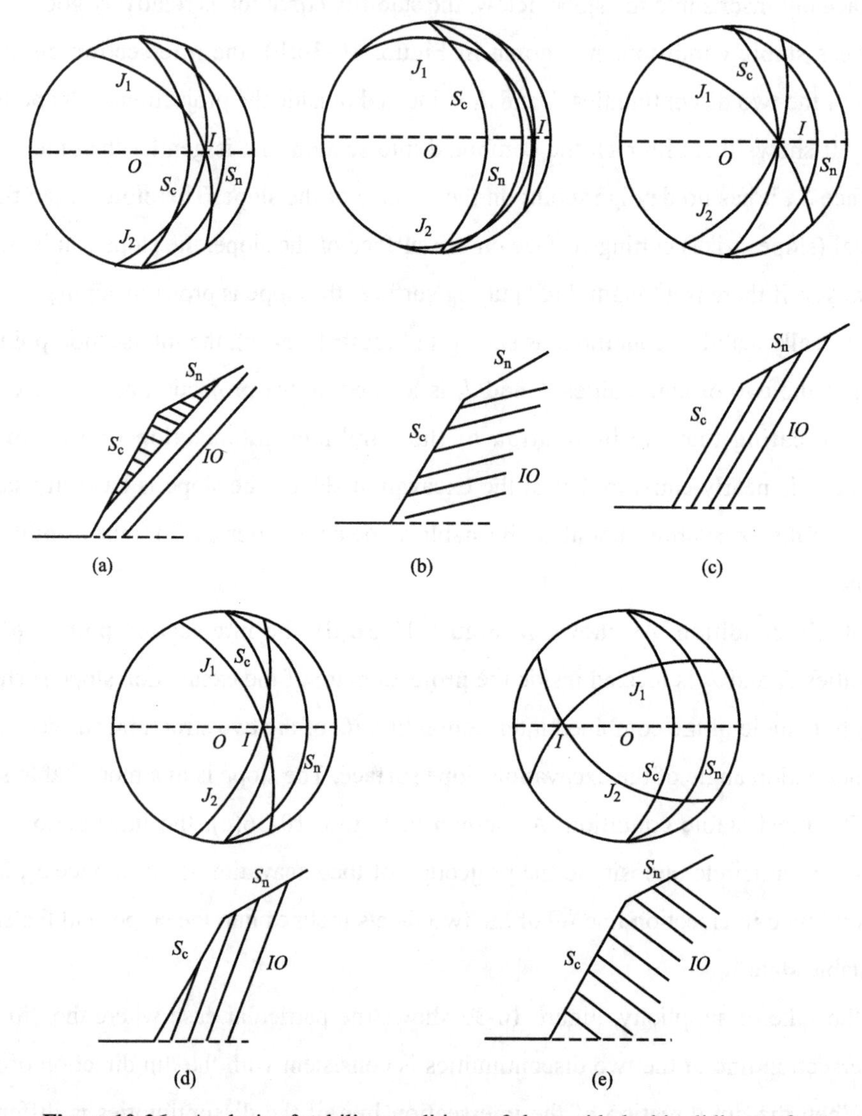

Figure 10-36 Analysis of stability conditions of slope with double sliding surfaces

10.6 Controls of deformation and destruction

For unstable slopes, to ensure the project's safety, some effective prevention and control measures must be taken. At present, the common methods used at home and abroad include: slowing down the slope and reducing the weight of the upper part; preventing the surface water from infiltrating into the rock mass; removing the groundwater in the unstable rock mass; building supporting structures; anchoring, etc. Before carrying out these treatments, the nature, type, and scale of the unstable slope damage and the factors that cause deformation, sliding, or collapse should be ascertained. Only by taking targeted measures can economic and effective results be achieved. Standard preventive measures include:

10.6.1 Load shedding and backpressure

Load shedding and back pressure are commonly used simple treatment methods, which can be used for various types of slopes such as rock or soil landslides, creeping, etc., which are generally not large in volume. The effect of load shedding is to reduce the sliding force of the sliding body. The primary method is to cut off a part of the rock and soil at the back of the landslide body. However, weight reduction alone cannot play the role of anti-sliding, and it is best to combine it with backpressure. Backpressure is to pile up rock and soil in anti-sliding areas outside and above the toe of the slope to play a back pressure and improve the anti-sliding force. Pile the soil and rock-cut by pressure reduction in the lower anti-slip part to make it have a backpressure effect so that the two are well combined to reduce the sliding force, increase the anti-sliding force, and stabilize the slope. This method can also be called head-cutting.

10.6.2 Seepage prevention and drainage

The main purpose of seepage prevention and drainage is to eliminate and reduce the hazards of surface water and groundwater (Figure 10-37).

Ground drainage, no matter it is a rock slope or a soil slope, the drainage of surface water often plays a vital role in the stability of the slope. On the one hand, the principle of ground drainage is to cut off the source of ditch spring and construction water flowing from the outside; on the other hand, to drain the precipitation on the slope to make it drain into the ditch as soon as possible to prevent it from infiltrating into the slope. Drainage ditches should be lined, and large cracks on the slope should be blocked simultaneously; in the vicinity of the building, sometimes use shotcrete protection. The distance between the drainage ditches along the slope is generally 50~60m.

Underground drainage reduce and maintain the groundwater level of unstable bodies to avoid groundwater infiltration in reservoirs or surrounding rock masses; it is of great significance to the stability of rock masses. The sliding surface composed of the weak interlayer can prevent the deterioration of long-term water immersion and reduce the shear strength. The discharge of groundwater can lower the water content of the slope and the pore water pressure in it, increase the sliding resistance and improve the stability of the slope. There are many prevention methods, mainly including interception ditch, blind ditch, water collection well, horizontal drilling, etc.

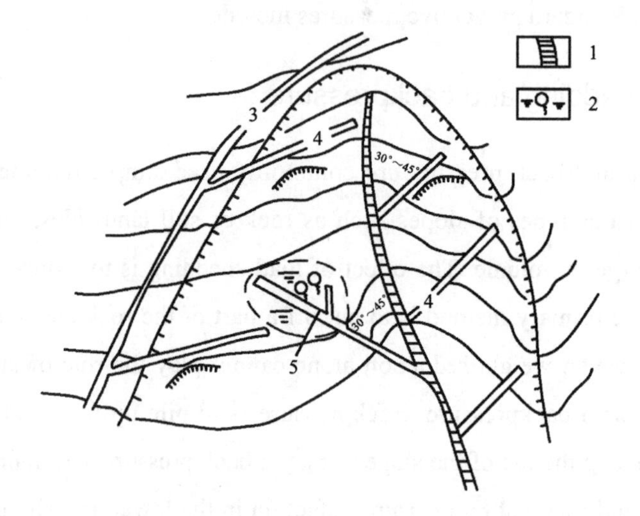

Figure 10-37 Surface drainage system

1—natural ditch paving; 2—spring water and wetland; 3—cut ditch; 4—drainage open ditch; 5—diversion ditch

10.6.3 Anchor system

Anchoring connects the rock masses on both sides of the sliding surface with anchor rods or cables to enhance the anti-sliding force of the sliding surface, thereby stabilizing the slope. It has sound effects when applied to prevent collapse and landslide. The anchor cable or steel cable can be better prestressed to increase the normal stress on the sliding surface, which is more significant for increasing the anti-sliding force. Anchoring is to drill holes first, then insert anchor rods or anchor cables (with the ends spread out) into the drill hole, apply prestress, and then perform cement grouting into the hole. Several parallel bolts can be used to form a bolt system (Figure 10-38).

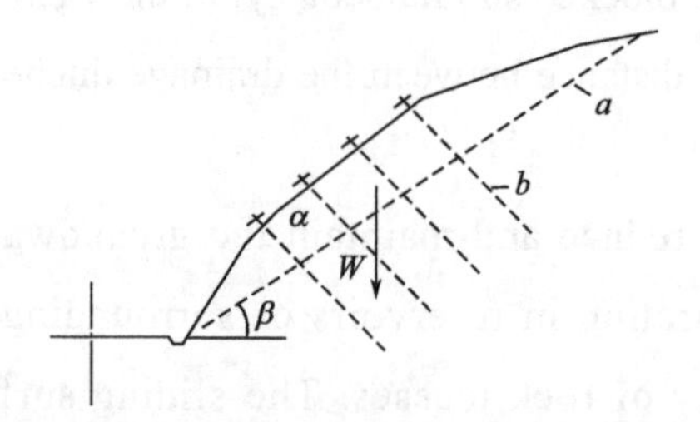

Figure 10-38 Schematic diagram of slope stabilization with pre-stressed bolts

a—rock layer (slip surface); *b*—pre-stressed bolts; *W*—weight of rock mass above the sliding surface

When anchor rods are used to reinforce slopes, they are often connected with retaining walls to become "anchor retaining walls," specifically composed of three parts: anchor rods, rib columns, and baffles. Its structure type is shown in Figure 10-39. The thrust of the landslide acts on the baffle, from the baffle to the rib column, from the rib column to the anchor rod, and finally to the stable ground below the sliding surface through the anchor rod. The anchoring force of the anchor rod sustains the whole stability of the structure.

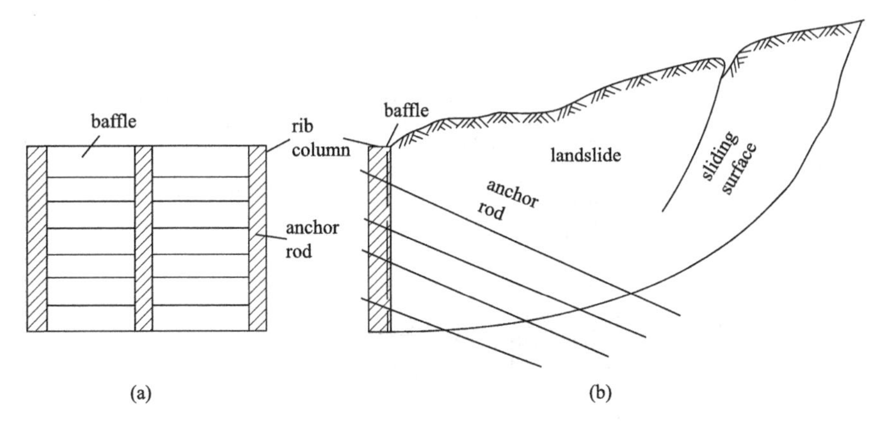

Figure 10-39 Anchor rod retaining wall structure

(a) front view of baffle plate; (b) section view

10.6.4 Anti-sliding engineering

The anti-sliding retaining wall is a kind of anti-sliding project commonly used at present. It is located at the front edge of the sliding body, with its weight to support the sliding force of the sliding body, and is used in conjunction with drainage measures. Different building materials and structural forms show anti-sliding stone stacks, anti-sliding stone bamboo cages, mortar-masonry anti-sliding retaining walls, concrete or reinforced concrete anti-slide retaining walls, etc. The advantage of the retaining wall is that the structure is relatively simple, the materials can be taken on the spot, and it can quickly stabilize the landslide. However, the foundation of the retaining wall must be set in a stable layer below the lowest sliding surface, and drainage holes should be reserved in the wall and continuous with the blind ditch behind the wall (Figure 10-40).

The anti-sliding pile is used to support the sliding force of the sliding body and fix it on the pile of the sliding bed. Its advantages are construction is safe, convenient, time-saving, labor-saving, material-saving, and less disturbance to the slope. So it is also a widely used retaining project (CAHE, 2019). Its materials include wood, steel, concrete, and reinforced concrete, and it can be poured or hammered during construction. Anti-slide piles are generally concentrated on the front edge of the landslide, and 1/3 to 1/4 of the total length of the pile is buried in a stable layer below the landslide surface (Figure 10-41).

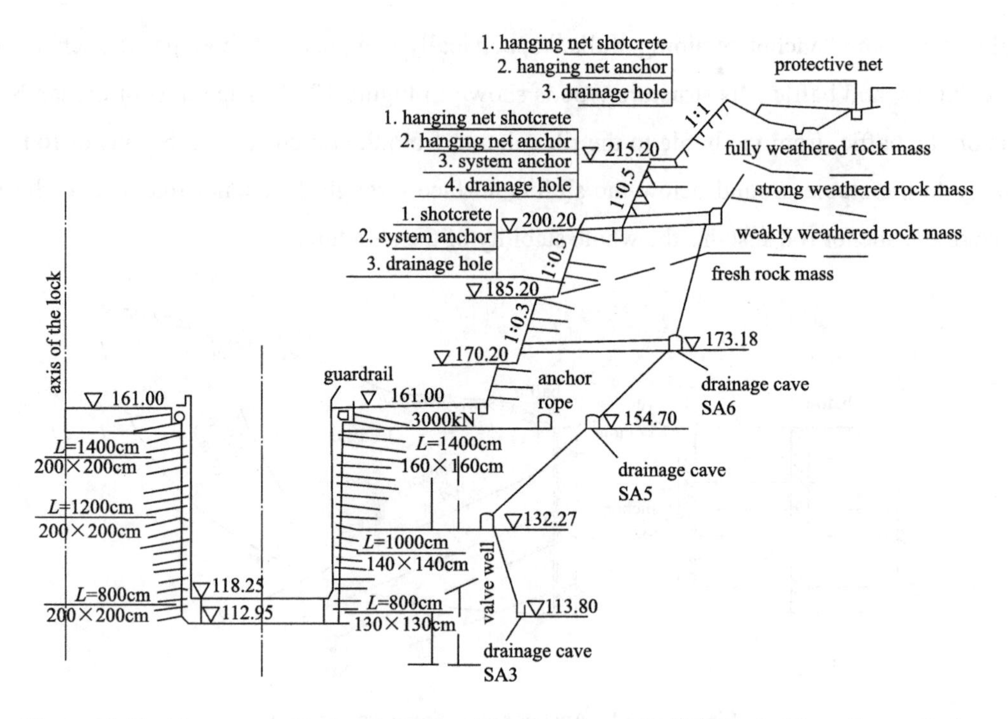

Figure 10-40 The systematic support system of the the permanent shiplock, the Three Gorges Project

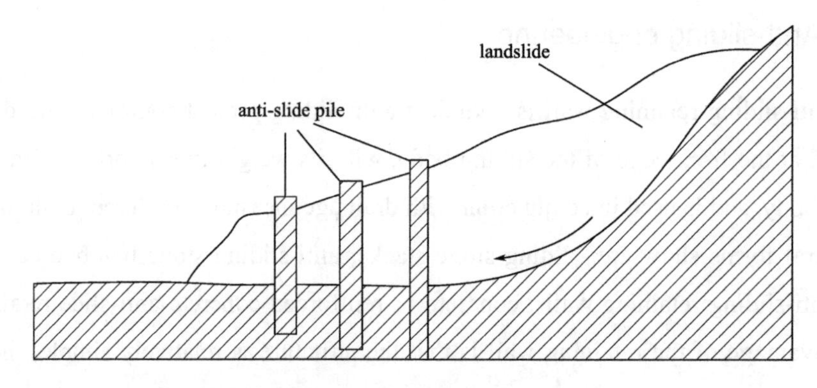

Figure 10-41 Anti-slide pile setting position

10.6.5 Other measures

Other measures refer to measures such as slope protection, improvement of geotechnical properties, and defensive evasion. Slope protection prevents the erosion or wave erosion of the slope by the current, and it can also avoid the weathering of the slope. To prevent river water erosion or sea, lake, and reservoir water wave erosion, water-retaining protection projects (such as retaining walls, breakwaters, stone masonries, and ripped-rock revetments, etc.) and water diversion projects (such as diversion dikes, spur dikes, waterproof guide walls, etc.) are generally built.

Vegetative techniques are most frequently used for aesthetic purposes, such as slope reclamation (Figure 10-42). However, many treatment methods use vegetation to improve the stability of a slope. Generally, these methods are most successful when minor or shallow instability (such as raveling or erosion) is involved, as is usually the case for soil slopes or highly fractured rock slopes (Buss et al. 1995). The establishment of vegetation on steep soil slopes or loose-rock slopes is often enhanced by constructing benches or stairstep terraces in the slope face. These arrangements act to hold the seed mix in place, encourage infiltration, and impede water flow to minimize erosion and sedimentation (Kliche,2009).

Figure 10-42 A slope with vegetative techniques

To prevent the weathering and erosion of the slope's surface composed of easily weathered rocks, slope protection measures such as shotcrete, mortar plastering, and mortar rubble can be used. The purpose of improving the properties of the rock and soil is to improve the anti-sliding ability of the rock and soil. It is also an effective measure to prevent the deformation and damage of the slope. Commonly used are the chemical grouting method, electroosmosis drainage method, and roasting method. They are mainly used to improve the properties of soil mass and can also be used to strengthen the weak interlayer in the rock mass.

Reference

[1] BUSS K, PRELLWITZ R, REINHART M A. Highway Rock Slope Reclamation And Stabilization, Black Hills Region, South Dakota, Part Ii, Guidelines. Final Report[R]. 1995.

[2] CAHE(Chinese Association of Hydraulic Engineering). Construction technology and measurement of hydraulic engineering[M]. Beijing: China Water & Power Press, 2019.

[3] COATES, D F. Rock mechanics principles[M]. Ottawa: Energy, Mines and Resources

Canada,1981.

[4] CRUDEN D M. Multilingual landslide glossary[M]. Richmond, B.C: Bitech Publishers Ltd,1993.

[5] DAWSON E M, ROTH W H, DRESCHER A. Slope stability analysis by strength reduction[J]. Geotechnique, 1999, 49(6): 835-840.

[6] DUNCAN J M, GOODMAN R E. Finite element analysis of slopes in jointed rock, Contract Report[J]. US Army Engineer Waterways Experiment Station, Corps of Engineers, No. S-68-3, 1968.

[7] FRÖHLICH O K. General theory of stability of slopes[J]. Geotechnique, 1955, 5(1): 37-44.

[8] GLOVER F, AUGUSTINE M, CLAR M. Grading and shaping for erosion control and rapid vegetative establishment in humid regions[M]. Madison, Wis: American Society of Agronomy, 1978.

[9] GOODMAN R E. Methods of geological engineering in discontinuous rocks[M]. Minnesota: West Group, 1976.

[10] GOODMAN R E. Introduction to rock mechanics[M]. New York: Wiley, 1989.

[11] GRIFFITHS D V, LANE P A. Slope stability analysis by finite elements[J]. Geotechnique, 1999, 49(3): 387-403.

[12] HIGHLAND L, BOBROWSKY P T. The landslide handbook: a guide to understanding landslides[M]. Reston: US Geological Survey, 2008.

[13] HOEK, E. & BRAY, J. W. Rock Slope Engineering[M]. 2nd ed. London: Institute of Mining and Metallurgy,1977.

[14] HUANG Z, LAW K T, LIU H, et al. The chaotic characteristics of landslide evolution: a case study of Xintan landslide[J/OL]. Environmental geology, 2009, 56(8): 1585-1591.

[15] Itasca Consulting Group Inc. 3DEC 3-Dimensional Distinct Element Code. Ver. 5.2 User's Manual[M]. USA: Minneapolis,2019.

[16] KLICHE, A.C. Rock Slope Stability[M]. Colorado: Society for Mining, Metallurgy, and Exploration,2009.

[17] LG (Learning Geology). Stereographic projection[EB/OL].2016-11-11.

[18] LI T F, CHEN H T, WANG R Q. Formation mechanism of Yanchihe landslide in Yichang city, Hubei province[J]. Journal of Engineering Geology, 2016, 24(4): 578-583.

[19] LIU R, HAN Y, XIAO J, WANG T. Failure mechanism of TRSS mode in landslides induced by earthquake[J/OL]. Scientific reports, 2020, 10(1): 1-11.

[20] PHILLIPS F C. The use of stereographic projection in structural geology[M]. 3rd ed.Arnold, London, 1972.

[21] PRIEST S D. Hemispherical projection methods in rock mechanics[M]. London:Allen & Unwin, 1985.

[22] ROWLAND S M, Duebendorfer E M, Schiefelbein I M. Structural analysis and synthesis: a laboratory course in structural geology[M]. John Wiley & Sons, 2013.

[23] STACEY T R. The stresses surrounding open pit mine slopes[J]. Planning open pit mines, 1970: 199-207.

[24] STACEY T R. A three-dimensional consideration of the stresses surrounding open-pit mine slopes[C]//International Journal of Rock Mechanics and Mining Sciences & Geomechanics Abstracts. Pergamon, 1973, 10(6): 523-533.

[25] STEWART A, WESSELS F, BIRD S. Design, Implementation, and Assessment of Open Pit Slopes at Palabora over the Last 20 Years[J]. Slope Stability in Surface Mining, 2000: 177-182.

[26] VARNES D J. Slope movement types and processes[M]. Washington, D.C: Transportation Research Board,1978:11-33.

[27] WALDRON J, SNYDER M. Geological Structures: a Practical Introduction[EB/OL]. 2020.

[28] WALDRON, J. EAS 233 Geologic Structures and Maps[EB/OL]. 2009.

[29] WANG T, WU H, LI Y, et al. Stability analysis of the slope around flood discharge tunnel under inner water exosmosis at Yangqu hydropower station[J/OL]. Computers and Geotechnics, 2013, 51: 1-11.

[30] WANG T, ZHOU Y, LV Q, et al. A safety assessment of the new Xiangyun phosphogypsum tailings pond[J]. Minerals Engineering, 2011, 24(10): 1084-1090.

[31] WINDSOR C R. Rock reinforcement systems[J/OL]. International journal of rock mechanics and mining sciences, 1997, 34(6): 919-951.

[32] WYLLIE D C, MAH C. Rock slope engineering[M]. 4th ed. CRC Press, 2004.

[33] YIN Y. Mechanism of apparent dip slide of inclined bedding rockslide-a case study of Jiweishan rockslide in Wulong, Chongqing[J]. Chinese Journal of Rock mechanics and Engineering, 2010, 29(2): 217-226. (In Chinese)

[34] ZÁRUBA Q., MENCL, V. Engineering geology[M]. Amsterdam: Elsevier,1976.

[35] ZHANG Z, WANG S, WANG L, et al. Principles of engineering Geological Analysis[M]. 4th ed. Beijing: Geological Publishing House(2016). (In Chinese)

[36] ZHANG Z, DONG X. Landslide terminology recommended by The International Geotechnical Societies' UNESCO Working Party on World Landslide Inventory[J]. Journal of Geological Hazards and Enviroment Preservation, 1995, 6(1):1-6. (In Chinese)

Exercises

Review Questions

(1) Describe the characteristics of the rock mass and the geological factors affecting the stability of the rock mass.

(2) What changes may occur to the gravity stress distribution in rock mass during slope formation?

(3) What are the factors affecting the stress distribution of slope rock mass?

(4) What are the types of slope rock mass deformation and failure? What are the main features?

(5) How many ways are landslides classified? What are the characteristics of various types of landslides?

(6) What are the main factors affecting slope stability? How is it affected?

(7) How to calculate the slope stability when the sliding plane is a broken line?

(8) How to make stereographic projection of point, line and plane?

(9) How to use stereographic projection to analyze slope stability?

(10) What are the main measures to prevent slope failure?

(11) When rockbolts are used for surface or underground excavations to reinforce a fractured rock mass, the rockbolts will be subjected to tension, shear and compressive forces. The diagram below (Windsor, 1997) indicates the types of forces on rockbolts *a* to *f* (Figure 1). Recognizing that rockbolts are designed to operate in tension with little shear force applied, rate these rockbolt configurations in terms of their suitability on this criterion alone.

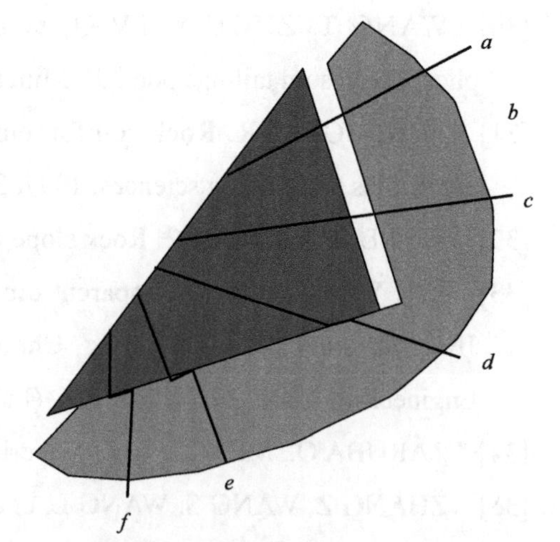

Figure 1 A rock block intersected by a number of reinforcing elements oriented to reinforce different release surfaces in the rock mass

(12) the rock slope is just on the right bank of the Yangtze River (Figure 2). The total volume of dangerous rock mass is approximately 3 million cubic meters, and it has been cut apart by more than ten cracks going deep into coal mine bottom.

Questions:

(a) What kind of failure can happen in the slope?

(b) How to control of deformation and destruction of the slope?

Figure 2 Lianzi cliff dangerous rock mass in Zigui, Yichang

11 Engineering geological study of underground caverns and tunnels

11.1 Introduction

The artificially excavated or naturally existing in rock and soil as structures for various purposes, collectively referred to as underground buildings. According to its purpose, it can be divided into traffic tunnels, hydraulic tunnels, mine tunnels, underground powerhouses, underground warehouses, and underground military projects. According to whether the inner wall has internal water pressure, it can be divided into two types: pressureless caverns and pressure caverns.

With the rapid development of China's water conservancy and hydropower construction industry, the number of underground buildings is increasing. The use is more and more extensive, and the scale is also getting bigger and bigger. For example, some diversion tunnels have a length of more than 10km and a diameter of more than 200m. An underground hydropower plant with an installed capacity of one million kilowatts usually has a span of up to 30m and a height of more than 60m. The total installed capacity of the Jinping Hydropower Project on the Yalong River (Figure 11-1) is 14 × 600MW (Feng et al., 2006; Chen, 2015). The diameter and length of its four diversion tunnels are 12m and 16.6km on average, respectively, with the maximum overburden depth up to 2525m, which raises a series of challenging problems such as high external water pressure (10MPa) and high tectonic stress (100MPa).

The construction of large-span, high-side wall underground powerhouses and long tunnels will inevitably encounter complex geological conditions and a large number of engineering geological problems. Among them, the most concerned is the stability of surrounding rock. Geology is the most critical factor that determines the nature, form, and cost of a tunnel. For example, the route, design, and construction of a tunnel primarily depend on geological considerations.

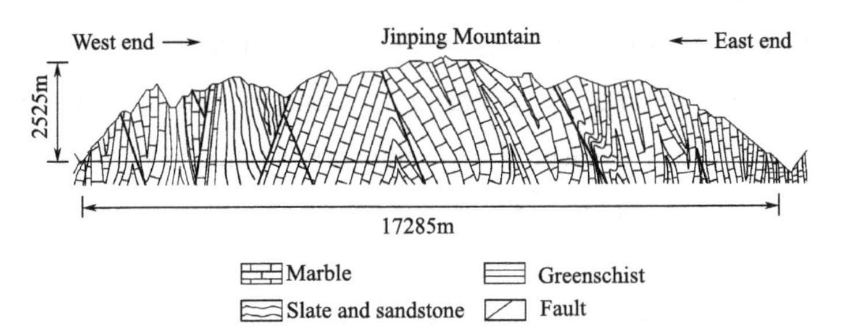

Figure 11-1 Geologgical profile along headrace tunnels in Jinping II hydropower station

11.2 Engineering geological research contents

The extent of engineering geological studies depends, as with other civil engineering structures, on the stage of the project and the importance of the work. They include surveys and detailed investigations, and cooperation during construction. Issues to be addressed include:

(1) Investigation of the geological conditions of the area and the properties of the rocks. The results are taken into consideration when choosing the overall alignment of the tunnel. Rock with high pressure will be undesirable for tunnel work in all cases. Since modern high-speed machines react sensitively to changes in the hardness of rocks, tunnels driven in rocks exerting a uniform medium pressure may be less expensive than those driven in strong rocks whose strength and hardness vary significantly over short distances.

(2) Hydrogeological investigation of the area along the tunnel route may be considered in which segments and in what amounts inflows of water into the tunnel can be anticipated. Hydrogeological conditions are essential in planning tunnels for water supply, underground, and all tunnels crossing areas where corrosive water is present. The occurrence of caustic water is so undesirable that, if possible, such sites should be evaded.

(3) When the general alignment of the tunnel has been chosen, features controlling its exact position are reviewed. The tunnel should not be located in rocks disrupted by weathering or faulting. If the fault zone cannot be avoided, aligning the tunnel to its right angles is desirable. The primary joint and fracture systems are also the controlling factors, particularly in underground openings of large diameter (e.g., for underground powerhouses).

(4) For the location of the tunnel entrance, care must be taken not to disturb the stability of slopes by approach cuttings. Sliding movements and pressures could delay the working progress.

(5) Study of the effect of tunnel working on the surrounding area. The drawdown of the groundwater table may result in the disappearance of the water in wells and springs. Subsidence

of the ground surface may destroy buildings, transportation routes, and others.

Detailed engineering geological investigation must be carried out to refine assumptions on the area's geological structure deduced from preliminary studies. A geological advance forecast is the most helpful tool for obtaining the necessary information. In excavating the diversion tunnel (13.5km long) of the river Isere (French Alps) in a complex of crystalline schists, the work had to be interrupted when the tunnel entered talc schists and anhydrite, and the tunneling method had to be changed. At first, horizontal test borings were used for investigation. Still, these proved unsuccessful because of small core recovery, so then an exploratory drift was drilled ahead to a length of 1km (Olivier-Martin & Kobilinski, 1955).

11.3 Engineering geological evaluation of location selection

After the general location of the tunnel or cavern has been proposed, especially concerning the rocks to be traversed and the hydrogeological situation, the more precise selection of the alignment is the next task. The engineers must carefully consider regional stability, mountain stability, topography, lithology, geological structure, groundwater, in-situ stress, etc.

11.3.1 Topography condition

In terms of topography, the mountain rock is required to be intact. It should be sufficient thickness of the mountain rock around the cavern, including the roof of the cave and the side of the mountain. The slope at the tunnel's entrance and exit should be steep downward and gentle upward, and there should be no landslides or collapses. The rock at the cave entrance should be exposed directly, or the diluvial layer should be thin. When excavating a hole on a high slope with steep terrain, it should be entered without cutting the slope or cutting the slope less. If necessary, the guide arch (Figure 11-2) or artificial auxiliary entrance (Figure 11-3) can be used to advance the tunnel to ensure the stability of the slope(Yuan et al., 2021). Both entrance and exit of the tunnel should not be selected in low-lying places where drainage is painful, nor should they be selected in areas that are easily scoured by water, such as gullies, riverside mountain outlets, and valley outlets.

Many cases exist where even deeply located tunnels have been afflicted by factors extending from the ground surface. The case of the Unterstein Tunnel in the Salzach river valley near Salzburg is instructive (Figure 11-4). Construction of this tunnel started in 1874 to cross a rock mass consisting of chlorite schists. Because the beds dipped into the mountain, their stability

Figure 11-2 Guide arch installation prior to commercing a drill-and-blast tunnel(Xichang,Sichuan) in a poor geological region

Figure 11-3 The artificial auxiliary entrance is used during the construction of the Zongyantou Tunnel in Zigui, Hubei

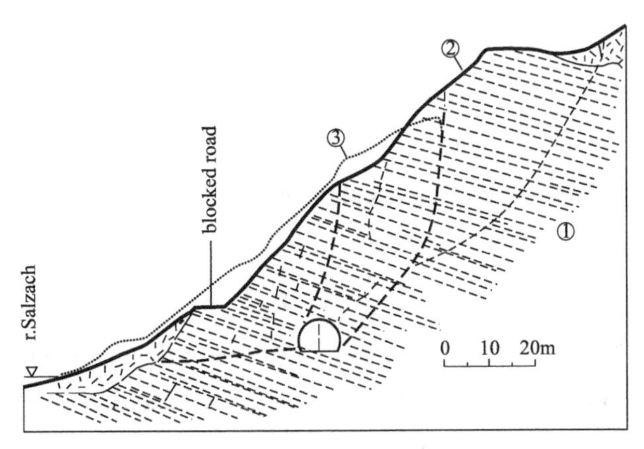

Figure 11-4 Cross-section of the collapsed slope above the Unterstein Tunnel (Wagner, 1884)

① —Chloritic schists; ② —original slope surface; ③ —slope surface after sliding

was not doubted, and the axis was situated near the slope surface. The work, in 8m long sections excavated one by one, continued without difficulty until February 1875, when the advance heading had been holed through for the entire length of 104m and 60m left. The lining had been finished over a length of 64m, 15m were under construction, and 25m were in excavation. In order to advance the operations, several sections were opened in the 15m length in the spring of 1875. This resulted in the caving-in of two sections.

The rock was found to be transected by fractures perpendicular to the foliation, i. e., dipping steeply into the valley. Immense pressure began to build up, the dolomite and conglomerate masonry lining of the finished sections was crushed and replaced by granite blocks. On June 10th, 1875, the operation had to be stopped, a total caving-in and collapse of the slope occurred. The highway was blocked and had to be relocated to the other bank of the river. The collapsed slope was investigated through galleries, and its displacements were measured. Results led to an abandonment of the proposed line, and a new tunnel, 400m long and located deep in the slope, was constructed over the years 1877~1888.

11.3.2 Formation lithology conditions

Formation lithology is one of the essential factors affecting the stability of surrounding rock. Such as hard and intact rock mass, surrounding rock is generally stable, can adapt to the shape of various sections of underground caverns. The weak rock mass, such as clay rock, broken and weathered rock mass, rock mass easy to expand with water, usually has low mechanical strength. It is easy to soften, collapse and expand when it meets water, which is not conducive to the stability of surrounding rock.

Therefore, the location of the cavern should be selected as far as possible in hard and intact rock. In the weak, broken, and loose rock, the roof is easy to collapse, sidewall and bottom plate are easy to cause swelling and extrusion deformation accidents. It usually needs to support while driving or advanced support, a long construction period, high cost.

Magmatic rock, a thick layer of hard sedimentary rock and metamorphic rock, the stability of surrounding rock is good, suitable for the construction of large-scale underground engineering. Tuff, clay rock, shale, poorly cemented sandy conglomerate, phyllite, and some schist have poor stability and are unsuitable for building large underground caverns. Loose and broken rock stability is very poor, so site selection should be avoided as far as possible. In addition, the combined characteristics of rock strata also have an important influence on the stability of surrounding rock. Generally, the rock mass with soft and hard interbedded or weak interbedded has poor strength. The more layers of layered rock, the thinner the thickness of a single layer, the worse the stability. Homogeneous thick layers and massive rock mass usually have good stability.

The Waggital water tunnel in Switzerland was driven under the Schwendibach valley (Figure 11-5). Because a moraine covered the valley, an exploratory shaft was deepened to the bedrock in the valley bottom. The results suggested that the gallery would be situated below at least 18m of rock, but the driving of the gallery showed that the lowest point of the bedrock does not lie under the valley axis, but about 50m farther to the north. Tunnel operations in this section proved to be very difficult owing to high water flows and considerable pressure of the cohesionless moraine material.

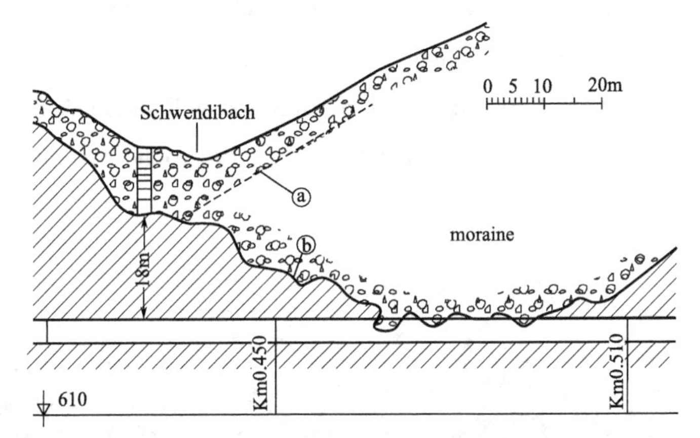

Figure 11-5 Waggital hydraulic structure; geological section along the gallery under an over-deepened valley shows that one exploration pit is not sufficient to determine the thickness of deposits in the valley (Schardt, 1924)

ⓐ—assumed; ⓑ—the actual surface of bedrock

11.3.3 Geological structure conditions

The geological structure is an essential factor in controlling the integrity and permeability of rock mass. When selecting a site, try to avoid areas with complex geological structures; otherwise, it will cause difficulties in construction. For example, the Simplon Tunnel in Italy is more than 20km long. Due to the severe folds and inversion of the ground and the accompanying large-scale reverse faults, the rock is broken, and collapses occurred many times during the construction. Rock bursts appeared in the Simplon Tunnel even after it was opened to railway traffic and caused an upheaval of the ballast, bending of the rails, and several derailments (Záruba & Mencl,1976). The following is a brief analysis of the influence of folds, faults, and rock formations on the stability of surrounding rocks.

1. The effect of folds

In areas with severe folds, faults are generally developed, especially in the core of the folds, where the rock formations have the worst integrity. The tunnel driven parallel to the bedding is a less favorable location, especially when the rock is not uniform (Figure 11-6a). Oblique crossing of beds may involve asymmetrical pressures on the supports and large over breaks. As shown in Figure 11-6(c), the rock layer is arched in the anticline core. Although the rock layer is broken, it is just like the stone arch structure, which can transfer the load of the overburden to the rock mass on both sides. It is conducive to the stability of the roof, but the excavation is not easy. Although the cave roof has developed cracks, the rock blocks are wide at the top and narrow at the bottom, making it difficult to fall off. However, as shown in Figure 11-6(b), the rock at the core of the syncline is inverted arched. The top of the rock-cut by the tension fissures is narrow in the top and wide in the bottom, which is easy to collapse. In addition, the syncline core is often a place for confined water storage, and groundwater will suddenly flood into the cavern when the underground cavern is excavated. Therefore, underground caverns should not be built in the syncline core.

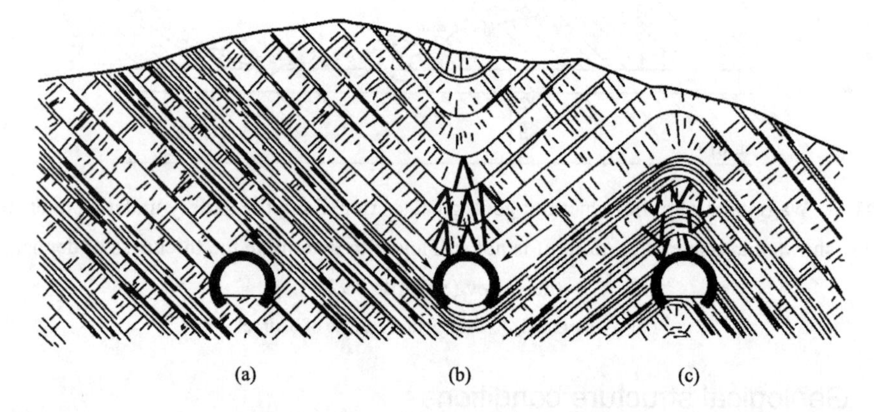

Figure 11-6 Schematic diagram of the tunnel at the core of the fold. The tunnel was driven through a folded rock complex parallel to the strike of beds

(a) asymmetrical pressure and large over breaks are to be expected; (b) along the syncline axis is accompanied by block gliding and water inflow; (c) when the tunnel axis coincides with the anticline axis, the local dip of beds is of great importance

However, the outer edge of the anticline core is in a tension zone, the inner edge is squeezed and weathered, and the rock is often very broken. Therefore, when arranging underground caverns, in principle, the fold core should be avoided. If it is necessary to build an underground project in a fold, the caverns can be placed on the two wings of the fold.

The most favorable case occurs when the tunnel axis runs at right angles to the strike of beds dipping in the driving direction. Somewhat less good is the case of crossing the beds which dip into the excavation (Figure 11-7a). Blasting is easy, but the danger of blocks sliding from the face arises. When crossing an anticline perpendicular to the strike of beds, the driving meets a wedged rock, obstructing the excavation. The danger that the blocks may fall increases due to the rock's dense fracturing (Figure 11-7b). Water flows can be expected when driving in a syncline (Figure 11-7c), especially when permeable and impermeable strata alternate.

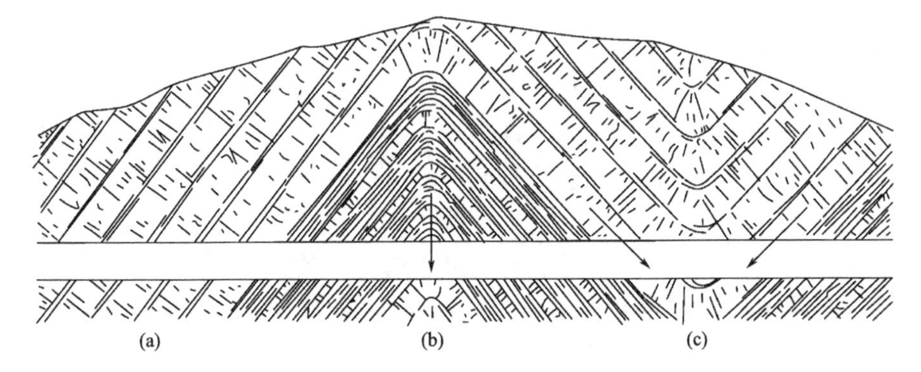

Figure 11-7 Tunnel driven through a folded rock complex perpendicular to the strike of beds
(a) beds dipping into the excavation make the blasting easy, but the danger of blocks sliding from the face arises; (b) when crossing an anticline, the driving meets a wedged, and fractured rock;(c) when driving in a syncline water flows can be expected

2. The effect of fault zone

The fault zone usually has poor stability. In case of large-scale faults, almost all of the underground excavation will collapse or even roof collapse. In general, the axis of the cavity should be avoided along the fault zone. If the axis of the cavern is perpendicular or close to the vertical fault zone, the unstable section that needs to be crossed is shorter. However, it can also produce collapse or a large amount of groundwater influx. Wherever possible, caverns should be orientated to avoid fault zones (Bell, 2007).It is necessary to avoid an orientation whereby a cavern' s long axis is parallel to steeply inclined major joint sets (Hoek and Brown, 1980). According to the data of GEODATA (YREC, 2012), four fault zones exist in the penstock area of Coca Codo Sinclair (CCS) in Ecuador, including two Andean tectonic zones in the NE–SW strike and two faults in the NW–SE strike; they are generally right-lateral strike-slip faults (Figure 11-8).

Underground openings of limited length provide the possibility to situate the axis normal to the major surfaces of weakness. This orientation is possible with the auxiliary caverns of surge chambers, access tunnels of underground power stations. The more comprehensive and higher the opening, the more urgent is this requirement, and it is fully respected during the design of caverns of power stations. Major fault zones should be crossed at right angles, even by longer tunnels, and the tunnel axis must be adapted to this requirement by inserting suitable bends (Záruba & Mencl,1976).

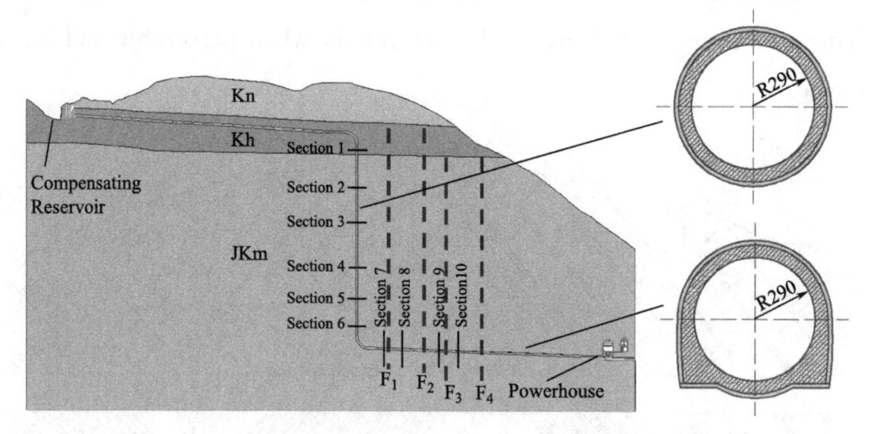

Figure 11-8 Geotechnical cross-section along a tunnel of CCS (Wang et al., 2018)

3. The effect of orientation of the bedding

The axis of the cavern is perpendicular to the strike of the rock formation. In this case, the surrounding rock has better properties, especially beneficial to the stability of the sidewall. The stability is best when the rock formation is steep (Figure 11-9a). When the rock strata dip is relatively flat, and the joints are well developed, local rock collapse is likely to occur on the roof of the cave, and stepped over-excavation often occurs on the roof of the cavern (Figure 11-9b).

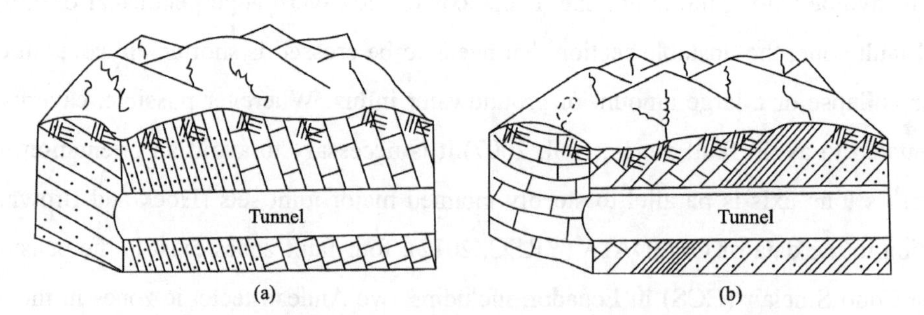

Figure 11-9 A cavern perpendicular to the strike of a monoclinal rock

The cavity strike is parallel to the rock strike. When the strata are nearly horizontal (dip angle $< 10°$), if the rock layer is thin and the connection between each other is poor, roof collapse often occurs in the excavation of caverns (especially large span caverns). Therefore, when arranging cavities in horizontal strata, the cavities should be located as far as possible in thick and hard strata (Figure 11-10a). If the cave must cut through different combinations of soft and hard rock strata, the hard rock strata should be used as the roof, avoiding the weak rock strata or weak interlayer on the top, which is easy to cause roof overhang or collapse (Figure 11-10b). Weak rock layers located on either side or at the bottom of the cavern (Figure 11-10c) are also unfavorable, which can easily cause side walls or floor plates to bulge and deform or be pushed out. For example, the underground powerhouse cavern of Shuicaozi hydropower station has a tuff interlayer in the rock mass, which affects the stability of the sidewall and top arch. Therefore, the cavern is moved 25m below the interlayer in site selection.

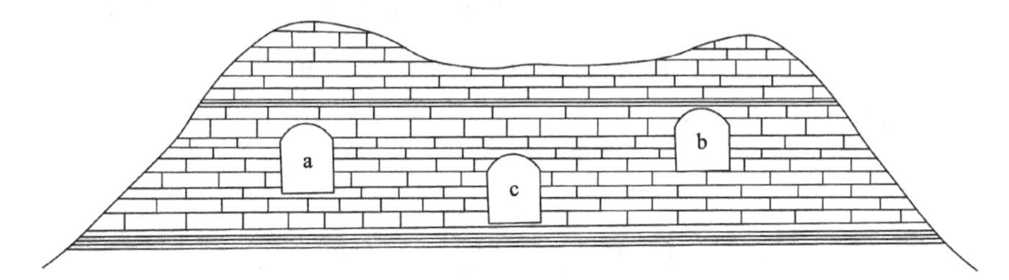

Figure 11-10 A tunnel arranged in the horizontal rock

In inclined strata, this is generally unfavorable. As shown in Figure 11-11 a&b, when the cave passes through the inclined rock with broken rocks or weak interlayer, the surrounding rock on the inclined side tends to deform or slide, resulting in great bias pressure. In contrast, the surrounding rock on the opposite side has slight lateral pressure, conducive to stability. Therefore, it is best to select the chamber in the homogeneous solid rock (Figure 11-11c). In addition, the dip angle of the rock layer also impacts the stability of surrounding rock, so the site selection should be considered comprehensively with other factors.

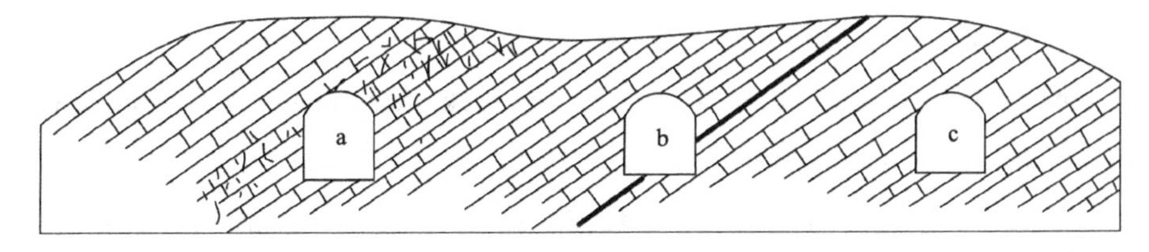

Figure 11-11 Bias pressure of tunnel in inclined rock formation

11.3.4 Groundwater conditions

Cave-in or roof collapse accidents in underground engineering construction are often related to groundwater activities. The so-called "water control first" is a meaningful experience. Therefore, it is best to choose the dry rock mass above the groundwater level or the rock mass with little underground water and no high-pressure aquifer. Figure 11-12 is a schematic diagram of the relationship between underground engineering and groundwater level. In the excavation of underground works in the aeration zone (Ⅰ), water may drip along the cracks in the rainy season and dry in the dry season. However, concentrated seepage may also be encountered when there is a large area of stable surface water. The groundwater table variable zone (Ⅱ), the amount of water inflow, and the external water pressure vary with the seasons. As the rock mass is saturated and dehydrated alternately, it can accelerate the deterioration of the weak and broken rock properties and cause landslides. For underground projects below the groundwater table (Ⅲ), there may be greater water gushing and seepage pressure at the beginning of the construction, so a waterproof and drainage design must be made.

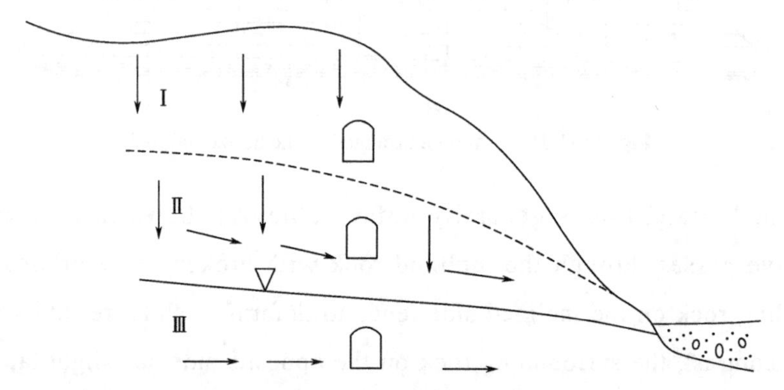

Figure 11-12 Relationship between underground engineering and groundwater level

Ⅰ—aeration zone; Ⅱ—variation zone of groundwater table; Ⅲ—perennial groundwater flow zone

11.3.5 In-situ stress conditions

In the process of tunnel design and construction, it is necessary to understand the distribution and variation of the initial stress field at the site of the project, and to obtain the characteristics of stress redistribution of surrounding rock after tunnel excavation, to choose the corresponding measures to maintain the stability of surrounding rocks.

The initial stress state is the main factor to determine the stress redistribution of surrounding rock. For example, for circular caverns, when the stress ratio coefficient $\lambda(\sigma_h/\sigma_v)=1$, the compressive stress distribution in surrounding rock is relatively uniform. Generally, the surrounding rock stability is good. When the $\lambda<1/3$ or $\lambda>3$, tensile stress will appear in the surrounding rock, and compressive stress concentration is also significant, which is unfavorable to the stability of the surrounding rock.

When the horizontal principal stress value of the initial stress field is large, the axis of the cavern had better be arranged parallel to the direction of the maximum horizontal principal stress. Otherwise, the sidewall will produce severe deformation and failure. The layout of the roadway in the Jinchuan mine is an example. In addition, the measured direction of the maximum horizontal principal stress of Ertan Hydropower station is about NE30°, and the difference between the two horizontal principal stresses is very large. If the axis of the tunnel is arranged parallel to the direction of the maximum horizontal principal stress, the lateral pressure on the sidewall is small. If the direction of the vertical maximum horizontal principal stress is arranged, the lateral pressure will increase by 1.57~3.63 times.

11.4 Analysis of induced stresses and displacement

When an underground opening is excavated into a stressed rock mass, the stresses in the vicinity of the new opening are redistributed.

11.4.1 Openings in Competent Rock

In rock stressed below its elastic limit, that is, below about one-half of the compressive strength, and in which joints are widely spaced and tightly precompressed or healed, it is often acceptable to consider an opening as a long hole of a constant cross-section in an infinite volume. It is the plane strain equivalent of a hole in a plate. We can use the problem of a circular hole in a biaxially loaded plate of homogeneous, isotropic, continuous, linearly elastic material solution (Kirsch, 1898). A point located at polar coordinate r, θ near an opening with radius a (Figure 11-13)has stresses σ_r, σ_θ, $\tau_{r\theta}$, given by

$$\begin{cases} \sigma_r = \dfrac{p_1+p_2}{2}\left(1-\dfrac{a^2}{r^2}\right)+\dfrac{p_1-p_2}{2}\left(1-\dfrac{4a^2}{r^2}+\dfrac{3a^4}{r^4}\right)\cos 2\theta \\ \sigma_\theta = \dfrac{p_1+p_2}{2}\left(1+\dfrac{a^2}{r^2}\right)-\dfrac{p_1-p_2}{2}\left(1+\dfrac{3a^4}{r^4}\right)\cos 2\theta \\ \tau_{r\theta} = \dfrac{p_1-p_2}{2}\left(1+\dfrac{2a^2}{r^2}-\dfrac{3a^4}{r^4}\right)\sin 2\theta \end{cases} \tag{11-1}$$

where σ_r, is the stress in the direction of changing r, and σ_θ is the stress in the direction of changing θ.

Substituting the value $r=a$ in Eq.(11-1)gives the variation of stresses on the walls of the opening.As shown in Eq.(11-2)

$$\begin{cases} \tau_{r\theta} = \sigma_r = 0 \\ \sigma_\theta = p_2(1+2\cos 2\theta)+p_1(1-2\cos 2\theta) \end{cases} \tag{11-2}$$

If $k=p_1/p_2$,then the Eq. (11-2) can be written as:

$$\begin{cases} \tau_{r\theta} = \sigma_r = 0 \\ \sigma_\theta = p_2\,[(1+k)+2(1-k)\cos 2\theta] \end{cases} \tag{11-3}$$

When θ=0° /180° , $\sigma_\theta = p_2(3k-1)$;when θ=90° /270° , $\sigma_\theta = p_2(3k-1)$.

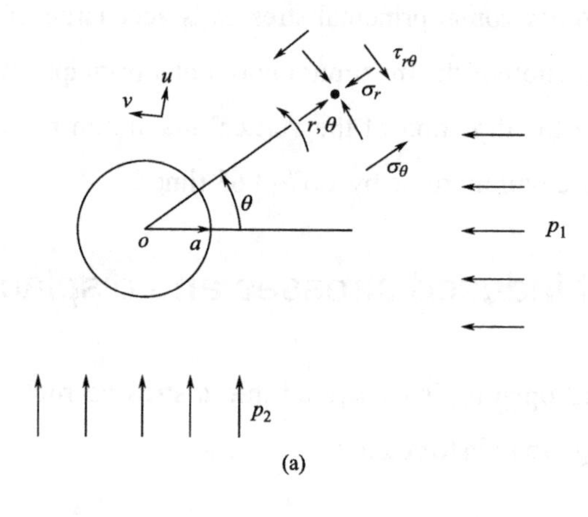

(a)

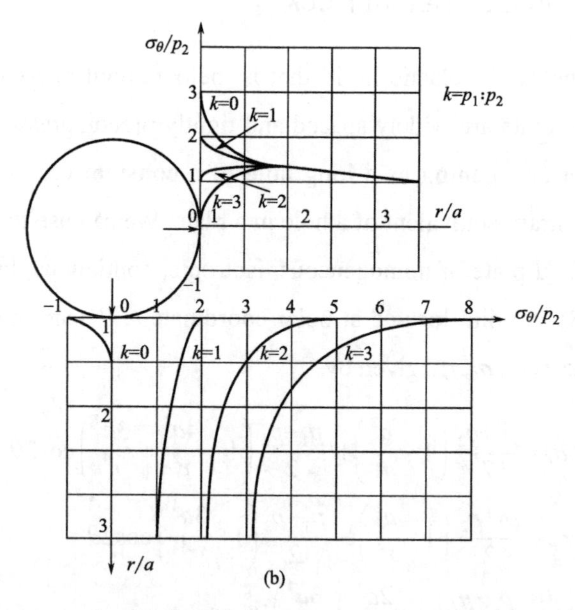

(b)

Figure 11-13 Stresses around a circular hole in an isotropic, linearly elastic, homogeneous continuum

The radial and shear stresses are zero since this is a free surface. The tangential stress σ_θ varies from a maximum of $3p_1-p_2$ at θ=90° to a mimmum of $3p_2-p_1$ at θ=0° . Away from the opening, the stress concentrations fall off quickly, as shown in Figure 11-13b and Table 11-1.

The Kirsch solution allows to calculate the possible influence of joints in the region of a tunnel. Assuming that a joint of giving the position and orientation does not introduce any change in the stress field, we can compare the shear and normal stresses along its surface. This

Table 11-1 stress concentration k in roof (θ=90°) and wall (θ=0°) accrding to kirsch solution

k	0		0.3		0.6		1.0	1.5		2.0		3.0	
r/a \ θ	90°	0°	90°	0°	90°	0°	All θ	90°	0°	90°	0°	90°	0°
1.0	−1.00	3.00	−0.10	2.70	0.80	2.40	2.00	3.50	1.50	5.00	1.00	8.00	0.00
1.0	−0.61	2.44	0.12	2.25	0.85	2.07	1.83	3.05	1.52	4.26	1.22	6.70	0.60
1.2	−0.38	2.07	0.25	1.96	0.87	1.84	1.69	2.73	1.51	3.77	1.32	5.84	0.94
1.3	−0.23	1.82	0.32	1.75	0.86	1.68	1.59	2.50	1.48	3.41	1.36	5.23	1.13
1.4	−0.14	1.65	0.36	1.60	0.85	1.56	1.51	2.33	1.44	3.16	1.37	4.80	1.24
1.5	−0.07	1.52	0.38	1.50	0.84	1.47	1.44	2.20	1.41	2.96	1.37	4.48	1.30
2.0	0.03	1.22	0.40	1.23	0.76	1.24	1.25	1.86	1.27	2.47	1.28	3.69	1.31
2.5	0.04	1.12	0.38	1.13	0.71	1.14	1.16	1.72	1.18	2.28	1.20	3.40	1.24
3.0	0.04	1.07	0.36	1.09	0.68	1.10	1.11	1.65	1.13	2.19	1.15	3.26	1.19
4.0	0.03	1.04	0.34	1.04	0.65	1.05	1.06	1.58	1.08	2.10	1.09	3.14	1.11

exercise defines an area of joint influence, which can be superimposed on the actual or assumed geological section to isolate potential problem areas.

In the hydrostatic case (k = 1), the stress concentration around the excavation boundary is always $2p$. The solution for stresses anywhere within the rock mass for this stress state is similarly simplified because there are no shear stresses: the terms $(1-k)$ are all zero. Hence the equations for radial and tangential stress reduce to:

$$\begin{cases} \sigma_r = p_v\left(1-\dfrac{a^2}{r^2}\right) \\ \sigma_\theta = p_v\left(1+\dfrac{a^2}{r^2}\right) \\ \tau_{r\theta}=0 \end{cases} \tag{11-4}$$

In Figure 11-14, there is a zone shown around the opening where the Mohr-Coulomb criterion for the intact rock has been satisfied. For the conditions of a hydrostatic field stress, as shown, this zone is circular and concentric with the centre of the opening.

Closed form solutions still possess great value for conceptual understanding of behaviour and for the testing and calibration of numerical models. For design purposes, however, these models are restricted to very simple geometries and material models. They are of limited practical value. Fortunately, with the development of computers, many powerful programs that provide numerical solutions to these problems are now readily available. A brief review of some of these numerical solutions is given below.

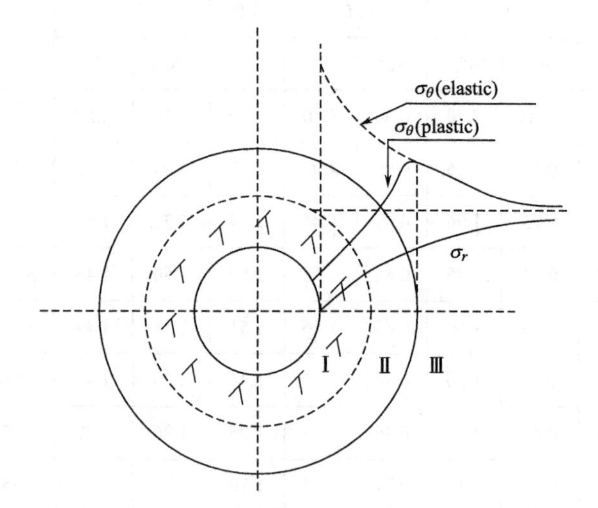

Figure 11-14 Elastic-plastic stress distribution in surrounding rock

I is the stress reduction zone, II is the stress increase zone, I and II form the plastic loosening zone, and III is the natural stress zone

11.4.2 Principal stresses induced in the rock surrounding a circular tunnel

Take the case of induced stresses in the rock surrounding a horizontal circular tunnel showing a normal vertical section at the tunnel axis. Before the tunnel is excavated, the in-situ stresses σ_v, σ_{h1} and σ_{h2} , and are evenly distributed in the given rock slice. Once the rock has been removed from the tunnel interior, the new stresses are induced. Three major stresses σ_1, σ_2 and σ_3 on a typical rock element are illustrated in Figure 11-15.

The convention used in rock engineering is that compressive stresses are always positive. The three principal stresses are numbered such that σ_1 is the largest compressive stress and σ_3 is the smallest compressive stress or the largest tensile stress of the three.

The three principal stresses are mutually perpendicular, but they may be inclined to the direction of the applied in situ stress. FLAC3D is used to calculate the induced stress. The material is assumed to be linearly elastic, perfectly plastic, with a failure surface defined by the Mohr-Coulomb criterion. For modeling purposes, the quarter-symmetry geometry is used. In the numerical simulation, the initial stress state is applied throughout the domain first, and then the hole is removed. It is evident in Figure 11-16 which shows the directions of the stresses in the rock surrounding a horizontal tunnel subjected to a isotropic in situ stress. The purple bars in this figure represent the directions of the maximum principal stress σ_1, while the green bars give the directions of the minimum principal stress at each element considered.

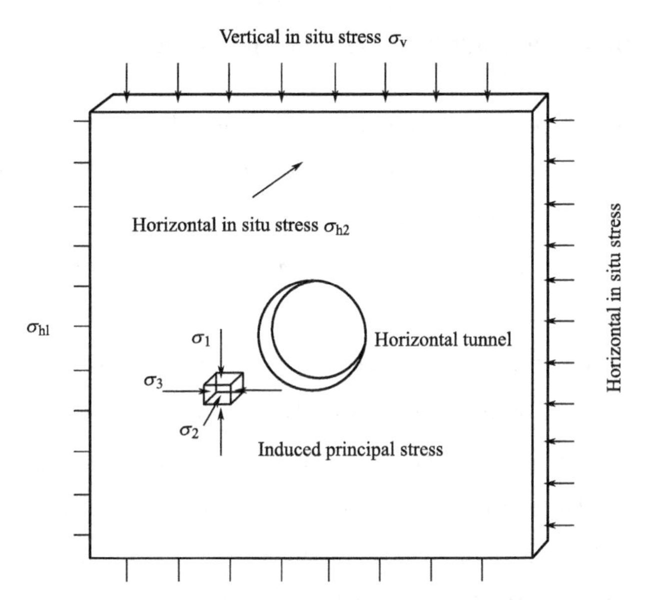

Figure 11-15 Illustration of principal stresses induced in an element of rock close to a horizontal tunnel subjected to a vertical in situ stress σ_v, a horizontal in situ stress σ_{h1} in a plane normal to the tunnel axis and a horizontal in situ stress σ_{h2} parallel to the tunnel axis

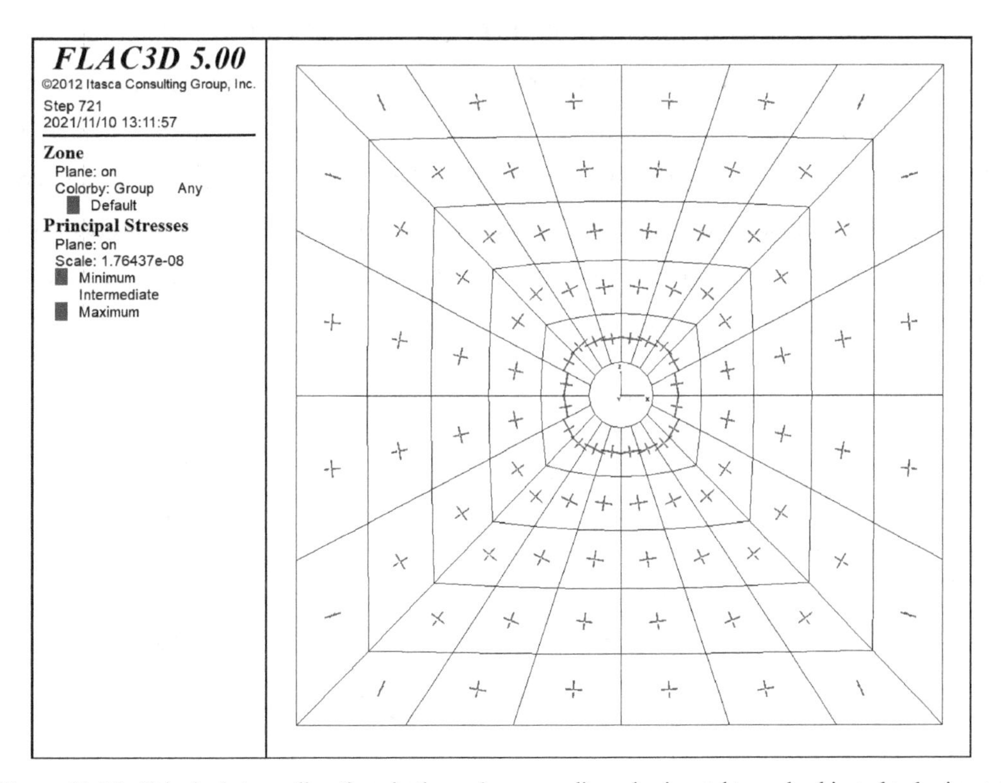

Figure 11-16 Principal stress directions in the rock surrounding a horizontal tunnel subjected to horizontal isotropic in-situ stress of a magnitude of 30MPa

11.4.3 Longitudinal displacement profile of Tunnel

The longitudinal displacement profile (LDP) describes the displacement at the monitoring section versus distance away from tunnel face and provides an analytical solution to establish a distance-convergence relationship. As shown in Figure 11-21, the origin of the x-axis is located at the monitoring section and the x-axis represents the relative location from monitoring section to tunnel face; the vertical axis represents the displacement of tunnel boundary at the monitoring section. When tunnel face proceeds along the x-axis to the coordinate value of x, the displacement at the section is denoted as u_x. Before the arrival of tunnel face, a portion of displacement has occurred. When tunnel face arrives at monitoring section, initial displacement u_0, approximate 30% of the ultimate displacement u_∞, has achieved, after which the convergence of surrounding rock will increase dramatically to the peak value and keep stable. In other words, the excavation disturbance effect by tunnel face will disappear until tunnel face is far away enough. It is commonly believed that the distance between the section and tunnel face is no less than three times opening diameter. In tunnel design, the appropriate location for the installation of the initial support can be determined by the LDP (Carranza and Fairhurst, 2000).

As shown in Figure 11-17(a), the displacement of the surrounding rock when tunnel face arrives at the monitoring section is denoted as u_0. When the distance away from tunnel face is x, the displacement at the monitoring section is u_x. When the tunnel face is far away enough, the ultimate displacement is denoted as u_∞. To reflect the propagation of the convergence during the tunnel face advancing, the dimensionless displacement, λ, is defined to normalize the displacement of surrounding rock as:

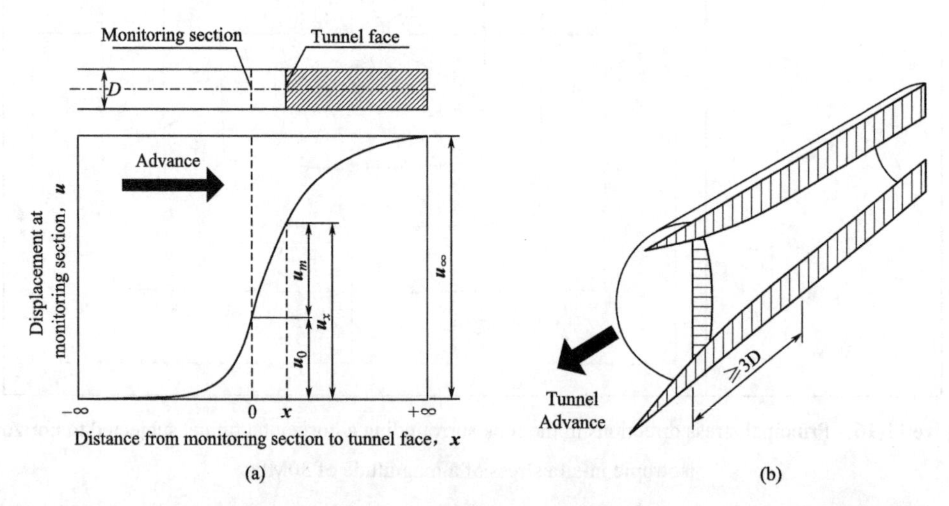

Figure 11-17 The longitudinal displacement profile of a tunnel (Vlachopoulos and Diederichs 2009)
(a) the meaning of LDP; (b) the face effect

$$\lambda=\frac{u_x}{u_\infty} \tag{11-5}$$

The longitudinal displacement profile can be obtained by the monitoring data. However, the field data usually just contain the excavated part and fail to monitor the unexcavated rock mass. To establish the intact relationship between the dimensionless displacement and distance to tunnel face, numerous authors have proposed alternative expressions. Sulem & Panet (1995) derived the relationship for the normalized LDP from elastic analysis as:

$$\lambda=0.25+0.75\left[1-\left(\frac{0.75}{0.75+x/R}\right)^2\right] \tag{11-6}$$

where R is the tunnel radius; x is the location away from the tunnel face and $x>0$ is behind the tunnel face whereas $x<0$ is ahead of the face.

Hoek proposed the empirical best-fit formula of the measured data in the Mingtam Power Cavern Project in 1999 (Carranza-Torres & Fairhurst, 2000) as follows,

$$\lambda=\left[1+\exp\left(\frac{-x/R}{1.10}\right)\right]^{-1.7} \tag{11-7}$$

The elastic theoretical expression was deduced by Zhao et al. (2016) as:

$$\lambda=\begin{cases}(1-\lambda_0)\left(1-e^{-\frac{x}{X}}\right)+\lambda_0 & x\geqslant 0\\ \lambda_0 e^{\frac{x}{X_1}} & x<0\end{cases} \tag{11-8}$$

where X and X_1 are constants, and moreover, $X_1=\frac{\lambda_0}{1-\lambda_0}X$.

And the expression based on the Eq. (11-9) was improved by the numerical analysis as (Zhao et al, 2016):

$$\lambda=\begin{cases}(1-\lambda_0)\left(1-e^{-\frac{(1-\lambda_0)x}{X}}\right)+\lambda_0 & x\geqslant 0\\ \lambda_0 e^{\frac{\lambda_0 x}{X_1}} & x<0\end{cases} \tag{11-9}$$

From the work that Carranza-Torres and Fairhurst (2000) compared the Eq. (11-6) and the Eq. (11-7) with the measured data of a tunnel in the Mingtam Power Cavern project, the Eq. (11-6) could illustrate the excavated part of LDP curve and this elastic approximation still had deviation with the field data. The Eq. (11-7) could meet the description requirement in the excavated part, but the error in the unexcavated part was obvious. Based on the theoretical analysis of elastic medium, the Eq. (11-8) had distinction with the rock material. The Eq. (11-8) is improved by the elastoplastic numerical simulation, and the Eq. (11-9) with a better effect is proposed now(Zhang et al., 2020).

11.4.4 Numerical methods of stress analysis

Most underground caverns are irregular in shape and are frequently grouped close to other excavations, and these groups of excavations can form a set of complex three-dimensional shapes. In addition, because of geological features such as faults and dykes, the rock properties are seldom uniform within the rock volume of interest. Consequently, closed-form solutions are of limited value in calculating the stresses, displacements, and failure of underground excavations' rock mass. Several computer-based numerical methods have been developed over the past few decades, and these methods provide the means for obtaining approximate solutions to these problems(Hoek,2003).

Numerical methods for analyzing stress-driven problems in rock mechanics can be divided into two classes: Boundary discretization methods, in which only the boundary of the excavation is divided into elements and the interior of the rock mass is represented mathematically as an infinite continuum. These methods are normally restricted to elastic analyses. Another type is domain discretization methods, in which the interior of the rock mass is divided into geometrical elements, each with assumed properties. The collective behavior and interaction of these simplified elements model the more complex overall behavior of the rock mass. In other words, domain methods allow consideration of more complex material models than boundary methods. Finite element and finite difference methods are domain techniques that treat the rock mass as a continuum. The distinct element method is also a domain method that models each block of rock as a unique element. The two classes of analysis can also be combined in the form of hybrid models to maximize the advantages and minimize the disadvantages of each method.

Before selecting the appropriate modelling technique for particular types of problems, it is necessary to understand the basic components of each technique.

1. Boundary Element Method

The boundary element method (BEM) is another numerical algorithm developed based on the integral equation theory(Wang et al.,2017& Zhou et al.,2020). Different from the finite element method, the boundary element method only needs to arrange elements at the boundary of the problem, that is, the solution domain of the problem is reduced by one dimension, which greatly reduces the workload of pre-processing.

From the process of establishing the boundary element method, one can see that it has the following characteristics:

(1) The dimension of the problem is reduced by one dimension. This can be derived from Green's formula and it is the cornerstone of the theory of the boundary element method. The

unknowns in the final system of equations can be greatly reduced since only the boundary of the domain is discretized.

(2) The accuracy is very high. This is reflected in the discretization process of the boundary element method. The boundary element method is not much different from the traditional finite element method, but it only discretizes the boundary. Therefore, the calculation error only comes from the boundary. Moreover, it also adopts the fundamental solution as the test function, which can be considered as a semi-numerical and semi-analytical method, which further improves the calculation accuracy.

(3) It is easy to be coupled with other methods such as the finite element method.

(4) The pre-processing is simple and easy to interface with computer modeling software, such as CAD.

(5) It is more suitable for solving infinite or semi-infinite domain problems, and has certain advantages in solving fracture problems, which are very common in geotechnical mechanics.

The boundary element method can be classified into two categories, i.e., the direct boundary element method and the indirect boundary element method.

(1) The direct boundary element method. In the direct boundary element method, unknown displacement and traction can be obtained directly by solving the final system of equations, according to the giving boundary conditions. It can be further classified into two types, i.e., the conventional displacement BEM (CBEM) and the traction (hyper singular) BEM (HBEM). The difference between these two kinds of BEM is that the results of the boundary integral for CBEM are displacement, while the results are traction in HBEM. Another type of direct BEM is the dual BEM (DBEM), which is a combination of the CBEM and HBEM, and is commonly applied for crack problems.

(2) The indirect boundary element method. In this type of boundary element method, the unknowns obtained by solving the final system of equations are not the real physical quantities and the displacements or tractions on the boundaries have to be calculated by further transformation. It can be further classified into the displacement discontinuity method (DDM) and fictitious stress method (FSM). These indirect BEMs can also be applied in crack problems, however, many researchers have found out the indirect BEMs are not stable and accuracy is very low sometimes, even using very fine meshes.

2. Finite element and finite difference methods

In practice, the finite element method is usually indistinguishable from the finite difference method; thus, they will be treated here as the same. For the boundary element method, it was seen that conditions on a domain boundary could be associated with the state at all points throughout the remaining rock, even to infinity. In comparison, the finite element method relates the conditions at a few points within the rock (nodal points) to the state within a finite closed area formed by these points (the element). In the finite element method, the physical problem is modeled numerically by dividing the entire problem region into elements.

The finite element method is well suited to solving problems involving heterogeneous or non-linear material properties since each element explicitly models the response of its contained material. However, finite elements are not well suited to modeling infinite boundaries, such as underground excavation problems. One technique for handling unlimited boundaries is to discretize beyond the excavation zone of influence and apply appropriate boundary conditions to the outer edges. In practice, efficient pre- and post-processors allow the user to perform parametric analyses and assess the impact of approximated far-field boundary conditions. The time required for this process is negligible compared to the total analysis time.

Joints can be represented explicitly using specific joint elements. Different techniques have been proposed for handling such elements, but no single method has found universal favor. Joint interfaces may be modeled using general constitutive relations, though possibly at increased computational expense depending on the solution technique.

Once the model has been divided into elements, material properties have been assigned, and loads have been prescribed, some technique must be used to redistribute any unbalanced loads and thus determine the solution to the new equilibrium state. Available solution techniques can be broadly divided into two classes - implicit and explicit. Implicit methods assemble systems of linear equations that are then solved using standard matrix reduction techniques. Any material nonlinearity is accounted for by modifying stiffness coefficients and adjusting prescribed variables. These changes are made in an iterative manner such that all constitutive and equilibrium equations are satisfied for the given load state.

The response of a non-linear system generally depends upon the sequence of loading. Thus, the load path modeled must be representative of the actual load path experienced by the body. It is achieved by breaking the total applied load into load increments, each increment being sufficiently small, and solution convergence for the increment is achieved after only a few iterations. However, as the system being modelled becomes increasingly non-linear and the load increment represents an ever-smaller portion of the total load, the incremental solution technique becomes similar to modeling the quasi-dynamic behavior of the body, as it responds to the gradual application of the whole load.

To overcome this, a "dynamic relaxation" solution technique was proposed (Otter et al., 1966) and first applied to geomechanics modeling by Cundall (1971). In this technique, no matrices are formed. Instead, the solution proceeds explicitly - unbalanced forces, acting at a material integration point, result in acceleration of the mass associated with the point. Applying Newton's law of motion expressed as a difference equation yields incremental displacements. Using the appropriate constitutive relation produces the new set of forces, marching in time, for each material integration point in the model. This solution technique has the advantage that both geometric and material non-linearities are accommodated, with relatively little additional computational effort compared to a corresponding linear analysis, and computational expense increases only linearly with the number of elements used. A further practical advantage is that numerical divergence usually results in the model predicting anomalous physical behavior. Thus, even relatively inexperienced users may recognize numerical deviation.

Most commercially available finite element packages use implicit (i.e., matrix) solution techniques. For linear problems and moderate non-linearity, implicit methods tend to perform faster than explicit solution techniques. However, as the degree of non-linearity of the system increases, imposed loads must be applied in smaller increments which imply a more significant number of matrix reformations and reductions, and hence increased computational expense. Therefore, highly non-linear problems are best handled by packages using an explicit solution technique.

3. Distinct Element Method

If ground conditions are conventionally described as blocky (i.e., where the spacing of the joints is of the same order of magnitude as the excavation dimensions), intersecting joints form wedges of rock that may be regarded as rigid bodies. That is, these individual pieces of rock may be free to rotate and translate, and the deformation that takes place at block contacts may be significantly more significant than the deformation of the intact rock. Hence, individual wedges may be considered rigid. For such conditions, it is usually necessary to model many joints explicitly. However, the behavior of such systems is so highly non-linear that even a jointed finite element code employing an explicit solution technique may perform relatively inefficiently.

An alternative modeling approach is to develop data structures that represent the blocky nature of the system being analyzed. Each block is considered a unique free body that may interact at contact locations with surrounding blocks. Contacts may be represented by the overlaps of adjacent blocks, thereby avoiding the necessity of individual joint elements. It has the added advantage that arbitrarily large relative displacements at the contact may occur, a situation not generally tractable in finite element codes.

Due to the high degree of non-linearity of the modeled systems, explicit solution techniques are favored for distinct element codes. The use of explicit solution techniques places fewer demands on the skills and experience than codes employing implicit solution techniques. Although the distinct element method has been used most extensively in academic environments to date, it is finding its way into the offices of consultants, planners, and designers. Further experience in applying this powerful modeling tool to practical design situations and subsequent documentation of these case histories is required to develop an understanding of where, when, and how the distinct element method is best applied.

4. Hybrid approaches

The objective of a hybrid method is to combine the above techniques to eliminate undesirable characteristics while retaining as many advantages as possible. For example, in modeling an underground excavation, most non-linearity will occur close to the excavation boundary, while the rock mass at some distance will behave elastically. Thus, the near-field rock mass might be modeled using a distinct element or finite element method, which is then linked at its outer limits to a boundary element model, so that the far-field boundary conditions are modeled precisely. The direct boundary element technique is favored in such an approach as it increases programming and solution efficiency.

Lorig and Brady (1984) used a hybrid model consisting of a discrete element model for the near field and a boundary element model for the far-field in a rock mass surrounding a circular tunnel. The FLAC & PFC coupled approach is to used to take advantage of each modeling scheme while at the same time minimizing the requirement for computational resources to simulate of Acoustic emission (AE) in large-scale underground excavations (Cai et al., 2007).

11.5 Measures to improve the stability of surrounding rockmass

It is very important to study the stability of surrounding rock of cavern, not only to carry out engineering design and construction correctly, but also to improve the stability of surrounding rock effectively. There are two ways to ensure the stability of surrounding rock: one is to protect the original stability of surrounding rock, so that it will not be reduced; Second, the rock mass is given a certain strength, so that its stability is increased. The former mainly adopts reasonable construction and supporting lining scheme, while the latter mainly strengthens surrounding rock.

11.5.1 Excavation schemes

Different construction schemes should be selected for different stability of surrounding rock. Reasonable selection of construction scheme is of great significance to protect the stability of surrounding rock. The principle to be followed is to dig the guide tunnel with smaller section size as far as possible, and to support or lining it in time after excavation. In this way, the loosening range of surrounding rock can be reduced or the early loosening of surrounding rock can be prevented. Prevent the surrounding rock loose, or the loose range is limited to the minimum. There are many construction schemes for the surrounding rock with different stability. It can be divided into three categories.

1. Partial excavation, partial lining, gradually expand the section

When the surrounding rock is not very stable, the top circumference is easy to collapse. Then, dig a guide tunnel at the upper part of the largest section of the cave (Figure 11-18a), immediately support, reach the required contour, and prepare the lining of the top arch. Then the section is enlarged under the protection of the top arch lining, and finally the side wall lining is made. This is the way to excavate the upper guide cave and arch the back wall first. In order to reduce construction interference and speed up transportation, the method of digging the upper and lower guide holes and arching the back wall first can also be used (Figure 11-18b). The surrounding rock is very unstable, collapse of the top circumference, side circumference slippery. In this way, a guide hole can be excavated at the side of the designed section (Figure 11-18c) and lined from bottom to top. To a certain elevation, then dig the top guide hole, make the top arch lining, finally dig the residual rock mass. This is the method of side guide tunnel excavation, first wall and then arch, or called core bracing method.

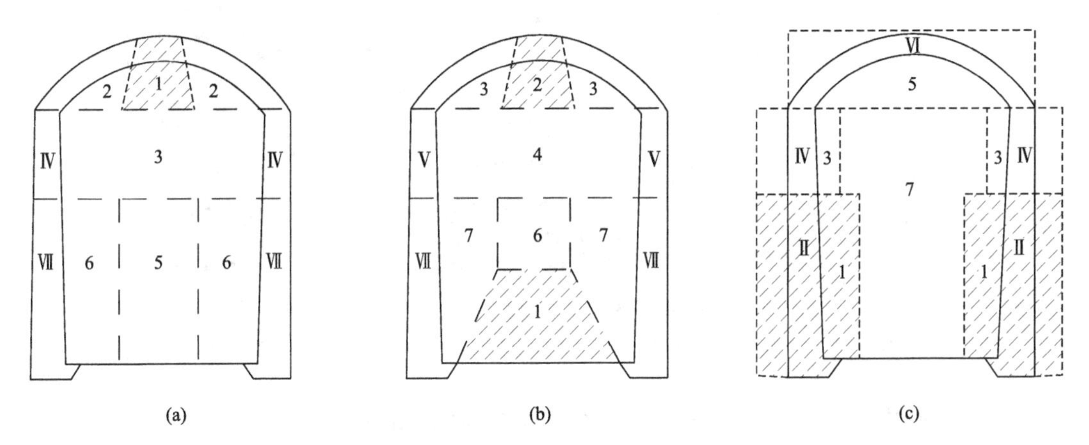

Figure 11-18 Schematic diagram of section of partial excavation and gradual expansion

(a) The upper guide first built the arch and then built the wall; (b) The upper and bottom guide first built the arch and then built the wall; (c) The side guide first built the wall and then built the arch. 1, 2, 3⋯⋯Excavating Sequence; Ⅳ, Ⅴ⋯⋯Lining Sequence

2. The pilot tunnel is fully excavated and lined continuously

The surrounding rock is relatively stable, so the guide tunnel can be constructed by comprehensive excavation and continuous lining. Either the upper and lower guide tunnels are fully excavated, or the lower guide tunnels are fully excavated, or the central guide tunnels are fully excavated. After the whole section is dug, the lining is lined from side wall to top arch. In this way, the towing speed is fast and the lining quality is high.

3. Full section excavation and TBM

When the surrounding rock is hard, the whole section can be excavated at one time. The construction speed is fast and the slag is convenient. TBM (Tunnel Boring Machine) is a comprehensive equipment that uses mechanical rock breaking, ballasting and support to implement full section continuous operation (Figure 11-19). It is developed from shield technology. TBM construction has been applied in more and more countries in the world due to its advantages such as fast construction speed, high efficiency, good tunnel formation, low impact on the surrounding environment and safe operation.

Figure 11-19 The world's first ultra-small turning radius hard rock TBM (Made by China Railway Group Limited) has been completed in Wendeng Pumping and storage Power station

11.5.2 Reinforcement measures for surrounding rock

1. Support

The support is the temporary measure to strengthen the cave wall, and the lining is the permanent measure to strengthen the cave wall. The support is a simple and feasible way to protect the stability of surrounding rock and prevent the early loosening of surrounding rock (Figure 11-20). The triumphant advance of the railways in the 19th century was mainly responsible for rapid developments in tunnel construction. In order to protect the excavated spaces from the weight of the rock or earth above, they must be secured during construction with temporary supports and by masonry when in permanent use.

Figure 11-20 The temporary supports of the Simplon railway tunnel traversing the Lepontine Alps between Switzerland and Italy

2. Lining

The lining serves the same function as the support, but is durable and makes the cave walls light. Brick and stone linings are cheaper, while reinforced concrete and steel plate linings cost more. The lining must be closely combined with the hole, and the gap can be filled with good effect. For the lining of the top arch, the grouting hole is usually reserved. After lining, backfill grouting, to achieve the purpose of tight, in the seepage area can also play a role in impermeability. Linings may be of different types according to the need for reinforcement(Figure 11-21).

3. Gunite and shotcrete

Gunite, shotcrete, and the lining mentioned above have many of the same functions, but the cost is much lower and can make full use of the strength of surrounding rock to achieve the purpose of protecting surrounding rock and making it stable. Gunite protection is economical and straightforward and has a good stability effect on protecting the easily weathered surrounding rock. After excavation, cement mortar is sprayed on the cave wall in time to form a protective layer to protect the original strength of the surrounding rock. The shotcrete is similar to the gunite methods. However, the concrete generally with the accelerating agent, sprayed to the cave wall in time, quickly solidified and has a greater strength (general 25 MPa above), can prevent the early loosening of the cave wall.

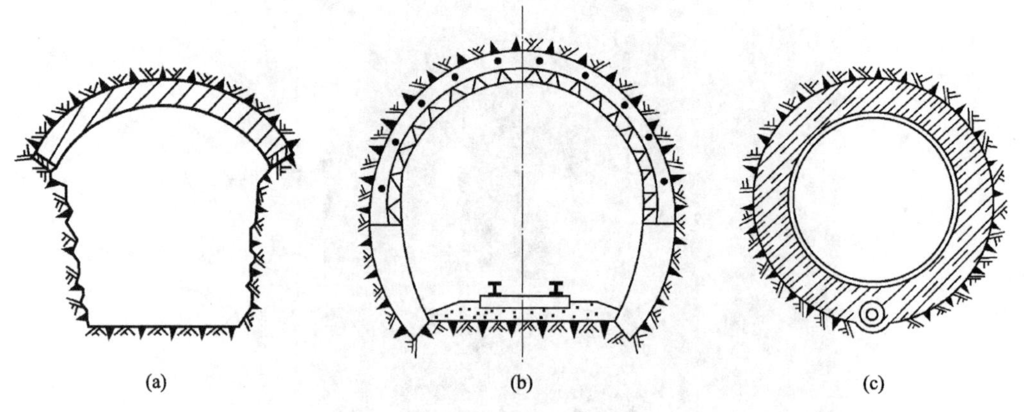

Figure 11-21 Schematic diagram of lining type
(a) semi-lined; (b) side-wall lined; (c) fully lined

4. Rock anchors/bolts

Anchor reinforcement, also known as bolt reinforcement, is widely used at present. The anchor bar was inserted into the surrounding rock to fix the loose surrounding rock and the stable surrounding rock around the hole, acting as a nail (Figure 11-22a). There are two commonly used wedge-head anchor bars (Figure 11-22b) and mortar anchor bars (Figure 11-22c). In mortar anchor bar, the steel bars is inserted into the drill hole of surrounding rock, and then fill the drill hole with cement mortar. The reinforcement is curved in the hole for better results. Wedge head anchor bar is to open a small seam at the inner end of the steel bar, put on the iron wedge, punch the steel bar into the manhole, and then hammer tight. Iron wedge force in the stool end open into a fork, the steel bar stuck in the hole. The outer end is screwed to fix potentially loose rocks to the rock mass. Sometimes mortar holes are also used to increase the anchoring effect. Anchor reinforcement is suitable for solid rock mass, and the anchor depth should be greater than the thickness of the loose zone of surrounding rock, or greater than the height of the balanced arch, or greater than the size of the separated body.

5. Grouting

In order to increase the stability of surrounding rock and reduce its permeability, it is often necessary to strengthen the excavation of caverns in fractured rock mass and unstable quaternary deposits. The most common reinforcement method is cement grouting, followed by asphalt grouting, sodium silicate (silicate) grouting, and freezing method, and so on. In this way, a roughly cylindrical or spherical consolidation layer is formed in the surrounding rock.

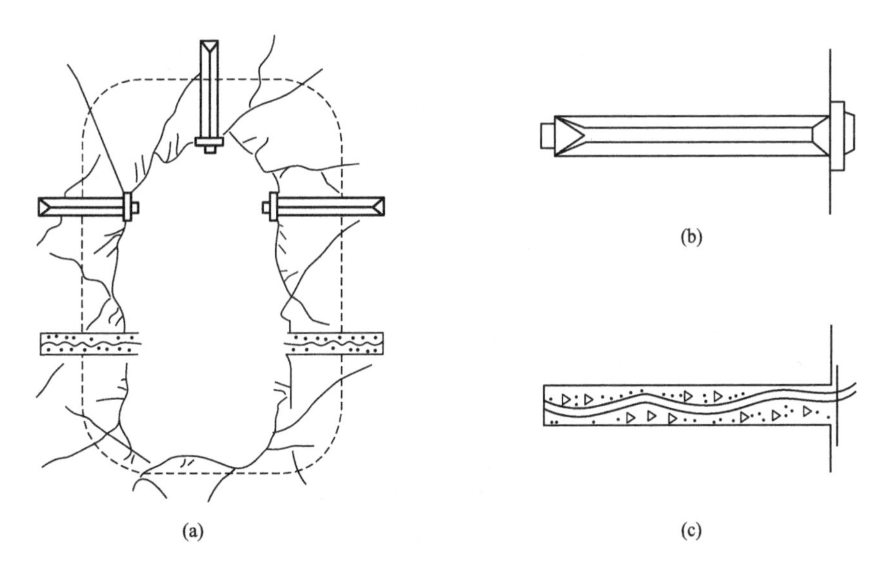

Figure 11-22 Schematic diagram of surrounding rock of tunnel reinforced by anchor bar

(a) Reinforcement section; (b) Wedge anchor bar; (c) Cement-And-Grouted Steel Bolt

11.5.3 Convergence-confinement methods

To describe the interaction between the rock and support system by an analytical method, the convergence-confinement method was developed in 1960s and 1970s by Sulem & Panet (1995) and has been employed as an industry standard for support design at present (Vlachopoulos & Diederichs, 2014).

There are three components of the method, as shown in Figure 11-23. The Longitudinal Displacement Profile (LDP) describes the relationship between the tunnel deformation and the distance away from the tunnel face. The Ground Reaction Curve (GRC) relates the inner pressure to the displacement in the tunnel wall. The Support Characteristic Curve (SCC) represents the stress-deformation relationship of the support system (Carranza-Torres & Fairhurst, 2000). This method emphasizes the synergy of the rock-support system to make full use of the rock's self-bearing capacity.

As shown in Figure 11-23, the rock-support system will reach equilibrium at the intersection of the GRC and SCC. Based on the modern support design theory, the moderate deformation of rock mass is permitted to form the bearing arch, and the large deformation should be restricted for the stability (Jenny et al., 1987). An early installation of the support will lead to excessive load in the support. Contrarily, a delayed installation of the support will lead to excessive tunnel deformation or collapse.

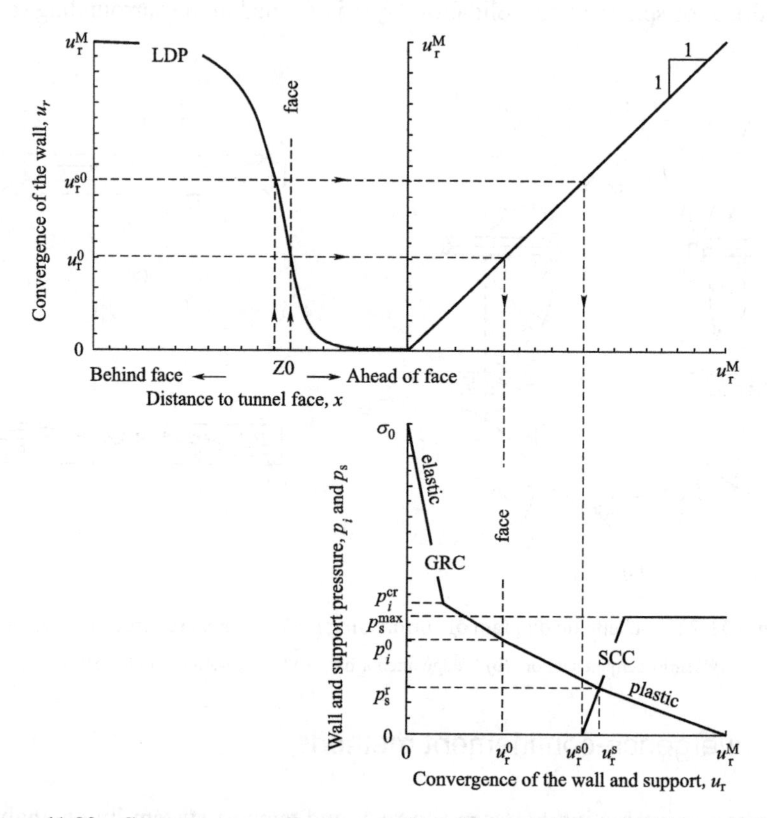

Figure 11-23 Convergence-confinement method (Carranza-Torres & Fairhurst, 2000)

In other words, the initial support should be installed when surrounding rock has released appropriate pressure or when surrounding rock has occurred appropriate deformation (Lee & Schubert, 2008). Accordingly, it is valuable to analyze the moment that the equilibrium between rock and support is achieved to ensure the safety of tunnel structure as well as utilize the full potential of rock mass.

11.5.4 *Q*-support chart

The traditional application of the six-parameter *Q* value in rock engineering involves selecting suitable combinations of bottom concrete and bolts for reinforcing and supporting the rock mass(8.6.3 in this textbook). In particular, it concerns the permanent estimation of the lining for tunnels or caverns in the rock, mainly for civil engineering projects.

The advancement of the *Q* system in the early 1970s followed a period of preoccupation with the shear strength of rock joints and clay-filled discontinuities. The original 212 case records of tunnels and caverns from 50 different rock types were analyzed several times during the six-month period needed to develop the Q parameters. This was to calibrate and re-calibrate the ratings to match the final Q value with the support and reinforcement needs. Parameter ratings needed successive fine-tuning to bring the "all-encompassing" Q value into reasonable correspondence with the necessary level of rock reinforcement (fully grouted rock bolts) and with the required level of shotcrete or concrete support for the excavated perimeters (arch, walls and sometimes the invert as well).

The early collection of 212 case records were obtained from a period (about 1960~1973) while plain shotcrete *S*, or steel-mesh reinforced shotcrete, termed *S*(*mr*), or cast concrete arches, termed CCA, were used for tunnel and cavern maintenance, together with different types of rock bolts. Since the 1993 update of support references with 1050 new case records, the superficial support has undergone an innovative enhancement to mostly steel-fibre reinforced sprayed concrete, termed *S*(*fr*) in place of *S*(*mr*). Cured cube strength qualities of 35~45MPa or more are now readily obtainable with the wet method, micro silica-bearing, and non-alkali accelerated sprayed concrete. The renewed Q support chart is shown in Figure 11-24.

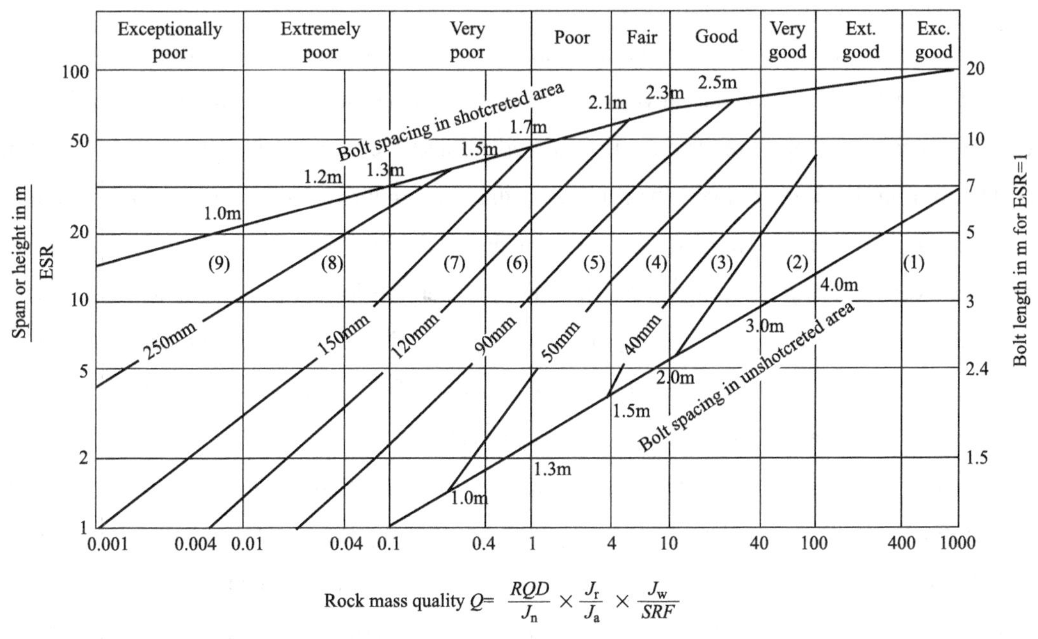

Figure 11-24 Estimated support categories based on the tunneling quality index Q (After Grimstad and Barton, 1993)

Despite the significant number of new case records, it was hardly found necessary to make any changes to the 20-year old *Q* parameter ratings. Just three of the strength/stress SRF ratings were increased to bring massive (high RQD/J_n) rock masses under extremely high stress sufficiently far "to the left" in the support chart to receive appropriate quantities of systematic bolting *B*, and *S(fr)*. Previously, such cases were supported entirely differently and were treated in a footnote before *S(fr)* development at the end of the 1970s.

During the period leading up to developing the updated support methods, Barton was involved in an increasing number of projects that required numerical verification of the empirical tunnel and cavern designs. Motorway tunnels in Norway, Hong Kong and Japan, caverns in Israel and England, and the 61m span Gjøvik cavern in Norway figured prominently in an identification of the apparent need for improved correlations between rock mass classification, general site investigation data, and the final input data for distinct element models developed principally by Cundall.

Norway built an underground Olympic ice hockey cavern (Gjøvik, 91m×61m×25m) with a capacity of 5600 spectators in 1993. It has long been regarded as the largest underground cavern open to the public in the world.

The rock type at the site is Precambrian gneiss varying in composition from granitic to quartz dioritic. The rock has developed a network of tectonic micro-joints, often filled or coated·with calcite or epidote. The result is a well-jointed rock mass with a mean *RQD* of about 70. The jointing is more frequent than in Norwegian basement rocks in general but is generally irregular, rough-walled, and has a significant dip and strike variations. The spacing of the more persistent jointing is often several meters. The general joint character is low persistence, moderate to marked roughness, and generally without clay filling, i.e., potentially positive characteristics for large spans. The *Q* value is typically 30 for the best and 1.1 for the poorest quality of the rock mass, with 12.2 as an average *Q* value.

Question number one was whether or not the rock mass in the small mountain hill had enough horizontal tectonic stresses to create sufficient compressive stresses in the roof of the hall (Broch et al., 1996). At an early stage, in-situ rock stress measurements with 3D overcoring technique were therefore carried out from one of the existing underground openings. The results showed dominating horizontal stresses in the order of 3~5MPa. The vertical stress (maximum ≈ 1MPa) is very low at a depth of only 25~50m. Later hydraulic fracturing measurements from vertical boreholes drilled from the surface verified the results. Based on these findings, it was decided to proceed with the investigations. Numerical modeling was carried out using BEM, FEM, UDEC, DDA (Scheldt & Lu, 2003), and FLAC codes. The thorough investigations concluded that a stable 61m span could be constructed under geological and rock mechanical conditions. Maximum roof deformation was estimated to be in the range of 5~10mm. Based on the readings from the leveling, the surface, and the cavern extensometers, it gives a maximum deformation of about 7mm. This is in full accordance with the predicted values from the different numerical models.

The use of grouted rock bolts combined with shotcrete (in later years, steel-fiber-reinforced) has a long tradition in Norwegian underground openings for various kinds of service. It was, therefore, the natural choice for the new large Mountain Hall in Gjøvik. More than two years after the support work was finished, there are no indications that this support does not work satisfactorily (Figure 11-25).

Observations and measurements have demonstrated that caverns with huge spans, i.e., greater than 50m, can be safely and economically excavated and supported even in rocks of just fair to good quality (Q = 4~40). The crucial question is the stress situation. Horizontal stresses of a certain magnitude are necessary to establish stable conditions for the arched roof. When this general stability is obtained, local stability in the roof can also be secured with rock bolts of limited length combined with steel-fiber-reinforced shotcrete.

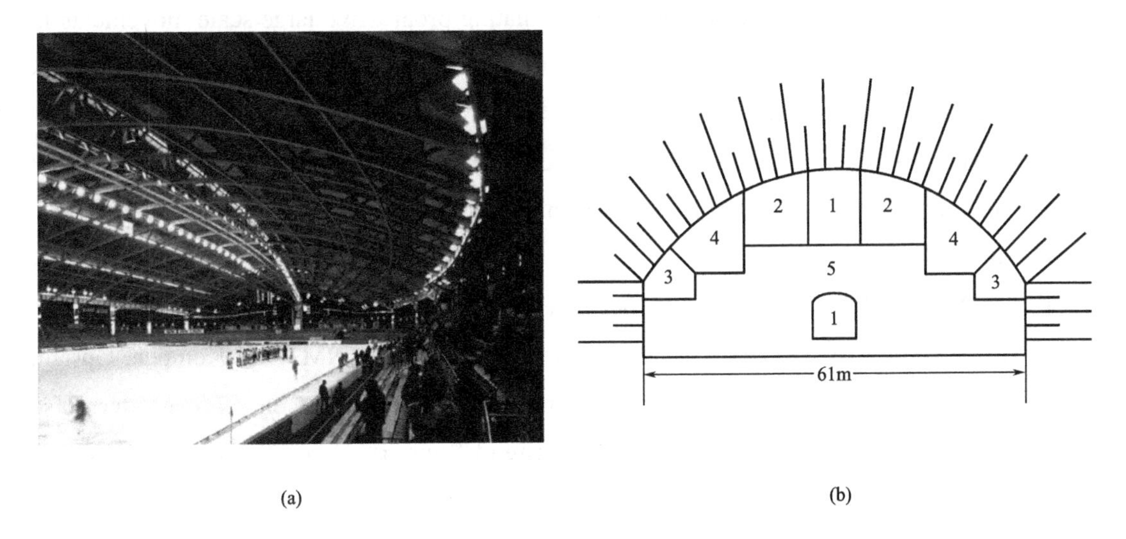

(a) (b)

Figure 11-25 The Gjøvik Olympic Cavern

(a) Inside the stadium after construction; (b) Excavation stages and bolting pattern

Reference

[1] BARTON N. Some new Q-value correlations to assist in site characterisation and tunnel design[J/OL]. International Journal of Rock Mechanics and Mining Sciences, 2002, 39(2):

185-216.

[2] BARTON N, BY T L, CHRYSSANTHAKIS P, et al. Predicted and measured performance of the 62 m span Norwegian olympic ice Hockey Cavern at Gjøvik[J/OL]. International Journal of Rock Mechanics and Mining Sciences & Geomechanics Abstracts, 1994, 31(6): 617-641.

[3] BELL F G. Engineering geology[M]. 2nd ed. Amsterdam: Elsevier, Butterworth-Heinemann, 2007.

[4] BROCH E, MYRVANG A M, STJERN G. Support of large rock caverns in No-rway[J/OL]. Tunnelling and Underground Space Technology, 1996, 11(1): 11-19.

[5] CAI M, KAISER P K, MORIOKA H, et al. FLAC/PFC coupled numerical simulation of AE in large-scale underground excavations[J/OL]. International Journal of Rock Mechanics and Mining Sciences, 2007, 44(4): 550-564.

[6] CARRANZA-TORRES C, FAIRHURST C. Application of the Convergence-Confinement method of tunnel design to rock masses that satisfy the Hoek-Brown failure criterion[J/OL]. Tunneling Underground Space Technology, 2000, 15(2): 187-213.

[7] CHEN S. Hydraulic structures[M/OL]. [S.l.]: Springer, 2015.

[8] CUNDALL P A. A computer model for simulating progressive large scale movements in blocky rock systems: Proceedings of the Symposium of the International Society for Rock Mechanics[C]. France, 1971.

[9] FENG X, ZHANG C, QIU S, et al. Dynamic design method for deep hard rock tunnels and its application[J/OL]. Journal of Rock Mechanics and Geotechnical Engineering, 2016, 8(4): 443-461.

[10] GOODMAN R E. Introduction to rock mechanics[M]. 2nd ed.Wiley, 1989.

[11] GRIMSTAD E, BARTON N. Updating of the Q-system for NMT[C]// Kompen, Opsahl, Berg. Proceedings of the International Symposium on Sprayed Concrete—Modern Use of Wet Mix Sprayed Concrete for Underground Support. Oslo: Norwegian Concrete Association, 1993.

[12] HOEK E. Practical rock engineering. Rocscience. https://www.rocscience.com/learning/hoeks-corner.

[13] HOEK E, BROWN E T. Underground Excavations in Rock[J/OL]. International Journal of Rock Mechanics and Mining Sciences & Geomechanics Abstracts, 1981, 18(2): 27.

[14] JENNY R J, DONDE P M, WAGNER H. New Austrian tunneling method used for design of soft-ground tunnels for Washington metro[J]. Transportation Research Record, 1987: 11-14.

[15] KIRSCH G. Die theorie der elastizitat und die bedurfnisse der festigkeitslehre[J].

Zantralblatt Verlin Deutscher Ingenieure, 1898, 42(28): 797-807.

[16] LEE Y Z, SCHUBERT W. Determination of the round length for tunnel excavation in weak rock[J/OL]. Tunneling Underground Space Technology, 2008, 23(3): 221-231.

[17] LORIG L J, BRADY B H G. A hybrid computational scheme for excavation and support design in jointed rock media[C/OL]// Brown E T, Hudson J A. Design and performance of underground excavations. London: Brit. Geotech. Soc., 1984.

[18] OLIVIER-MARTIN M, KOBILINSKY M. L' exécution d' un grand souterrain pour l'amén-agement hydroélectrique d'Isère-Arc[J]. Construction, 1995, X(4): 145-156.

[19] OTTER J R H, CASSELL A C, HOBBS R E. Dynamic relaxation[J/OL]. Proce-edings of the Institution of Civil Engineers, 1996, 35(4): 633-656.

[20] SCHARDT H V. Die geologischen Verhältnisse des Stauund Kraftwerkes Wäggitai[J]. Eclogae Geol. Eclogae Geologicae Helvetiae, 1924, 18: 525-543.

[21] SCHELDT T, LU M. Large span cavern stability analysis–practical use of DDA[J/OL]. Development and Application of Discontinuous Modelling for Rock Engineering, 2003: 103-109.

[22] SULEM J, PANET M. Le Calcul des Tunnels par la Méthode Convergence-Confinement[J]. Press de l'école Nationale des Ponts et Chaussées, Paris, France, 1995.

[23] VLACHOPOULOS N, DIEDERICHS M S. Improved Longitudinal Displacement Profiles for Convergence Confinement Analysis of Deep Tunnels[J/OL]. Rock Mechanics and Rock Engineering, 2009, 42(2): 131-146.

[24] VLACHOPOULOS N, DIEDERICHS M S. Appropriate Uses and Practical Limitations of 2D Numerical Analysis of Tunnels and Tunnel Support Response[J/OL]. Geotechnical and Geological Engineering, 2014, 32(2): 469-488.

[25] WAGNER C J. Die Beziehungen der Geologie zu den Ingenieur-Wissenschaften[M/OL]. Spielhagen und Schurich, Vienna: Spielhagen & Schurich, 1884: 88.

[26] WANG Q, ZHOU W, CHENG Y G, et al. A line integration method for the treatment of 3D domain integrals and accelerated by the fast multipole method in the BEM[J]. Computational Mechanics, 2017, 59(4): 611-624.

[27] WANG T, ZHU H. The application of FLAC and FLAC3D to the support design of underground cavern[C]. Proceedings of the Fourth International FLAC Symposium on Numerical Modeling in Geomechanics-2006 (Madrid, Spain): Itasca. Minneapolis, MN, 2006: 37-40.

[28] WANG T, HU W, WU H, et al. Seepage analysis of a diversion tunnel with high pressure in different periods: a case study[J/OL]. European Journal of Enviro-nmental and Civil

Engineering, 2018, 22(4): 386-404.

[29] YREC (Yellow River Engineering Consulting Co. Ltd.). Coca Codo Sinclair hydroelectric project Design Report[R], 2012.

[30] YUAN P, WANG S, WANG T, et al. Construction technology and particle flow code modelling of diversion tunnel support in Xigeda stratum[J]. Engineering Journal of Wuhan University, 2021, 54(03): 205-211.

[31] ZÁRUBA Q, MENCL V. Engineering geology[J]. Elsevier, 1976.

[32] ZHANG Y, SU K, QIAN Z, et al. Improved Longitudinal Displacement Profile and Initial Support for Tunnel Excavation[J]. KSCE Journal of Civil Engineering, 2019, 23(6): 2746-2755.

[33] ZHAO D, JIA L, WANG M, et al. Displacement prediction of tunnels based on a generalised Kelvin constitutive model and its application in a subsea tunnel[J]. Tunneling Underground Space Technology, 2016, 54: 29-36.

[34] ZHOU W, LIU B, WANG Q, et al. Formulations of displacement discontinuity method for crack problems based on boundary element method[J]. Engineering Analysis with Boundary Elements, 2020, 115: 86-95.

Exercises

Review Questions

(1) What are the effects of topography, lithology, geological structure, groundwater, and ground stress on the site selection of underground caverns?

(2) Describe the characteristics of secondary stress distribution after excavation of underground cavern and concept of surrounding rock.

(3) What are the engineering measures to improve the stability of surrounding rock?

(4) What is the Longitudinal Displacement Profile (LDP)? How to establish the intact relationship between the dimensionless displacement and distance to tunnel face?

(5) A cavern (165m long, 22m wide, and 15m high) is excavated in chalk strata beneath the sea. The crown of the cavern will be 35m below the seabed. What is the primary geological information you would like to have before proceeding with the excavation?